Future-Proofing Emerging Technologies for Business Transformation

Ravichandran Krishnamoorthy
Jain University, India

Easwaramoorthy Rangaswamy
Amity Global Institute, Singapore

IGI Global
Scientific Publishing
Publishing Tomorrow's Research Today

Vice President of Editorial Melissa Wagner
Director of Acquisitions Mikaela Felty
Director of Book Development Jocelynn Hessler
Production Manager Mike Brehm
Cover Design Phillip Shickler

Published in the United States of America by
IGI Global Scientific Publishing
701 East Chocolate Avenue
Hershey, PA, 17033, USA
Tel: 717-533-8845
Fax: 717-533-7115
Website: https://www.igi-global.com E-mail: cust@igi-global.com

Library of Congress Cataloging-in-Publication Data

Names: Krishnamoorthy, Ravichandran, 1969- editor | Rangaswamy, Easwaramoorthy editor
Title: Future-proofing emerging technologies for business transformation / edited by: Ravichandran Krishnamoorthy, Easwaramoorthy Rangaswamy.
Description: Hershey, PA : IGI Global Scientific Publishing, [2025] | Includes bibliographical references and index. | Summary: "The main goal of the book is to help organizations to build strategies based on relevant skills and frameworks that will enable them to excel and to be future ready when facing disruption. It underlines the concept of linking the digital process change agendas with organizational objectives, customer needs, and market trends, while offering a long-term guide"-- Provided by publisher.
Identifiers: LCCN 2025001386 (print) | LCCN 2025001387 (ebook) | ISBN 9798337313207 hardcover | ISBN 9798337313214 paperback | ISBN 9798337313221 ebook
Subjects: LCSH: Organizational change | Business enterprises--Technological innovations
Classification: LCC HD58.8 .F89 2025 (print) | LCC HD58.8 (ebook) | DDC 658.4/06--dc23/eng/20250425
LC record available at https://lccn.loc.gov/2025001386
LC ebook record available at https://lccn.loc.gov/2025001387

British Cataloguing in Publication Data
A Cataloguing in Publication record for this book is available from the British Library.

Table of Contents

Detailed Table of Contents

Chapter One defines digital disruption and grounds it on disruptive innovation theories. They detail how such technologies as AI, blockchain, IoT disrupted industries and produced new winners. The chapter also outlines who the major digital disruptors are and what forces are driving their successes, noting that today's consumer is an essential piece in the business model puzzle that is today's globalised environment. Primarily, through concrete examples of major companies, such as Kodak and Blockbuster, the book describes negative lessons of industries' failure to address disruption and underlines the importance of adaptability to change.

In the second chapter, the author examines the compelling reason why organizations must go digital. It compares digital tools and techniques with traditional business management methodologies and reveals how early adaptors benefit from a wide range of benefits when compared to slow adaptors. The chapter explains the expensive price of failure especially when business environments are unpredictable and shows how modern organizations can predict hurdles in advance. Even through examples from Amazon and Tesla, it effectively argues the need to adapt because failure puts companies out of business

Chapter three explores change management in organisations. It analyzes the psychological part of change management as well as the aspects of fear and resistance during a transition to digital transformation. The cultivation of transformation culture is one of the most important strategic shifts that require leadership at the sub unit and corporate level. Chapter 8 looks at ways of increasing creativity and innovation within groups as well as stressing the need to adopt an agile approach. It also examines the creation of sustainable strategic plans to transform the environment to digital.

Dinesh Nilkant, Jain University, India
S. Hemanth Kumar, Jain University, India
Prabha Kiran, Westminster International University in Tashkent, Uzbekistan
Sumithra Sreenath, Independent Researcher, India

The chapter explores the critical role of leadership in navigating the complexities of digital transformation, emphasizing the integration of technology, culture, and strategy to drive sustainable organizational change. Through comprehensive case studies, it highlights successful transformation journeys at Microsoft, Tesla, and Walmart, alongside challenges faced by organizations like General Electric. The discussion delves into frameworks such as Transformational Leadership Theory, Agile methodologies, and Digital Maturity Models, underscoring their importance in fostering adaptability, innovation, and inclusivity. Key themes include addressing resistance to change, bridging digital skills gaps, and embedding ethical practices in leadership strategies. The chapter concludes by providing actionable recommendations for future leaders to harness emerging technologies, balance innovation with sustainability, and build resilient organizations in an interconnected global economy.

Divya Tiwari, Amity Global Institute, Singapore
Easwaramoorthy Rangaswamy, Amity Global Institute, Singapore
Adrian Raj Angappan, Amity Global Institute, Singapore

In the rapidly evolving world of today, the digital economy is changing sectors, occupations, and skill needs. The need for new skills has increased dramatically as technologies like big data and artificial intelligence continue to develop. Businesses must concentrate on gaining the necessary talent and abilities if they want to be competitive in the digital age. Establishing a culture of ongoing learning is essential for both businesses and educational institutions in the quickly changing global economy. Fostering creativity, flexibility, and resilience in the quickly changing world of today requires a learning culture. To help people and organizations achieve continual improvement, this study examines the idea of a learning culture and highlights its significance. The chapter incorporates theoretical frameworks including social learning, andragogy, and transformational leadership. This paper explores the principles, strategies, and outcomes of creating a learning culture that prioritizes continuous development.

This chapter explores the key factors contributing to resistance to digital transformation and provides strategies to overcome them effectively. It underscores the importance of stakeholder involvement in fostering trust, collaboration, and alignment with organizational goals, ensuring smoother transitions and greater commitment to change. Effective communication is highlighted as a vital tool in addressing uncertainties, enhancing transparency, and building confidence among employees. The chapter also investigates the critical role of organizational culture in enabling transformation by promoting adaptability, continuous learning, and cross-functional collaboration. Through case studies of organizations such as General Electric (GE) and IBM, practical insights are provided on how leadership-driven communication, targeted upskilling programs, and participatory approaches can successfully mitigate resistance. These insights offer actionable strategies to enhance change readiness and drive sustainable transformation.

Data has emerged as a critical strategic asset in today's digital economy, and in driving innovation, it has become the source of competitive advantage. Companies are under flood studies with data coming from diverse sources, and effective use of such data is an indispensable activity for market differentiation. Value capture refers to converting created value into profit by strategies such as data monetization, dynamic pricing, and customer segmentation. Examples include Google's targeted ads and Netflix's personalized recommendations. Companies like Amazon and Coca-Cola show how it is done. AI, machine learning, and cloud computing give captured data additional value. On the other hand, ethical considerations and conformity with regulations like GDPR are extremely critical. In sum, organizations have to invest in a robust data infrastructure and a culture of responsible use of data to succeed in this data-driven world.

The chapter tracks the historical development of AI together with Blockchain and IoT technologies which are now consumer-oriented digital disruptors that modify customer experiences. Throughout their development this chapter shows that empowerment, personalization, trust, and connectivity emerged as central features. AI establishes smarter automated systems which help the users with their consumer needs. The secure transparent procedures of Blockchain enable everyone to trust its reliable systems. Through IoT the components of our world maintain automatic real-time connections to perform harmonious continuous operations. New advances create market disruption that enables customers to enjoy expanded streaming, crypto and smart home system options. Better customer experience exists in the market but privacy as well as scalability and ethical factors need resolution. The model understanding between consumers and technology provides essential tactical knowledge to both businesses and digital market regulators.

Businesses are increasingly using predictive analytics to get a better understanding of and forecast customer behaviour within the context of Civilization 5.0, a human-centric civilization that combines the internet and physical space. This research investigates the critical role of predictive analytics in the evolution of marketing innovations, emphasizing on marketers' ability to forecast trends, tailor experiences, and develop data-driven marketing programs. This study aims to look at how predictive analytics technologies are changing marketing methods in Society 5.0 by providing deeper consumer insights and enabling more precise targeting. This study is unique in that it investigates the confluence of AI-driven analytics and marketing approaches within a Society 5.0 framework, which seeks to balance economic growth with societal wellbeing.

Preface

In an era defined by rapid technological evolution, the convergence of emerging technologies has reshaped the landscape of business, creating both opportunities and challenges. *Future-Proofing Emerging Technologies for Business Transformation* delves deep into the transformative power of digital innovation, offering a comprehensive exploration of how businesses can harness cutting-edge technologies to adapt, thrive, and sustain competitive advantage in an increasingly disruptive world.

This edited volume brings together insights from leading experts across diverse fields, including artificial intelligence (AI), blockchain, data analytics, and IoT, to illuminate how these technologies are not just enhancing business operations but also defining the future of industries and entire economies. The contributors to this book offer a holistic approach to understanding digital disruption, business transformation, leadership, change management, and the essential skills needed in the digital age.

The opening chapters lay the foundation by exploring the very concept of digital disruption, providing historical context through case studies of iconic business failures such as Kodak and Blockbuster. These early lessons set the tone for understanding the urgency of adaptation and transformation. As the book progresses, readers will gain an in-depth understanding of the imperative for digital business transformation, the role of leadership in driving change, and the challenges organizations face in building resilient, future-ready teams.

One of the key themes throughout the book is the crucial role of leadership and organizational culture in driving successful transformations. Chapters explore the psychological and strategic elements of change management and how a culture of continuous learning and adaptability is essential for fostering innovation and resilience. From managing resistance to change to creating a learning culture, the contributions highlight the human aspect of digital transformation, a perspective often overlooked in more technically focused discussions.

As businesses face an unprecedented pace of technological advancement, data has emerged as a central asset, with its potential to drive strategic decisions and create competitive advantages. Several chapters focus on the strategic importance of data analytics, from its use in personalized marketing to its potential in shaping smarter, more efficient business operations. The integration of AI and machine learning into decision-making frameworks provides businesses with the ability to predict trends, optimize resources, and better meet customer expectations.

A particularly innovative contribution is the exploration of blockchain technology, not only in enhancing security and transparency in business transactions but also in redefining entire processes, such as recruitment, through smart contracts and decentralized arbitration systems. As companies strive to adopt these transformative technologies, the book emphasizes the importance of balancing innovation with ethical practices and sustainability.

This volume concludes by offering a future-oriented perspective on the role of emerging technologies in shaping the business landscape. Whether discussing the implications of Society 5.0 for marketing, the integration of AI and blockchain into financial services, or the potential of open digital networks in emerging economies, the chapters provide a roadmap for businesses seeking to future-proof their operations in an ever-evolving digital ecosystem.

CHAPTER OVERVIEW

Chapter 1: The Digital Disruption Revolution

In this foundational chapter, Ravichandran Krishnamoorthy explores the concept of digital disruption, tracing its roots to disruptive innovation theories. Through real-world examples of companies like Kodak and Blockbuster, the chapter demonstrates the profound impact that emerging technologies—AI, blockchain, and IoT—have had on industries, highlighting the importance of adaptability in a rapidly evolving business environment.

Chapter 2: The Imperative for Business Transformation

Selvi S. and Nilesh G. delve into why organizations must embrace digital tools to survive and thrive. Comparing traditional business methods with digital advancements, they argue that early adoption leads to substantial competitive benefits, while failure to act can result in dire consequences, as exemplified by Amazon and Tesla.

Chapter 3: Transforming: Growing the Right Way

Selvi S. and Suman Mishra focus on the human side of business transformation, analyzing the psychological challenges of change management and the cultivation of a transformation culture. The chapter emphasizes leadership at all organizational levels and the importance of fostering creativity, agility, and sustainable strategic plans.

Chapter 4: Leadership for Digital Transformation

Dinesh Nilkant, Hemanth S., Prabha Kiran, and Sumithra Sreenath examine the pivotal role of leadership in managing digital transformation. They explore leadership frameworks, case studies of successful companies like Microsoft and Walmart, and discuss strategies for overcoming resistance to change and fostering innovation.

Chapter 5: The Right Talent and Skills for the Digital Economy

Divya Tiwari, Easwaramoorthy Rangaswamy, and Adrian Angappan argue that the digital economy demands new skills. This chapter explores how businesses must invest in continuous learning and fostering a learning culture to remain competitive, while also examining frameworks like social learning and transformational leadership.

Chapter 6: Managing Resistance to Change

Keai Lim, Chen Bao, Ya Li, Ruizhe Wang, and Xiang Zhou explore the key factors behind resistance to digital transformation and offer practical strategies to overcome them. The chapter stresses the importance of stakeholder involvement, effective communication, and cultivating an organizational culture that embraces adaptability.

Chapter 7: Data as a Strategic Asset: A Study

Shrinivas Patil, Pujari Reddy, and Premalatha P. highlight the strategic value of data in today's digital economy. The chapter discusses how companies can leverage data for competitive advantage, using case studies from Google, Netflix, and others, while also addressing the ethical considerations involved in data use.

Chapter 8: Digital Analytics in Action: From Data to Impact

Ajitha S., Reshma K. J., Geevarathna, and Huxley S. explore how digital analytics transforms raw data into actionable business insights. The chapter covers applications in customer behavior analysis, conversion rate optimization, and advertising performance tracking, while emphasizing ethical and strategic data handling.

Chapter 9: A Blockchain-Enabled Arbitrable Smart Hiring System (BEASH)

Jayanta Banerjee, Ajit Danti, Ayman Nassoura, and Dileep R. present a blockchain-enabled solution to modernize the hiring process. This innovative approach uses blockchain, smart contracts, and decentralized arbitration to reduce fraud and increase efficiency, with a focus on scalability and regulatory compliance.

Chapter 10: Customer-Centric Change

Ajit Kumar, Ravichandran Krishnamoorthy and Melvin Goh examines how digital transformation has shifted business focus toward customer-centric models. The chapter explores how businesses can leverage AI, data analytics, and personalization to enhance customer experience and build loyalty in an increasingly digital world.

Chapter 11: Marketing Innovations in Society 5.0: A Role of Predictive Analytics in Analyzing Consumer Behavior

Parihar Dahake, Nihar Dahake, Saket Bansod, Prachita Patil, and Akhilesh Ingle explore the intersection of predictive analytics and marketing within the context of Society 5.0, focusing on how AI and data-driven approaches can revolutionize consumer insights and enable more precise marketing targeting.

Chapter 12: Future-Proofing Financial Institutions: Strategies for Digital Transformation and Operational Excellence

Manjit Kour discusses how financial institutions can future-proof themselves in the face of digital disruption. By adopting AI, blockchain, and advanced analytics, financial institutions can optimize workflows, reduce risk, and stay competitive, with an emphasis on customer-centric and sustainable strategies.

Chapter 13: Future Perspectives of Business in the Context of Digitalization

Poornima K. explores the future of business in a digitalized world, focusing on how emerging technologies like AI, cloud computing, and IoT are reshaping industries. The chapter emphasizes the importance of open digital networks and embedded financial services in fostering business growth, particularly in emerging economies like India.

Chapter 14: Future-Proofing Emerging Technologies

Viknesh Ponnusamy and Hans Ravi Aruldas provide an overview of the emerging technologies that are shaping the future of business, including AI, big data, blockchain, IoT, and automation. The chapter explores the risks and rewards of adopting these technologies and provides real-life examples of businesses leveraging these tools for growth and disruption.

This volume serves as a vital resource for scholars, business leaders, and practitioners, offering actionable insights and strategic frameworks to navigate the evolving landscape of digital transformation and future-proof businesses against the challenges and opportunities posed by emerging technologies.

Chapter 1
The Digital Disruption Revolution

Ravichandran Krishnamoorthy
https://orcid.org/0000-0002-9092-480X
Jain University, India

ABSTRACT

Chapter One defines digital disruption and grounds it on disruptive innovation theories. They detail how such technologies as AI, blockchain, IoT disrupted industries and produced new winners. The chapter also outlines who the major digital disruptors are and what forces are driving their successes, noting that today's consumer is an essential piece in the business model puzzle that is today's globalised environment. Primarily, through concrete examples of major companies, such as Kodak and Blockbuster, the book describes negative lessons of industries' failure to address disruption and underlines the importance of adaptability to change.

1.1 INTRODUCTION:

It can be defined as a digital generation, because it alters the nature of work and competition as well as how value is delivered. The Digital Disruption Revolution, the inaugural chapter of Disrupt to Thrive: Explaining how these innovations affected the general stream of the global economy, the concepts represented in the Mastering Business Transformation in a Digital World offer the necessary background information. AI or Machine learning, Blockchain, Automation and other related

DOI: 10.4018/979-8-3373-1320-7.ch001

technologies are not an evolution rather they are revolutionary which transform sectors and tap the traditional existing business model.

In this chapter, the authors present the concept of digital disruption which is more than merely the newest phenomenon. This is revolutionary in how business organizations consider innovation activities, customer feedbacks and organizational efficiency. This research analyses how new entrants like Amazon, Netflix and Tesla benefited from using digital technologies while traditional giants like Blockbuster and Kodak perished.

1.2 Disruptive Innovation

Disruptive innovation has formed as a dominant theme in analyzing process and how technology and new markets tem change industries. Used by us for the first time in this paper, this term was first coined by Clayton Christensen in his book titled The Innovator's Dilemma published in 1997, and refers to a process whereby new entrants into an industry that are often of a small size and modest resources, undermine incumbants through developing simpler, cheaper or more innovative technologies. These disruptors start off with offering their product or service in moderately desirable or untapped markets and then over time, enhance their value proposition in such a way that they substantially replace existing players with new benchmarks. From the shift of mainframe computers to personal computers, to the overthrow of traditional taxi services by ride hailing applications Disruptive innovation has been the order of the day across many industries.

Fundamentally, disruptive innovation seeks to bring together the technology proposition, the consumer demand, and the value network proposition. Present day traditional firms despite being dominant during the initial stages of disruption rarely identify or address them. Due to their emphasis on sustaining innovations – routines that the firms use to refine and improve existing products and services to meet needs of their most lucrative customers – they prematurely ignore the opportunities arising from disruptive innovations. It is in this environment that disruptors can sneak in to grab those that are not served and then inch by inch creep into other territories.

Another feature of disruptive innovation has something to do with the process of democratizing of innovation. What was earlier only owned by the big firms or the well of financially endowed customers is now available to scale-down firms and the general populace. For example, advances in the Internet and cloud computing, artificial intelligence (AI) have made it possible for most startups to achieve what was almost impossible for large companies. Likewise, the technology continue to evolve in the renewable energy field, which has posed tremendous threats to industries that deal in fossils.

Disruptive innovation is different from merely replacing one technology with another and applies to business models and customer behavior. Netflix and Amazon are ideal illustrations that knowledge-based strategies can unseat well-entrenched industries such as motion picture and retail. Behind such approaches, these firms have revolutionized the contours of convenience, accessibility, and customization making a hornet's nest of imitable strategies for other players in the market.

Despite disruptive innovations being rich opportunities, they create flavors for incumbent and entrant firms. Sustaining and disruptive innovations are especially challenging for incumbent competitors since a company must continue to invest in the technology that made it successful while at the same time create disruption that will ensure the company's continued relevance. In contrast, disruptors are confronted with the problem of how to expand their businesses and how to sustain competitive advantage as they expand.

1.3 Disruptive Technologies- Definition

Disruptive technologies are those which reshape industries and organizations, eliminate traditionally recognized market offerings and offer simpler, often superior substitutes. Originally defined by Christensen (1997) disruptive technologies emerged with subsets of a specific market or a low performing segment and incrementally developed to revolutionise and unseat incumbent solutions.

About Disruptive Technologies Disruptive technologies display the following characteristics:

1. **Affordability:** In many cases, disruptive technologies cause the costs of complex solutions to decrease, thereby reaching the masses. For example, cloud computing introduced more numerous and diverse persons to high level computational resources that were once an exclusive property of large companies (Armbrust et al., 2010).
2. **Simplicity and Accessibility:** These technologies, in particular, aim at satisfying known user needs with the particular emphasis on the best and most convenient tools. For instance, the Smartphones made mobile computing what it was; a viable tool for use in society (West & Mace, 2010).
3. **Scalability:** Disruptive technologies allow organisations to expand as activity necessary for corporate growth occurs while costs do not have to escalate at a similar pace. This scalability is evidenced in supply chain context where blockchain guarantees security of decentralized transactions across global networks (Tapscott & Tapscott, 2016).

4. **Addressing Unmet or Unrecognized Needs:** Disruptive innovations bring to light or develop new opportunities in the marketplace. The IoT has created new applications in smart city application and connective appliances civilization pattern transforming (Atzori et al., 2010).

1.4 List Disruptive Distribution Channels

- **Artificial Intelligence (AI):** Big data helped advanced industries like health care, and play central role in finance, where it introduced predictive, automated, and improved customer experience (Brynjolfsson & McAfee, 2017).
- **E-commerce Platforms:** Examples of new business models include the contingency establishments such as Amazon and Alibaba, which revolutionized traditional commerce by providing e-shop, anchored by substantially differentiated variety, value and convenience (Parker et al., 2016).
- **Electric Vehicles (EVs):** Tesla's electric cars think of sustainability issues in the automobile sector and brought innovations, such as self-driving capability into the market (Christensen, 1997).

1.5 The Broader Implications

Cognitive technologies redraw the environments in which economies, industries, and societies exist. They tend to produce waves that ensure industries that are in contact with them transform or else become obsolete. For example, with the introduction of new technologies such as the PayPal and Stripe in the processing of payment, the conventional banking industry has been tweaked (Arthur, 2009).

But with disruption there are difficulties as well. Current firms have to ensure that they maintain innovations whilst at the same time, trying to explore disruptive paths – which most of the time may call for drastic changes in the overall culture, and operational methods. In addition, governments and regulatory authorities cannot leave these concerns unaddressed and ignore the ethical, economic, and social consequences of such technologies.

1.6 Theories on Disruption:

1.6.1. Concept of disruption

The idea of disruption, especially in management think and planning has received a lot of attention particularly in the last couple of decades owing to the part played by technology in such processes. Disruption is the overturn of an industry by 'game changing' technologies or strategies that replace the dominant incumbents and create new forms of competition. Various understandings explain nature and impact of disruption particularly in business and technology fields.

1.6.2. Clayton Christensen: Disruptive Innovation Theory.

Can be considered to originate with the book published by the author in 1997 It explains that disruption occurs when new entrants with a weaker business model than existing companies ... These disruptors first enter by going after small, niche markets or through offering scaled down, lower cost products that large competitors do not bother to address because they are too busy working on sustaining technologies—inventions or innovations that provide incremental improvements to the existing products to the company's best customers.

They make new entrants provide better product or service, and replace the incumbent players with their good enough product or service for the masses. The best example here is the disruption of the mainframe and minicomputers market share by personal computers which although initially created for the small enterprises and home users offered cheaper and simpler solutions that progressively supplanted all the other classes of machines. Chrisheory going further by suggesting that for large organisations there few opportunities to recognise or respond to disruptive innovation, because at first they are often worse than the existing solutions. Outsider and hidden high-protein disruption is preferable to insiders and overt disruption from firms that have become established in the well, known prime focus area.

1.6.3. Technetium and Technological Determinism

Technology disruption is therefore the process of applying technology to reverse the orientation in the market and break the mold in which industries exist. This supports the thesis part of Technological deterministic perspective which has it that technology is the main cause of change within the society and the economy. From this perspective, the internet, smartphones, artificial intelligence, blockchain and so on are in practice viewed as disruptive technologies which bring novel competencies to transform business models, customers' behaviors and even the value creation system.

For example, the internet altered the manners of trade and communication because newspapers, clothes, anything could be bought online and television was available online. In this theory we are going to talk about the fact that technote are not neutral; they govern markets and sometimes they can eject initial business models.

1.6.4. Value Network Theory

The other theoretical lens that assists in explaining disruption is the(Value Network Theory that highlights how organisations can create values alongside suppliers, partners and buyers. In this theory disruption is described as the process of using technology or innovation to establish new mechanisms of creating value within an industry in a way that the prior method of creating value becomes irrelevant. Such value networks may be more effective, less costly, or merely offer more unique value and indeed replace the other networks in the market.

For instance Uber and Lyft disrupted the established model of taxi hailing business and adopted pragmatic and efficient model of application hailing which is more effective than the normal mode of taxi hailing services. This theory helps scholars explain why incumbents operating in existing orks have the powers to resist disruptive forces from outside their orbits but lack the capacity to respond.

1.6.5. Blue Ocean Strategy

Although it is neither a theory of disruption itself, the Blue Ocean Strategy created by W. Chan Kim and Renée Mauborgne helps to use a different perspective on disruption . This strategy aims at developing new markets instead of competing in the existing markets – "Red Oceans," – whes little or no competition exist. The most promising innovation mode is actually building completely new value propositions that are not related to the existing market incumbents at all.

What is known as disruption therefore happens when a firm finds out there are still latent customer requirements and markets, or threats, which the incumbent firms cannot perceive. For example, Nintendo Wii shifted the Video Game Industry dynamics of the worth buying since it targeted the casual and family audience something that although Microsoft and Sony had failed to capture in their obsession of the hard-core gamers.

1.6.6. Schumpeter's Creative Destruction

Other classical ideas are commented by Schumpeter, known as Creative Destruction which means that a new form of production is created with destruction of the former. Schumpeter held a simple but firmly established view that the capitalist world economy is cyclical New technologies and forms of enterprise act destructively and overthrow the old patterns of industrial circulation and create opportunities for fresh sources of profitable investment.

It is observed that disruption theories are concerned with the possibility of a given technology, the associated market as well the business model. Among all the current disruptive innovation theories, the most credible theory is that developed by Clayton Christensen because this theory provides the most information pertaining to how new entrants displace incumbents. Other related theories include Technological Determinism, Value Network Theory, Blue Ocean Strategy and Creative Destruction that complements the proposed theory in explaining how business and industries transfigure itself to be in harmony with technological and market forces. Taken together the theory of change, resilience, innovative disruption and the strategy of sustainability depicted organisational survival and success in wake of or because of change accumulation as complex process involving much more than mere practices and right conditions.

2.THE RISE OF DISRUPTIVE TECHNOLOGIES

In the modern and fast evolving world and continuously evolving business arena, industries are observing shifts in Technology disruptions on how organizations competed and even survived. These technologies introduce changes that displace one market whilst bringing new market players and concepts. Disruptive technologies are not just refinements on existing value frameworks, nor are they an augmentation of an existing value proposition with an additional layer of value into the current consumer experience; rather, they open up new value propositions, alter consumer behaviour/psychology and most often make the existing approaches of delivering goods and services obsolete.

Disruptive technologies can thus be defined as being those which transform business or industries. The worst kind of competitive pressures is that they are most often effective because they are normally capable of introducing solutions that are cheaper, more easily to acquire or are simply superior to all the other existing ones. For example, cloud computing revolutionized the software and it market that for most organizations needed local servers to host their organizational data. This change ensured that far less supermarket firms benefited from the application-ideal IT infrastructure thus eliminating the stark fact of massive capital intensity (Arthur, 2009).

The disruptive technologies are emerging as new normative for industries, providing complete transformations of conventional business models carrying opportunities of innovation at their best. Term disruptive technologies was coined by Clayton Christensen in The Innovator's Dilemma in 1997, disruptive technologies are innovations that disrupt existing systems and offer simpler, cheaper, and more accessible solutions. Such technologies normally start by targeting certain specific market segments or by neglecting them altogether, only to first transform these particular segments into the most strategic prize in today's markets. They are not only a technological revolution but also a social and economic process which pervades all spheres of the contemporary world.

Another characteristic of disruptive innovations is the way they make powerful tools and services, initially accessible only to the select few, available, to a much wider population. For example, cloud computing has provided alternative ways through which startups, and small businesses can be able to acquire computing power previously accessible to large scale corporations (Armbrust et al., 2010). Likewise, technologies including solar power and wind power have made available green energy technologies to homes and small businesses (IRENA, 2019).

Of the disruptive technologies, AI and machine learning appears to be most disruptive with wide ranging impacts. AI is now disrupting industries from healthcare to finances because it can automate these tasks, make prognosis, and even improve clients' experiences. For instance, the diagnostics of diseases using Artificial intelligence has been proven to have immense accuracy hence will improve patient's care while at the same time decreasing costs (Topol, 2019). In the same way, blockchain is applying disruptive techniques in the financial systems by providing secure, decentralized and transparent transaction means. Cryptocurrencies such as BitCoin and Ethereum demonstrate how block chain can disrupt the traditional banking systems (Tapscott & Tapscott, 2016).

However, the disruptions caused by such technologies give rise to our last challenge. Its application typically distrupts local markets for labor, marginalizing some skills while creating a demand for fresh competencies. However, with issues like data privacy, algorithms' bias, and impacts on the environment people discussed how these smart technologies should be managed, or regulated (Brynjolfsson & McAfee, 2017).

In the case of businesses, disruptive technologies' innovations are both blessings and curses. Those organisations that adopt them stand to gain improved market opportunities, productivity, and competitive advantage. On the other hand, those that fail to adopt new models are likely to f , this is evident by the overrun giants such as Kodak and Blockbuster.

In the present world where change technologies are becoming the order of the day it is important to study the process and effects of disruptive technologies. The thought patterns set up in this introduction provide the base from which to understand how these innovations reconfigure industries, economies, and societies, for better and for worse.

2.1 Seven Features Defining Disruptive Technologies

It is possible to list a few characteristics that are attributable to disruptive technologies, and which, while not entirely universal in this instance, remain largely relevant. First, they made it affordable through solutions that could only formerly exist as exclusive to the larger firm or the high-end clientele. Subjects like artificial intelligence (AI) or automation are things that were previously available only to large corporations, while now start-ups, or relatively young companies, can leverage those kinds of technologies and apply them to their work. For instance, embracing of Artificial Intelligence is revolutionizing industries such as the medical, marketing industries through the aspects of Decision making, Data analysis and Customer relation (Kelly, 2016).

Second, disruptive technology means that operations are scalable, or organizations can grow operations significantly without a proportional increase in overhead expenses. One of the characteristics of the modern electronic business environment is the use of new technologies, and bitcoin, and other forms of virtual currencies, embody this characteristic, as they are based on the blockchain. Blockchain was established initially for the use of fostering the sale transaction and most importantly for facilitating the financial transaction using the cryptocurrency but has extended its field to areas such as for supply chain and health care. That is therefore progressive due to its ability to record and even effect transactions across 'borders,' and does not require other parties or an intermediary, thereby eliminating costs and achieving more effectiveness (Parker et al., 2016).

Third, such technologies always solve unacknowledged needs or create new demands hence altering consumer expectation. Smart home, Smart cities and IoT running in manufactory etc are some key disruptive technology of many sectors as devices connected to internet. The Internet of Things has transformed and enriched the supply chain and management by using tracking technologies to track the products and monitor the utilization and effective managing of the production chain. These have resulted to so called smart cities, that have IoT data driven infrastructures in sectors such as transport, energy and security systems (Arthur, 2009).

2.2 Impact Across Industries

Technological advancement disrupts one or the other field of human endeavor affecting industries and modifying conventional practices. By bringing changes that make processes more effective, less costly, and more effective these technologies are revolutionizing how companies work and services are provided. This section analyses how disruptive technologies are reshaping industries including health care, production, financial services as well as the retail sector.

2.3 Healthcare: Precision Medicine and AI

Technologies like AI and Machine learning for instance are negatively disrupting precision medicine in the healthcare industry. Is used in diagnostics; it increases the disease detection rate and improves the corresponding clinical results. For instance, AI in medicine can diagnose diseases from images better than normal methods by analyzing the medical images of patients with diseases like cancer at an early stage. In the words of Topol (2019), the above technologies not only enhance the ability to correctly diagnose patients' ailments but also enhance the choice of the right treatment procedures that will suit different patients.

AI is not only in diagnostics but in drug discovery and in monitoring of patients as well. Based on big data, machine learning algorithms can quickly scan out possible drugs within the shortest time and the least cost instead of relying traditional product development. Smart wearables connecting to the IoT platform read people's biometrics continually and alert users of developing ailments, if any, and help patients regain control of their health.

2.4 Manufacturing: The Rise of Smart Production

The manufacturing industry is going through an evolution as computerization and automation have become a trends. Much of the supply chain production paradigms are becoming obsolete to be substituted with more resilient, accurate and efficient supply chain management technologies like robotics, IoT and add-on manufacturing like 3D printing. Manufacturing facilities characterized as Smart Factory or Industry 4.0 integrate real-time data to improve production efficiency, minimize costs and maximize quality.

To illustrate this, connected devices in the industry scan the actual condition of machines and advise on when they should be serviced. This concept is used to reduce the amount of lost production time and to also increase the life of equipment. Likewise, the implementation of 3D printing has transformed the practice of prototyping and small-batch manufacturing through which manufacturers can make intricate designs with relative efficiency and little material wastage. This paper has further demonstrated how Tesla automates its production lines, a direct application of disruptive technologies as emphasized by Christensen (1997) within the automobile industry.

2.5 Financial Services:

Blockchain is the technology that make it based on the decentralised system. The world has seen disruption in a number of ways particularly in the financial services industry due to emergence of blockchain. Transaction processing is made secure, transparent, and decentralized using blockchain, hence the user does not need to use a third party such as banks. This innovation also contributes to decreasing of the transaction costs and increasing of security and anti-fraud measures within the financial value chain (Tapscott & Tapscott, 2016). Digital currencies such as Bitcoin and Ethereum have threatened the conventional financial sector by offering a new way of managing value. Further, the DeFi application also uses blockchain for various services like lending, borrowing, and trading without involving any financial institution. Other than cryptocurrencies though, the use of Blockchain in supply chain finance and international payments are transforming world trade by making it transparent as well as efficient.

2.6 Retail: Advanced systems and Customer Individualisation

In the retail industry AI is acting as a driver in how companies are engaging their customers as well as how they are managing their business. Consumers find it enjoyable, and retailers utilize artificial intelligence to enhance the customer experience: its primary applications are product recommendations and specific promotions based on the consumer's past purchases and interests. The recommendation tool in Amazon perfectly illustrates how the utilization of AI improves the organization's customer service and revenue.

AI is also widely used to monitor supply chain processes and forecasts demand, manage inventory, and detect disruptions. For example, the use of decision support systems helps retailers to make accurate forecast of which products to supply and when to supply them to avoid have high stock levels and low stock levels simultaneously. Automated checkout systems and, for example, AI-based chatbots help to change the speed of interactions and make the customer experience quicker and more convenient.

2.7 Potential Uses of AI across Industries

The use of disruptive technologies in various industries offers the world immense values and significant opportunities but at the same time it contains some difficulties. On one hand, these technologies result in time and cost-saving, eradicating waste, and consequent generation of new gross income. While they increase industrial output and create new demand channels, they interfere with employment relations and make companies rewrite the rules of workforce management.

For instance, when AI is used in the completion of monotonous tasks such as production or writing, there are likely going to be mass layoffs as companies embrace the new technology. But on the other hand, it opens opportunities for those people who will be assigned to create, maintain and analyse AI systems – high-skilled positions. Likewise, the changes that have happened in financial services due to blockchain have forced those institutions to invest in their digital transformation strategies to stay afloat.

Disruptive technologies are the only way industries are able to spearhead new opportunities and adapt to consumer needs in the future. With time, these innovations will inarguably effect further areas of the human endeavour in the future, the work environment, the market square and the society.

2.8 The Current Need for Business to Change

He still points out, however, that in the future traditional technologies will be non-existent given that they are not sustainable for business given that disruptive improved technologies are in the market and any business that does not tap into the technologies will be closed. The organisations that embrace such virtues are well placed to change their processes and develop a competitive advantage through great operational efficiency, a perfect customer satisfaction, and ways of generating income that were non-existent before. However, any organisation which is unable to change might find that competition with new firms that are leaner and faster in response to change is out of their reach. Similarly, cloud computing and A.I driven recommendation has shaped the world's largest e-commerce company Amazon and it further expands throughout new domain (Christensen, 1997).

This chapter reviews disruption as a threat and opportunity for growth for all industries and firms engaged in disruption technology. only those who can maintain a track on these technologies which can be adopted will be advantaged in an ever-changing society. In such possible characteristics of these innovations as availability, modularity, and opportunity, such organizations can reveal potential for subsequent advancement. History also shows that disruptive technologies are also able to recreate industries but most importantly the rules of the game .

3. EXPLORING DIGITAL DISRUPTORS AND THEIR DRIVING FORCES

He still points out, however, that in the future traditional technologies will be non-existent given that they are not sustainable for business given that disruptive improved technologies are in the market and any business that does not tap into the technologies will be closed. The organisations that embrace such virtues are well placed to change their processes and develop a competitive advantage through great operational efficiency, a perfect customer satisfaction, and ways of generating income that were non-existent before. However, any organisation which is unable to change might find that competition with new firms that are leaner and faster in response to change is out of their reach.

One potential general implementation area is that attaining successful adaptation also involves the institution-wide cultivation of the successful innovation culture to effectively capture the technologies. For instance, the emerging models, such as the use of artificial intelligence and automation technology, have challenged the construction of electric cars manufactured by Tesla to a level where conventional auto makers cannot match. Similarly, cloud computing and A.I driven recommendation has shaped the world's largest e-commerce company Amazon and it further expands throughout new domain (Christensen, 1997).

This chapter reviews disruption as a threat and opportunity for growth for all industries and firms engaged in disruption technology. only those who can maintain a track on these technologies which can be adopted will be advantaged in an ever-changing society. In such possible characteristics of these innovations as availability, modularity, and opportunity, such organizations can reveal potential for subsequent advancement. History also shows that disruptive technologies are also able to recreate industries but most importantly the rules of the game.

3.1. Technological Advancements

The second set of factors is therefore closely related to what we have been discussing so far and these are the rates of technology creation which appears to be one of the main motors behind digital disruption. Three popular business enablers include; AI because it lowered the barrier to entry for struggling players and enabled disruptors to grow fast and achieve economies of scale, Blockchain because it provided the disrupting players with a commendable tool that helped to reduce the number of entrants in the market, and Cloud computing since it provided an efficient tool to the disrupting players to enable them establish themselves, achieve more growth, and reduce barriers to entry. For instance, the cloud computing based offers an environment where organizations facilitate storage, computation and processing of big data at a lower cost than developing a structure within an organization resulting in freeing up business energies for inventiveness instead of energy to support (Arthur 2009). AI also assists digital disruptors in enhancing their general capacity for un-complication of engaging in process and personalized unique services provision to their users (Kelly, 2016).

Technology has been named one of the biggest agents of digital disruption as it has facilitated fast growth of new industries and new ways of doing business. Increased generational speeds of creating and adopting new technologies have helped to radically open the entrance to new gamers; innovations threaten traditional incumbents. Of these advances, AI, blockchain, and cloud are apparent enablers that enable disruptors to grow quickly, scale effectively and transform industries' rules of engagement.

3.2 Cloud Computing: Empowering Agility and Creativity

It is worth understanding that cloud computing has transformed rather dramatically the way companies work with data and computations. Cloud platforms have greatly made it possible for organizations to make considerable reductions in internal infrastructure investment by making their solutions affordable and easily scalable. This has led to the decentralization of supports that would have otherwise taken much resources to be maintained thus allowing businesses to channel there efforts towards creativeness. As pointed out by Arthur (2009), cloud computing creates a setting through which a firm can process, analyse, warehouse and deploy big data, analytics and applications at a much cheaper fashion as compared to other possible techniques.

For instance, IT firms and small businesses can leverage on such service providers such as AWS, Azure, and Google Cloud to acquire enterprise-level computing. Such environments help to host applications, analyze large data, and deploy AI using machine learning without the need for largescale investments. Further, this has led to democratization of technology thus enabling disruptors to work on the same system of fairness with the consolidated firms, leading to innovations particularly in evolving sectors of retail, healthcare and finance.

3.3 Artificial Intelligence: Enhancing Personalization and Efficiency

AI is now a fundamental part of the digital revolution and is capable of providing organisations with tools to improve numerous processes, enhance the customer experience and automate tasks. It entails the use of computers in learning from data to make high-precision predictions and consequently advice business organizations. AI applications such as chatbots and virtual assistants for example, help organizations offer round the clock client support boosting customer satisfaction and ensuring reduced organizational expenses.

The first primary way that reflects the positive impact of AI is that it offers highly personalized services. Thus, it becomes easier for the consumers and AI systems to recommend products and services, advertise products based on individual consumer needs and interests, and post content that will be consumed by the audience. Such feature is demonstrated by Netflix and Spotify where the latter uses Artificial Intelligence to recommend and create customized playlist that would make the client stick around (Kelly, 2016). As well, industries have benefited from holding AI applications, including health care whereby machine learning algorithms are used in identifying diseases, predicting patient's outcomes, and evaluating treatment plans.

3.4 Blockchain: Redefining Transparency and Security

Technology has challenged traditional business activities through the use of secure, decentralised and; transparent transactions through application of blockchain technology. Hence Blockchain ensures reduction of transaction costs and improvement in efficiency across the chain in various industries. For instance, in supply chain, blockchain eliminates any form of fraud by offering an avenue to track goods at a real-time basis. In the same way, DeFi which is blockchain-based provides an extended set of services, including lending, trading, and others, without the use of traditional financial organizations (Tapscott and Tapscott 2016).

Cryptocurrencies such as Bitcoin and Ethereum are yet another example of blockchain technology, disrupting the old banking models and delivering new perspectives for storing and transferring value. Except for the provision of financial services, the ability of the distributed ledger can be useful in other fields like storing of records such as patients and real estate in facilitating property deals.

3.5 Internet of Things (IoT): Connecting the Physical and Digital Worlds

Internet of Things has facilitated an integration of physical and cyber worlds to allow for proper connections and proper sharing of information between devices. It is done using sensors, actuators and other communication technologies that make IoT systems gather real time data and transform it into information on how to improve processes or even the user experiences. In smart cities for example, IoT devices track traffic conditions, regulate energy usage, as well as enhance security. In the same way, smart manufacturing IoT applications optimize working efficiency and minimize breakdown time in factories.

Closely associated with this is change in the consumer behaviour due to proliferation of IoT devices such as the Amazon Echo and Google Nest. These devices employ IoT to offer custom, voice-based interactions; from managing home appliances to looking for information. The technological provided by IoT has made it possible to help organizations explore avenues of supporting customers and grow business.

3.6 Data analytics as the Key Facilitator of Innovation

Data analytics is another crucial step in the technological process, whereby businesses are capable of drawing beneficial information from masses of information. Big data can reveal certain patterns and facilitate work with a better understanding of the key factors for further business evolution. For example, predictive analytics to forecast the demand and use it to manage the stock, and prescriptive analytics, which can tell an organization how to go about getting the desired results.

Some sectors like the finance and the health sector have been most affected by the analytics of data. From financial to health care industries, algorithms are used to make market determination for investments in the same manner that analytics tools are used to determine risk factors and patient care. With the help of data analytics with the help of Artificial Intelligence and Internet of Things the process increases it impact at an optimum level as it provides query answer in the direction of the business at proper time.

3.7 Ethical and Regulatory Challenges

Our society benefits greatly from today's technology, but this progress brings many ethical and regulatory problems that we need to tackle right now. As technology grows to support daily life it creates problems with privacy issues as well as risks from biased algorithms and cybersecurity vulnerabilities. The community needs to understand and resolve these problems to make sure new technologies serve society's rules and help everyone live better lives.

3.7.1 Privacy Breaches

The growing use of technology creates major risks to our private information. The constant tracking and handling of personal data by artificial intelligence systems makes people worry about losing their privacy. Academic studies show that weak data controls allow unauthorized tracking activities while exposing user data to misuse and unauthorized public exposure. In "The Age of Surveillance Capitalism"

Zuboff illustrates the impact of technology on our privacy rights demonstrating why governments must create strong data protection rules.

3.7.2 Algorithmic Bias

Although computers process decisions efficiently, they replicate the biases hidden inside their basic programming. Noble's research in "Algorithms of Oppression" shows that algorithms make unfair decisions which harm minorities more often than other groups according to data studies from "Algorithms of Oppression." Data sets without diversity combined with insufficient oversight when creating systems result in built-in biased performance. Ethical AI demands total AI process visibility so teams can find and fix discrimination in AI workings.

3.7.3 Cybersecurity Concerns

Digital technology's growing importance creates more opportunities for hackers to attack personal and systemic targets with threats to user safety and the security of entire nations. According to World Economic Forum research cyber resilience creates global risk that demands both strong security systems and multination collaboration to handle cyber threats effectively. Systems data needs ethical rules and regulations to keep user information safe though reliable access and true results.3.8. Influence of Consumer Expectation Changes

3.8 The Decentralization System Poses Governance Problems

The governance and compliance challenges of decentralized systems become clear through blockchain technology. The benefits of decentralized systems include clear organization and limited reliance on main control bodies, yet their rules become harder to enforce. In their study "Blockchain and the Law" De Filippi and Wright share how blockchain tech lets criminals use pseudonymous transactions across borders to hide money laundering and tax crimes. The rules that govern decentralized systems should protect citizens from harms while letting them benefit from decentralized technology.

3.8.1 Our Choices Create Important Ethical Problems

The decision to let AI handle choices creates important problems about who should take blame when something goes wrong. Although AI systems help make choices in work areas and police work human oversight becomes harder which creates ethical issues. Floridi and Cowls explain through "The Ethics of Artificial

Intelligence" that standard procedures must define machine activities while preserving and strengthening human decision-making power.

3.8.2 Government Organizations Guard Ethics

Government organizations and oversight agencies lead the fight against these industry problems. Through their rule making and enforcement they create safeguards that protect companies from technological changes. Regulatory systems need to change quickly as technology develops at high speed. The GDPR of the European Union shows how rules protect privacy and encourage technology progress as a data protection best practice.

3.9 The Rise of Personalization

Personalization is an outstanding example of how consumers anticipate the strategic use of their information. Many of today's customers are in a position where they want products and services, specific experiences, tailored just for them. Such companies as Netflix are leading trends of approach, acting on the principles of artificial intelligence and machine learning to observe the consumers' behavior and offer them appropriate content. Thus the intelligent recommendation system guarantees that by continuous observation of viewing habits, preferences and patterns, only programs that would sustain the viewer's interest are recommended at any one time. Such a high level of personalization has changed the approach to consuming the material and continues to act as a valuable precedent for most industries (Kelly, 2016).

There is another sector that today's competition considers the possibility of personalization as an essential element – it is retail. Shops like Amazon have incorporated AI and machine learning to make suggestions of other products for a buyer to buy based on the buyer's browsing and purchasing history or based on suggestions from other users. It is a proven marketing strategy, which apart from providing an added-value and a friendly shopping experience also increases sales and customer loyalty. Likewise, clothing merchants such as Stitch Fix utilizes Artificial Intelligence alongside with professional stylists in order to offer specific clothing and apparel choices based on the customer's aesthetics. Compared to conventional cab hailing services whose cost structures and supply time are obscure, Uber's price transparency and supply availability serve as key factors that revolutionize consumers' demands (Parker et al., 2016).

3.10 Flawless Integration

The customer today requires a consistent experience across all touch points and screens. Consumers have demanded seamless appropriate regardless of the touch point and this has pushed businesses to envelop omnichannel strategies hence creating a consistent experience. For instance, e-commerce players have synchronised the physical and the online storefront through an integrated online and offline fulfilment system where customers are able to shop, buy and return products seamlessly in the physical outlets without going online.

3.11 Data Analytics and AI in the Context of Customer Journey

Hence, data analytics and AI are key factors in responding to new customer expectations. In huge data mines, opportunities are available to obtain essential information regarding customer behaviour, their tendencies, and their requirements. This assists the company in understanding future customer needs and delivering a products and services that meet the customer needs, hence improving their satisfaction.

For example, Spotify uses AI in order to select optimal songs to be suggested to a user based on his or her behavior in terms of listening to the music. In this way, not only does it enhance users' satisfaction but also distinguish Spotify from rivals at this level of the interface. In the travel industry of today, companies like Airbnb put into practice data analysis to give customers the best of what they are looking for and improve the customer experience throughout the entire journey.

Customer service has been highly innovated using chatbots and virtual assistants that powered by artificial intelligence. Because of instant response and replies given by these tools, questions are answered as and when they are asked, without having to wait for a long time as is often the case with traditional approaches. For instance, intelligent virtual assistants that engage customers on Facebook via customer support desk make it possible for business to engage customers face and build trust.

3.12 Transforming Traditional Industries

The shift in consumer expectations has had a profound impact on traditional industries, forcing them to adapt to remain competitive. Retail, transportation, and entertainment are among the sectors that have undergone significant transformations.

In retail, brick-and-mortar stores are integrating digital technologies to enhance the shopping experience. Smart mirrors in fitting rooms, mobile payment options, and augmented reality (AR) tools that allow customers to visualize products in their environment are just a few examples of how retailers are meeting modern demands.

Transportation, as exemplified by Uber, has shifted toward app-based solutions that prioritize convenience and transparency.

3.13. Challenges and Opportunities

Just like restaurant food delivery, Food delivery platforms such as DoorDash and Uber Eats meet this demand of instant satisfaction. These platforms integrate real time tracking, and artificial intelligence to enhance delivery routes of the orders. This element of immediacy has determined new consumer habits and has made convenience a necessary factor for contemporary commerce.

3.14. Integration and Seamlessness

The customer today requires a consistent experience across all touch points and screens. Consumers have demanded seamless appropriate regardless of the touch point and this has pushed businesses to envelop omnichannel strategies, hence creating a consistent experience. For instance, e-commerce players have synchronised the physical and the online storefront through an integrated online and offline fulfilment system where customers are able to shop, buy and return products seamlessly in the physical outlets without going online.

And the banking industry has also followed this trend. Particularly, digital players like PayPal and Square innovations provide answers for customers to pay seamlessly from one device to another. New generation traditional banks have embraced many more providing banking services through mobile apps and online platforms that can handle account management to loan applications.

3.15. Data Analytics and AI in the Context of Customer Journey

Hence, data analytics and AI are key factors in responding to new customer expectations. In huge data mines, opportunities are available to obtain essential information regarding customer behaviour, their tendencies, and their requirements. This assists the company in understanding future customer needs and delivering a products and services that meet the customer needs hence improving their satisfaction.

For example, Spotify uses AI in order to select optimal songs to be suggested to a user based on his or her behavior in terms of listening to the music. In this way, not only does it enhance users' satisfaction but also distinguish Spotify from rivals at this level of the interface. In the travel industry of today, companies like Airbnb put into practice data analysis to give customers the best of what they are looking for and improve the customer experience throughout the entire journey.

OPPORTUNITIES

3.16 Technological Innovation

As much as there are challenges present by these technologies there are great opportunities that come with them. AI and machine learning can help business entities to collect and analyze lots of data, forecast trends, and draw applicable conclusions and make pertinent decisions. Blockchain innovation establishes improved sincerity and protection of business transactions, and IoT establishes connectivity in addition to real-time monitoring among industries. The usage of these tools can help companies be innovative and build value for the customer.

For instance, to advance the diagnosis and overall health of its patients, healthcare engages AI and smart production through IoT-enabled smart manufacturing in industries and manufacturing. Leaders of corporations can advice their organization to adopt these emerging technologies since doing so can be a strategic move for an enterprise to create a competitive edge in the market.

3.17. Enhanced Customer Experiences

Changing customers' preferences provide chances for the provision of individualized, integrated, and exciting solutions. Companies and organizations that are able to analyze customer behavior through data analytIcs and artificial intelligence are able to develop products and services that will find fertile ground among the consumers. Examples of Amazon and Netflix have proved that applications of personalization are very effective in making customer strategies that will help to increase loyalty and revenues.

3.18 Sustainable development as a marketing tool or as a competitive advantage

The aspect of sustainability has emerged as an influential area that customer considers while purchasing products, giving companies an opportunity of differentiation. It enables notably companies that embrace ecological and sustainable gestures and provide clear supply chains to generate attention among ethical consumers and investors. The example of such companies as Patagonia and Tesla also vividly show that sustainability acts as a stimulus for brand loyalty and commercial success.

Also, green innovations, including in the generation of electricity, in consumption, and on the premise of circular economy, create new revenue sources. The research also established that companies that embrace environmental objectives enjoy government incentives and partnerships for their business and in the long run minimize costs.

3.19 Workforce Transformation

The trend of generating new automation and AI leads to challenges that can be used to transform the company's workers, promoting the corporate culture of innovation. If the reskilling programs are invested, the employee can easily take up the challenge of competing various high value tasks, which in effect will enhance productivity and engage the employees. The organizations that invest in the development of the workforce can help establish competitive Human Capital Strategy that attracts the best and brightest and creates a sustainable workforce of the future.

3.20. Global Market Expansion

Technological advancement has led to the destruction of barriers to geographical locations; organisations can freely and easily access global markets. Today's technologies such as e-commerce, digital marketing, and also distance communication technologies help the companies to address the audiences of diverse segments and enter the emerging markets. For example, independent merchants on Shopify are able to sell goods across borders and global companies use big data to develop strategies on how to market goods in given countries.

4. CONSUMER IN THE MODERN BUSINESS WORLD

This chapter establishes consumer as one of the most important models of change in the current business context. Today's consumer is more informed, interconnected, and properly conscious of values as compared to the consumers of the pre-digital technology ad globalization era. It is no longer a war of who controls the other through price and quality but how a company delivers great solutions in relation to a new model of customer. This has been true across all the functions in marketing and communication, products' design, and service delivery.

4.1 Empowered by Technology

The usage of digital technologies replaced the traditional methods of how consumers engage themselves with companies. These major advancements in technology from Smartphones, social media platforms to e commerce platforms make consumers informed individuals. They can search for specific products or services, look up for the best offer price or offer, and check out the customers' feedback quickly. In the view of Chaffey (2020), a staggering number of consumers, amounting to 81% of

Internet users, rely on information found on the Web when making their purchase decisions, stressing the role of trust and openness in terms of brand followership.

In addition, social media revolution has escalated consumer voices in various companies. Instagram and TikTok give consumers a platform to voice and express and share their experience, co-create consumer trends, and even report the wrong-doings from brands. Organizations need to be involved in the social media platforms where they post, enhance relationships with the consumers.

4.2 Demand for Personalization

Personalization is no longer a great innovative concept it is the new customer expectation. Today's customers want brands that are unique but cater for their own personal classification. In the context of more traditional retail, businesses like Amazon or Netflix use artificial intelligence (AI), or data analysis for recommendations, so that each touchpoint will be different and seem more relevant. A survey carried by Accenture (2018) found that majority of the consumers, (specifically 91%) said they are willing to shop with brands that engage them through personalization. Major players such as Patagonia and Tesla have adopted sustainability as their organizational values, which are relatively similar to those of the consumers. It also creates loyalty and has a competitive advantage in markets that are highly competed for the multinationals. Ethical consumerism is not a 'flavor of the month' but a wake-up wake up call for companies to embrace practices that will add value to society and the environment.

4.3 Implications for Businesses

It can be asserted that the current consumer expectations put emphasis on agility, innovation, and customers needs. There is the need for companies to procure technology to provide customer offers and cater to ethical, using multiple channels to create lasting relationships.

To promote strategic customer involvement, there exists the need to meet the emerging shifts in customer demands that can enhance the business and customer relations while delivering competitive advantage. Here I claim that the consumer is not only an actor in the contemporary business environment but also an active agent of change and change-maker.

4.4. Empowered by Technology

However, the most noteworthy and consequential change of resistance is driven by information, and technology. The consumers are more selective in their purchasing behaviors today than they used to be earlier because they can search for reviews, ratings or social media pages, or comparison. It has introduced the "democratization of information" whereby majority of the consumers is likely to have a clue on which brands or platforms offer what they need in their comparison (Kelly, 2016). They also point that they provide information on the fact that, customer are far more fickle meaning can easily switch their business from firm to firm if their demands are not met.

4.5. Demand for Personalization

This practice was common in the past, nowadays it is a tradition for customers to have special conditions for a particular product or service they wish to buy. Accenture also noted that the analysis was done and it found out that 91% of the consumers want to engage with the brands that suggest and promote products that they would be interested in. Some of the industries that use big data analytics and artificial intelligence to understand customers' behavior and satisfaction levels are the ones floating in this environment. Three examples that were mentioned include Netflix, Amazon and Apple which are usually very close to the consumer because the company is actively observing and following the consumer's behavior in a bid to anticipate their needs (Kelly, 2016).

4.6. Ethical consumerism.

Questions of Ethics have also been involved in the current society concerning the various commodities in the market. Today's consumers, therefore, are all aware on the environment as well as the kind of society organization that they want to support through their money. Consumer trends are highly in favor of products supplied by firms that exercise high levels of corporate social responsibility, use sustainable methods of production, and adhere to proper labor standard. Ethical consumption is also on the rise as this places pressure on the companies to be mindful of their material and manufacturing in the market place (Kim & Mauborgne, 2005).

4.7 The Growth of On-Demand Services

Instant availability has become the major requirement for consumers, and on-demand services fully meet this requirement. Startups such as Uber, DoorDash and Netflix are all good examples of this trend since they provide solutions that remove old-world constraints to entry. Lunch can be delivered, transport called and entertainment obtained all through a smartphone and a few taps.

The demands from the transport sector have been met by on-demand services. Uber and Lyft displaced taxi hailing techniques by offering the riders an easy to use application where they can order for a ride. Availability of fares, GPS navigation and payments through digital wallets have created high levels of mobility that make traditional taxis to look quite rigid.

4.8 Implications for Businesses

Social media is an incredible tool which demonstrates a weak in the culture of consumers to buy something impulsive and have it instantly delivered to their doorstep. On one hand, such trends are valuable to companies as they show that by implementing such approaches as velocity, convenience and customization companies will benefit.

5. THE FUTURE OF THE DIGITAL DISRUPTION REVOLUTION (2025-2040)

Through digital advance technology the years between 2025 and 2040 will create major transformations in every part of our world. Over the next decade and a half technical progress in AI blockchain quantum computers and IoT systems will rapidly reshape industries economies and social structures. Your research will help you seize digital growth while ensuring responsible and environmentally safe transformations.

5.1 Technology is blending together making new innovations possible

Technologies that disrupt businesses will combine together to build fresh ways to solve problems. Quantum computing and AI developments will enhance how society solves challenges across healthcare treatment systems and supply chains but especially for climate change predictions. Machine learning-powered autonomous systems will transform our infrastructure to run better and smarter across transportation systems and city designs while making agricultural production more productive.

5.2 The Future of Work

Digital transformation will keep reshaping how work happens and how companies use temporary workers. Automation and robotics will complete normal tasks so people can concentrate on imaginative and effective work. The workforce needs to learn new abilities since digital change is gaining momentum. Experts agree that employees must continuously learn new skills through academic studies which show growing employer demand for professionals who connect technical knowledge with design thinking.

5.3 Economic Transformations

As digital platforms continue to take over more markets the entire global economy will change. Blockchain technology will improve economic systems by building DeFi networks and showing supply chain details clearly. Digital currencies especially CBDCs will create new ways to pay and organize money policies. The new digital business models will help new businesses enter the market but create tougher competition between market sectors.

5.4 Societal Implications

Technology becoming more essential in our lives will lead us to new social patterns. Smart cities built with connected Internet of Things devices and artificial intelligence technology improve urban quality of life through better public systems and resource management. New developments will invite important privacy and moral discussions. Protecting data privacy against technology gains will create an essential battle for our future society.

5.5 Environmental Sustainability

Technology leads the way to climate change resolution while making our world more sustainable. By improving energy storage technology as well as solar and farming methods we will lessen our impact on nature. Digital supply chain solutions and predictive analytics help us run our resources better to combine economic development with environmental preservation.

5.7 The Path Forward

The digital disruption revolution between 2025 and 2040 will bring numerous opportunities together with tough problems to solve. Businesses governments and individuals must work together actively to overcome this digital disruption era successfully. Spending money on education programs and developing facilities plus exploring technology science will bring out digital progress and keep dangers under control. Through technology advancements and open access programs the digital revolution creates transformational success while spreading benefits to everyone.

REFERENCES

Accenture. (2018). *Personalization Pulse Check*. Accenture.

Armbrust, M., Fox, A., Griffith, R., Joseph, A. D., Katz, R., Konwinski, A., Lee, G., Patterson, D., Rabkin, A., Stoica, I., & Zaharia, M. (2010). A view of cloud computing. *Communications of the ACM, 53*(4), 50–58. DOI: 10.1145/1721654.1721672

Armbrust, M., Fox, A., Griffith, R., Joseph, A. D., Katz, R., Konwinski, A., Lee, G., Patterson, D., Rabkin, A., Stoica, I., & Zaharia, M. (2010). A view of cloud computing. *Communications of the ACM, 53*(4), 50–58. DOI: 10.1145/1721654.1721672

Arthur, W. B. (2009). *The nature of technology: What it is and how it evolves*. Free Press.

Atzori, L., Iera, A., & Morabito, G. (2010). The Internet of Things: A survey. *Computer Networks, 54*(15), 2787–2805. DOI: 10.1016/j.comnet.2010.05.010

Atzori, L., Iera, A., & Morabito, G. (2010). The Internet of Things: A survey. *Computer Networks, 54*(15), 2787–2805. DOI: 10.1016/j.comnet.2010.05.010

Brian Arthur. (2009). *The Nature of Technology: What It Is and How It Evolves*. Free Press.

Brynjolfsson, E., & McAfee, A. (2017). *Machine, platform, crowd: Harnessing our digital future*. W. W. Norton & Company.

Chaffey, D. (2020). *Global social media statistics*. Smart Insights.

W. Chan Kim and Renée Mauborgne, *Blue Ocean Strategy: How to Create Uncontested Market Space and Make the Competition Irrelevant* (Boston: Harvard)

Christensen, C. M. (1997). *The innovator's dilemma: When new technologies cause great firms to fail*. Harvard Business Review Press.

Christensen, *The Innovator's Dilemma*, pp. 24-27.

Clayton, M. (1997). *Christensen, The Innovator's Dilemma: When New Technologies Cause Great Firms to Fail*. Harvard Business Review Press.

Geoffrey, G. (2016). *Parker, Marshall W. Van Alstyne, and Sangeet Paul Choudary, Platform Revolution: How Networked Markets Are Transforming the Economy—and How to Make Them Work for You*. W.W. Norton & Company.

Ibid., pp. 89-92.

IRENA. (2019). *Renewable energy statistics 2019*. International Renewable Energy Agency.

James, F. (1996). *Moore, The Death of Competition: Leadership and Strategy in the Age of Business Ecosystems*. HarperBusiness.

Joseph, A. (1942). *Schumpeter, Capitalism, Socialism, and Democracy*. Harper & Row.

Kelly, K. (2016). *The inevitable: Understanding the 12 technological forces that will shape our future*. Viking Press.

Marshall McLuhan. (1964). *Understanding Media: The Extensions of Man*. McGraw-Hill.

Nielsen. (2015). Global Sustainability Report.

Parker, G. G., Van Alstyne, M. W., & Choudary, S. P. (2016). *Platform revolution: How networked markets are transforming the economy and how to make them work for you*. W. W. Norton & Company.

Tapscott, D., & Tapscott, A. (2016). *Blockchain revolution: How the technology behind bitcoin is changing money, business, and the world*. Portfolio Penguin.

Topol, E. (2019). *Deep medicine: How artificial intelligence can make healthcare human again*. Basic Books.

West, J., & Mace, M. (2010). Browsing as the killer app: Explaining the rapid success of Apple's iPhone. *Telecommunications Policy, 34*(5-6), 270–286. DOI: 10.1016/j.telpol.2009.12.002

Chapter 2
The Imperative for Business Transformation

S. Selvi
https://orcid.org/0000-0001-7954-9976
Jain University, India

G. Nilesh
https://orcid.org/0000-0001-7954-9976
DataMatics, India

ABSTRACT

In the second chapter, the author examines the compelling reason why organizations must go digital. It compares digital tools and techniques with traditional business management methodologies and reveals how early adaptors benefit from a wide range of benefits when compared to slow adaptors. The chapter explains the expensive price of failure especially when business environments are unpredictable and shows how modern organizations can predict hurdles in advance. Even through examples from Amazon and Tesla, it effectively argues the need to adapt because failure puts companies out of business

THE IMPERATIVE FOR BUSINESS TRANSFORMATION

As the consumers transmute rapidly into netizens, the need for change has become more important in business today than ever before (Accenture, 2018). Globalization, every changing consumer and technological innovation are factors that are forcing firms to adopt new strategies and operations (McKinsey & Company, 2020). Many companies that used to rely on conventional strategies are experiencing tremendous pressure, and dynamism is now a key to success (Christensen, 1997).

DOI: 10.4018/979-8-3373-1320-7.ch002

Companies that do not evolve are at the peril of being left behind, as disruptive players and nimble incumbents take advantage of advances in technology to eat the market and reshape verticals (Parker et al., 2016). It is therefore not a simple matter of implementing a new technology in an organisation but a question of a complete and dramatic transformation of the organisation together with its modes of interacting with customers and/or delivering customer value (Deloitte, 2020). From redesigning operations inside the business to cultivating an environment that enables the successful imposition of change, business transformation is a process of reinvention in the context of disruption (Kelly, 2016).

The key being change and that change is needed in every industry; whether that change involves transitioning from traditional bank branches to online services, reducing the negative impact of industries on the environment and society, or changing the type of business model to become competitive (McKinsey & Company, 2020). In forward-leaning economies, the companies that continue to thrive will be those that evolve, are innovative, and open to changes within the emerging digital marketplace (Statista, 2021).

The Necessity of Business Transformation

The need for businesses to adapt is greater than ever as the world's consumers transition from traditional to digital-first netizens (Accenture, 2018). Globalization, changing customer needs, and the speed at which technology is developing are some of the factors forcing businesses to reconsider their operational frameworks and strategies (Deloitte, 2020). Agility is becoming the key to success, and companies that previously prospered on traditional patterns are under tremendous pressure to innovate (Christensen, 1997). Companies that don't change risk being surpassed by nimble rivals and disruptive newcomers that are quick to use technology to take over markets and completely change sectors (Parker et al., 2016). Therefore, the problem involves more than just implementing new technologies; it involves a profound and extensive change in the way companies function, interact with their clientele, and generate value (Kelly, 2016).

Beyond merely implementing new technology, this change necessitates a thorough redesign of the company's internal procedures as well as its approach to customer care and contact (McKinsey & Company, 2020). Whether it's switching from physical to digital shopping, implementing sustainable practices to lessen the impact on the environment, or reevaluating business models to remain competitive in the market, change is necessary in all sectors of the economy (Statista, 2021). Businesses that embrace change, support innovation, and maintain flexibility in the face of the digital revolution will be the ones who thrive in progressive economies (Accenture, 2018).

Figure 1. Enterprise transformation

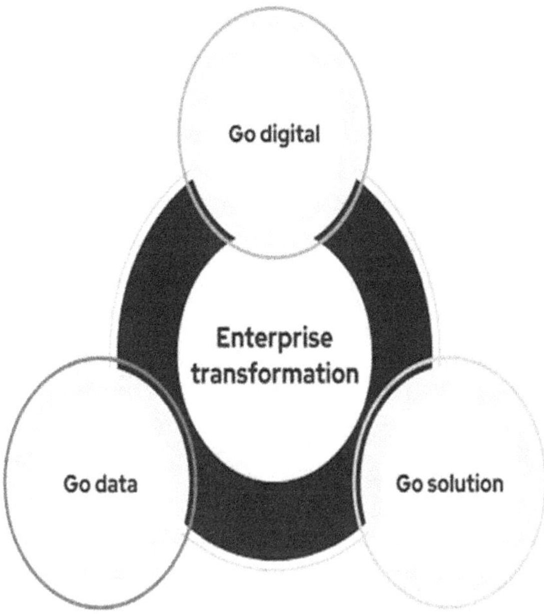

Source (https://www.aha.io/roadmapping/guide/enterprise-transformation/what-are-the-types-of-business -transformation)

This picture is an illustration of an enterprise transformation framework, which is a comprehensive method for improving and updating an organization's strategy, operations, and culture. Three interrelated components form the core of the diagram:

Make the switch to digital: Refers to implementing and utilizing digital technology to change how companies run, provide value to clients, and maintain competitiveness.

Cloud computing, AI, IoT, automation, and digital platforms are a few examples.

Get Data: Draws attention to how crucial data-driven decision-making, focuses on using big data, analytics, and insights to enhance procedures, client interactions, and results.
Go Solution: Focuses on creating and executing customized solutions that tackle particular corporate objectives or obstacles.

Entails using cutting-edge tools, procedures, and technologies to efficiently handle issues.

Importance of Businesses to Transform Right

Figure 2. Business process transformation

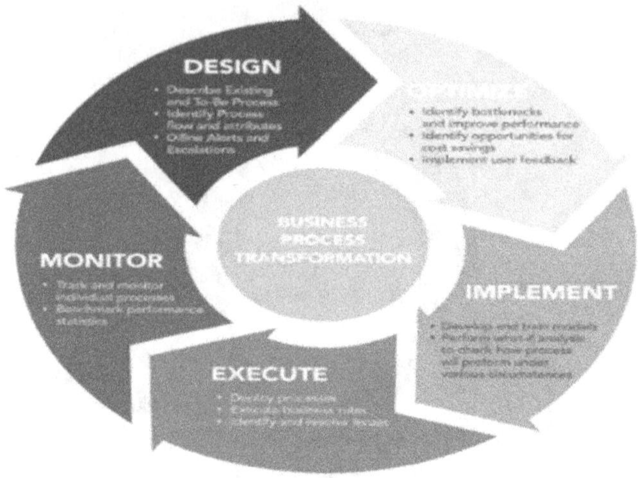

Source (https://modernizenow.com/business-process-transformation/)

1. **Design:** entails examining and comprehending current procedures.
   Design: A Critical Step in Business Transformation

In the process of business transformation, the design phase is crucial since it entails a careful analysis and comprehension of existing workflows and procedures (Deloitte, 2020). This stage ensures that the company can successfully adjust to new possibilities and challenges by laying the groundwork for the shift from current practices to future operations that are optimized (McKinsey & Company, 2020).

Key Activities in the Design Phase

a) Analyzing and Documenting Current ("As-Is") Processes

A thorough assessment of the current procedures is necessary. This entails recognizing dependencies, identifying workflows, and locating inefficiencies. Organizations obtain important insights into their existing operational environment by recording these "As-Is" procedures, which act as a foundation for upcoming enhancements.

b) Defining the Desired ("To-Be") Processes

The next stage is to imagine the ideal situation, or what the company hopes to accomplish. This entails creating streamlined, effective procedures that complement customer expectations, technology developments, and business objectives. The "To-Be" procedures should incorporate creative ways to improve efficiency and value delivery while addressing issues with the current system.

c) Identifying Key Features and Process Flow

It is essential to provide a concise and thorough description of the main characteristics and workflow of the new procedures. In addition to integrating tools or technology that will facilitate smooth execution, this entails outlining procedures, roles, and duties. The intended workflows can be made concrete and intelligible for all stakeholders by using visual aids like process maps or flowcharts.

d) Establishing Escalation Points and Notifications

Maintaining seamless operations depends on the early detection and fixing of issues. To guarantee that any problems that arise during the process flow are quickly identified and forwarded to the proper teams for resolution, escalation points must be precisely specified. Automated notification systems can improve responsiveness even more by facilitating speedier decision-making and reducing interruptions.

Importance of the Design Phase

Through a comprehensive analysis of current procedures and the establishment of future workflows, businesses establish a strong foundation for executing change (Deloitte, 2020). By ensuring congruence between operational realities, technology capabilities, and business goals, this phase lowers risks and prepares the way for a smooth transition (McKinsey & Company, 2020). Furthermore, an effective process not only increases productivity but also promotes a proactive, continuous improvement and problem-solving culture (Christensen, 1997).

In the end, design serves as a roadmap for innovation and sustainable growth, giving businesses the foundation they need to handle change and stay competitive in a market that is changing quickly (Parker et al., 2016).

2. Optimize: Enhancing Efficiency and Driving Excellence

An essential part of company transformation is the optimization phase, which focuses on finding inefficiencies and putting plans in place to enhance procedures (McKinsey & Company, 2020). In order to increase productivity, cost-effectiveness, and user pleasure, this stage makes sure that processes are not just simplified but also in line with company goals (Deloitte, 2020). Businesses can develop procedures that are effective and flexible enough to meet changing demands by addressing pain points and embracing continual improvement (Christensen, 1997).

Key Activities in the Optimization Phase

a) Identifying Bottlenecks and Enhancing Performance

Analyzing processes carefully to identify bottlenecks—places where delays, redundancies, or inefficiencies impair overall performance—is the first step in optimization. Outdated procedures, a lack of team coordination, or budget limitations could be the cause of these problems. Once found, specific solutions can be put into place to increase throughput and operational efficiency, such as automating repetitive operations or reallocating resources.

b) Exploring Cost-Reduction Opportunities

Process optimization also entails paying close attention to financial efficiency. This entails finding underutilized resource areas, cutting waste, and combining tasks to get rid of duplication. For instance, companies can save a lot of money while keeping or even improving operational effectiveness by implementing technology-driven solutions like process automation or cloud computing.

c) Incorporating User Feedback to Refine Processes

Without taking into account input from stakeholders, such as consumers or employees, who have direct interaction with the processes, no optimization effort is complete. Finding pain spots, identifying opportunities for development, and making sure the revised procedures satisfy practical requirements are all made possible by their insightful opinions. Businesses may develop user-friendly, effective solutions that meet the expectations of all stakeholders by proactively integrating this feedback.

d) Enhancing Agility and Scalability

Making sure that procedures are not just effective now but can also adjust to changing needs is a crucial component of optimization. Workflows must be made flexible in order for firms to scale up or down activities as needed without sacrificing performance. Using agile approaches or integrating modular solutions, for example, might help businesses react quickly to opportunities for expansion or changes in the market.

The Value of Optimization

Optimization is a continuous commitment to improvement rather than a one-time activity. Businesses can gain a great deal by fixing inefficiencies, from improved customer satisfaction to quicker response times. Additionally, a culture of excellence is fostered by streamlined procedures, which inspire teams to embrace innovation and always look for better methods to operate.

Optimization gives firms the advantage they need to succeed in a competitive environment. Organizations can attain long-term success and establish themselves as leaders in their respective industries by cutting expenses, enhancing performance, and developing reliable and adaptable processes.

Implement: Turning Optimized Processes into Reality

The planning and optimization efforts are realized during the implementation phase. The goal of this step is to ensure that the changes are implemented efficiently and successfully by incorporating the newly streamlined procedures into the organization's regular operations. Achieving success in this stage is essential because it opens the door for observable outcomes by bridging the gap between theoretical advancements and real-world implementation.

Key Activities in the Implementation Phase

a) Building and Preparing Teams for Deployment

Making sure the teams in charge of carrying out the new processes are properly prepared and in line with the transformation objectives is the first stage in the implementation process. This involves holding thorough training sessions to acquaint team members with the most recent systems, tools, and workflows. In order to gain support and reduce opposition to change, it is crucial to communicate clearly about the goals, advantages, and expectations of the new procedures. Giving staff members the information and abilities they require promotes self-assurance and guarantees a more seamless adoption process.

b) Conducting Rigorous Testing of the Updated Processes

It is essential to test the improved procedures in controlled settings prior to full-scale implementation. Often called a "pilot testing" or "beta testing" step, this phase aids in locating any unexpected problems or design flaws. Teams may verify the efficacy of the modifications through testing, which guarantees that the procedures produce the desired results without interfering with ongoing business activities. Additionally, it offers a chance to get user input, which can be utilized to improve the system prior to its final implementation.

c) Simulating Real-World Scenarios

A key component of testing is exposing the new processes to a range of real-world situations. This guarantees that the procedures are strong enough to manage various edge cases and operational difficulties. To assess the system's performance under various circumstances, simulations can incorporate high-volume processes, unforeseen interruptions, or different user involvement levels. Organizations can find any weaknesses and take proactive measures to fix them by stress-testing the procedures.

d) Establishing Feedback Loops and Support Mechanisms

It is imperative to establish systems for continuous feedback from stakeholders and end users during the implementation phase. This reduces downtime and raises user satisfaction by guaranteeing that any early problems are found and fixed quickly. The transition can also be made easier by establishing support teams or help desks to give staff members and clients the help they require while they adjust to the new procedures.

Importance of the Implementation Phase

The key to a successful change is effective implementation. If not implemented properly, even the best-designed procedures may not work as intended. Organizations can reduce risks and increase the return on their investment by meticulously preparing teams, extensively testing systems, and making sure that procedures are robust in a variety of scenarios.

A successful implementation also increases organizational confidence, which promotes an innovative and flexible culture. Teams are more inclined to support upcoming transformation projects when they observe the benefits of the changes in action, starting a circle of ongoing development and expansion.

4. Implement: Turning Optimized Processes into Reality

The planning and optimization efforts are realized during the implementation phase. The goal of this step is to ensure that the changes are implemented efficiently and successfully by incorporating the newly streamlined procedures into the organization's regular operations. Achieving this stage is essential because it opens the door for observable outcomes by bridging the gap between theoretical advancements and real-world implementation.

Key Activities in the Implementation Phase

a) Building and Preparing Teams for Deployment

Making sure the teams in charge of carrying out the new processes are properly prepared and in line with the transformation objectives is the first stage in the implementation process. This involves holding thorough training sessions to acquaint team members with the most recent systems, tools, and workflows. In order to gain support and reduce opposition to change, it is crucial to communicate clearly about the goals, advantages, and expectations of the new procedures. Giving staff members the information and abilities they require promotes self-assurance and guarantees a more seamless adoption process.

b) Conducting Rigorous Testing of the Updated Processes

It is essential to test the improved procedures in controlled settings prior to full-scale implementation. Often called a "pilot testing" or "beta testing" step, this phase aids in locating any unexpected problems or design flaws. Teams may verify the efficacy of the modifications through testing, which guarantees that the procedures produce the desired results without interfering with ongoing business activities. Additionally, it offers a chance to get user input, which can be utilized to improve the system prior to its final implementation.

c) Simulating Real-World Scenarios

A key component of testing is exposing the new processes to a range of real-world situations. This guarantees that the procedures are strong enough to manage various edge cases and operational difficulties. To assess the system's performance under various circumstances, simulations can incorporate high-volume processes, unforeseen interruptions, or different user involvement levels. Organizations can find any weaknesses and take proactive measures to fix them by stress-testing the procedures.

d) Establishing Feedback Loops and Support Mechanisms

It is imperative to establish systems for continuous feedback from stakeholders and end users during the implementation phase. This reduces downtime and raises user satisfaction by guaranteeing that any early problems are found and fixed quickly. The transition can also be made easier by establishing support teams or help desks to give staff members and clients the help they require while they adjust to the new procedures.

Importance of the Implementation Phase

The key to a successful change is effective implementation. If not implemented properly, even the best-designed procedures may not work as intended. Organizations can reduce risks and increase the return on their investment by meticulously preparing teams, extensively testing systems, and making sure that procedures are robust in a variety of scenarios.

A successful implementation also increases organizational confidence, which promotes an innovative and flexible culture. Teams are more inclined to support upcoming transformation projects when they observe the benefits of the changes in action, starting a circle of ongoing development and expansion.

In conclusion, strategy and execution collide at the implementation phase. Businesses may convert their improved processes into actionable results and secure long-term success and competitive advantage by concentrating on careful planning, thorough testing, and all-encompassing support.

MONITOR: ENSURING PROCESSES PERFORM AS INTENDED

The monitoring phase, which focuses on confirming the efficacy of implemented procedures and identifying chances for further refinement, is an essential part of the business transformation path. Organizations can make sure their procedures meet expectations, stay effective, and change to meet changing demands by regularly evaluating performance. This phase ensures long-term sustainability and relevance by measuring success and establishing the framework for continuous improvement.

Key Activities in the Monitoring Phase

a) Tracking and Observing Individual Procedures

The first step in monitoring is a thorough assessment of each procedure's performance under actual circumstances. This entails monitoring workflows methodically, evaluating team performance, and spotting any departures from the planned process design. Real-time analytics and performance dashboards are examples of sophisticated technologies that may be used to collect data and produce insights that can be put to use. Frequent monitoring guarantees that the procedures are operating as intended and contributes to transparency.

b) Comparing Performance Against Predefined Benchmarks

Comparing actual performance to predefined criteria or objectives is an essential part of monitoring. Metrics including punctuality, quality, efficiency, and customer satisfaction may be included in these standards. Organizations can assess if the new processes are producing the intended results by looking at these key performance indicators (KPIs). Any discrepancies between goals and actual performance can point to areas that need improvement.

c) Identifying Bottlenecks and Areas for Improvement

Inefficiencies or difficulties that were not noticeable during the planning or execution stages are frequently discovered during monitoring. These could include unanticipated setbacks, resource shortages, or problems brought on by unanticipated outside influences. Organizations can ensure smoother operations and improved overall performance by proactively addressing bottlenecks of this kind.

d) Soliciting Feedback from Stakeholders

An essential component of the monitoring phase is getting feedback from those who are directly involved with or impacted by the processes, such as partners, customers, or workers. Stakeholder feedback offers a comprehensive picture of how the processes are operating by supplementing quantitative data with qualitative observations. This input might help identify subtleties that could otherwise go overlooked and direct future improvements to increase efficacy.

e) Leveraging Technology for Real-Time Monitoring

The way that monitoring is done has been completely transformed by cutting-edge technology like artificial intelligence, process automation tools, and Internet of Things (IoT) sensors. These solutions give businesses the ability to monitor procedures in real time, allowing for prompt remedial action when deviations occur and instant alerts. This guarantees that procedures stay on course, decreases downtime, and increases accuracy.

The Value of Monitoring in Business Transformation

To make sure that the organization's efforts result in observable results, the monitoring phase is crucial. It offers a feedback loop that enables companies to evaluate the success of their changes and make data-driven choices for ongoing development. Additionally, monitoring promotes responsibility by keeping teams focused on the aims and objectives of the company.

Additionally, organizations can reduce risks and stop minor issues from growing into more significant ones by seeing possible problems early. The knowledge acquired during this stage not only improves ongoing procedures but also provides guidance for upcoming transformation projects, fostering an innovative and flexible culture.

In conclusion, the monitoring phase protects the company by making sure that its procedures continue to be successful, efficient, and in line with its strategic goals. Businesses may remain ahead of the curve and prosper in a constantly changing environment by being vigilant in their observation, benchmarking, and improvement.

Conclusion: An essential component of organizational agility and growth, this cycle is iterative, meaning that it repeats itself to continuously improve and adjust to shifting business needs.

In the modern business environment, the ability to **transform correctly** is critical to a company's long-term success and survival. The rapid pace of technological advancements, globalization, and shifting consumer expectations have created a highly competitive and disruptive landscape. To remain relevant, businesses must not only adapt but also ensure that their transformation efforts are aligned with strategic goals, sustainable growth, and customer needs. Proper transformation, when done right, offers significant advantages, including increased agility, improved customer experience, and sustained competitive advantage.

Adapting to Technological Change

One of the primary drivers of the need for businesses to transform is the rapid advancement of technology. Emerging technologies like artificial intelligence (AI), blockchain, and automation are disrupting industries by introducing new efficiencies and capabilities. Companies that fail to integrate these technologies risk falling behind competitors who leverage them to streamline operations and improve customer engagement. According to a McKinsey & Company report, businesses that embrace digital technologies as part of their transformation see significant increases in profitability and market share (McKinsey & Company, 2020). However, the key is to not just adopt technology for its own sake, but to implement it in ways that align with the company's long-term vision and customer needs.

Figure 3. global automation adoption scenarios

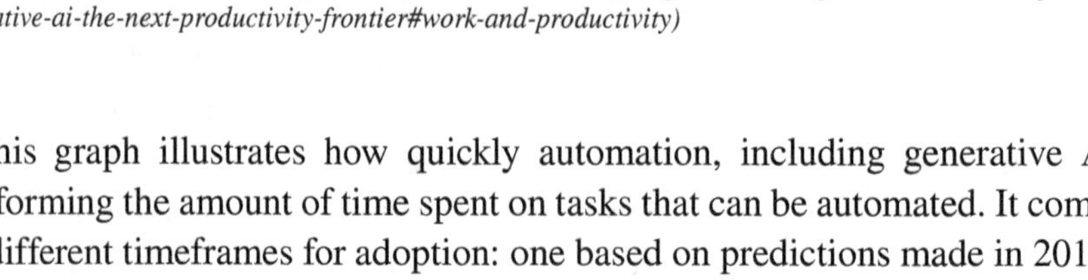

Source (https://www.mckinsey.com/capabilities/mckinsey-digital/our-insights/the-economic-potential-of-generative-ai-the-next-productivity-frontier#work-and-productivity)

This graph illustrates how quickly automation, including generative AI, is transforming the amount of time spent on tasks that can be automated. It compares two different timeframes for adoption: one based on predictions made in 2017 and another updated to include modern advancements like generative AI.

Key Points:

Faster Adoption with Generative AI:

- The updated scenarios show that automation is happening much faster than predicted in 2017.
- For example, reaching 50% automation is now expected about 10 years earlier than before.

Four Scenarios:

- The Updated Early Scenario (yellow line) represents an aggressive approach where automation technologies are adopted quickly.
- The Updated Late Scenario (orange line) assumes slower adoption but still faster than earlier predictions.
- The 2017 Early Scenario (red dashed line) and 2017 Late Scenario (pink dashed line) show older predictions, which now appear slower compared to current advancements.

50% Milestone:

- The dotted horizontal line marks the point where half of the time spent on work activities could be automated.
- In updated scenarios, this milestone is achieved around the 2030s, while 2017 scenarios predicted it closer to the 2050s.

Conclusion:

Generative AI and related technologies are accelerating the adoption of automation, reshaping work patterns much faster than earlier projections. This reflects the growing impact of AI on how businesses and industries operate globally.

2.1.3. Meeting Evolving Consumer Expectations

Consumer expectations are evolving rapidly, with demands for personalized, convenient, and seamless experiences increasing. Digital disruptors like Amazon and Netflix have set new standards for customer experience, pushing traditional businesses to adapt. Customers now expect brands to understand their preferences, provide instant gratification, and offer smooth digital interactions. Failure to meet these expectations can result in loss of customer loyalty. As Accenture notes, 75% of consumers are more likely to purchase from a company that recognizes them by name and offers personalized experiences (Accenture, 2018). Therefore, transforming customer engagement strategies is crucial for businesses to stay competitive.

Figure 4. enterprise resource planning based core

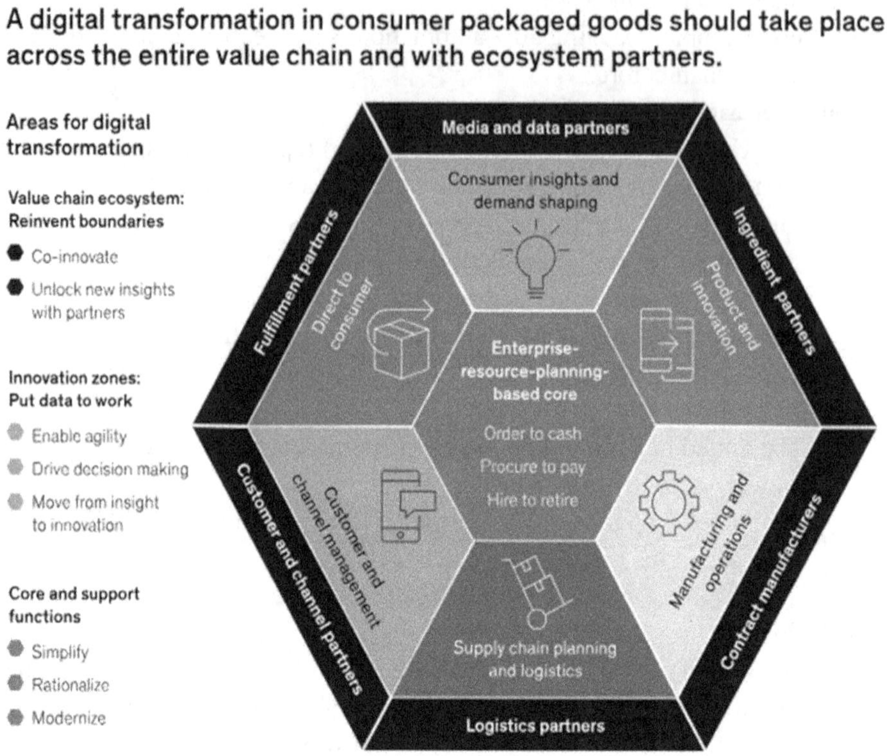

Source (https://www.mckinsey.com/industries/consumer-packaged-goods/our-insights/fortune-or-fiction-the-real-value-of-a-digital-and-ai-transformation-in-cpg)

This image highlights the importance of digital transformation in the consumer packaged goods (CPG) industry. It emphasizes that this change should involve the entire value chain and collaboration with partners. Here's a simplified explanation:

1. **Core Digital Hub**: At the center is the enterprise resource planning (ERP) system, which serves as the backbone for managing key processes like ordering, hiring, and procurement.
2. **Ecosystem Partners**: Surrounding the ERP system are six groups of partners who play a crucial role:
    - **Media and Data Partners**: Help with understanding consumer behavior and shaping demand.
    - **Ingredient Partners**: Provide the raw materials for production.
    - **Contract Manufacturers**: Assist in producing goods.
    - **Logistics Partners**: Manage the transportation and delivery of products.

- **Supply Chain Planning and Logistics**: Ensure smooth operations and coordination across the supply chain.
- **Fulfillment Partners**: Deliver products to customers efficiently.

3. **Focus Areas for Transformation**:
    - **Value Chain Ecosystem**: Redefine boundaries by collaborating and innovating with partners.
    - **Innovation Zones**: Leverage data to make smarter decisions, increase agility, and drive creativity.
    - **Core Functions**: Simplify, streamline, and modernize key operations.

The diagram underscores the need for companies to integrate technology across all areas and work closely with their ecosystem to succeed in today's competitive market.

2.1.4. Sustaining Competitive Advantage

Effective transformation enables businesses to build long-term competitive advantage. Companies that transform successfully can differentiate themselves by creating value in ways that competitors cannot easily replicate. For example, Tesla's transformation of the automotive industry through electric vehicles and autonomous driving technologies has given it a distinct advantage over legacy automakers (Christensen, 1997). By focusing on innovation, sustainable practices, and future-oriented strategies, businesses can carve out new market positions that insulate them from disruption.

The importance of businesses transforming correctly cannot be overstated. In a world of constant change, the ability to adapt thoughtfully and strategically is essential for long-term success. Businesses that integrate new technologies, meet evolving consumer expectations, and develop sustainable competitive advantages are more likely to thrive in the face of disruption. Proper transformation is not just about survival; it's about positioning for future growth and innovation.

Figure 5. This image outlines the digital transformation efforts of the Greater Boston Food Bank (GBFB) to enhance food security in eastern Massachusetts. Here's a simplified explanation:

Source (https://www.mckinsey.com/featured-insights/lifting-europes-ambition/videos-and-podcasts/how-advanced-are-european-companies-in-leveraging-digital-transformation)

Vision:

GBFB aims to lead in food distribution by using digital innovation to end hunger in the Greater Boston area.

Intended Goals:

1. **Using Technology and Data**: They plan to harness technology and treat data as a valuable resource to integrate and improve their operations.
2. **Collaborative Efforts**: By working together with stakeholders, they aim to predict needs and engage support effectively.
3. **Feeding More People**: They aim to expand their ability to provide nutritious food to individuals and families in over 190 cities and towns in eastern Massachusetts.
4. **Sustainability**: They aim to reduce their environmental impact while working towards ending hunger.

Key Ecosystem Stakeholders:

1. **People Facing Hunger**: These are individuals and families who rely on food assistance.
2. **Internal Stakeholders**: This includes government officials, corporate and individual volunteers, national organizations, consultants, and the GBFB team.
3. **External Stakeholders**: These are food vendors, donors (both food and monetary), and suppliers.

Figure 6. This image explains how the Greater Boston Food Bank (GBFB) is investing $5 million over three years to enhance its digital systems and better serve food-insecure communities.

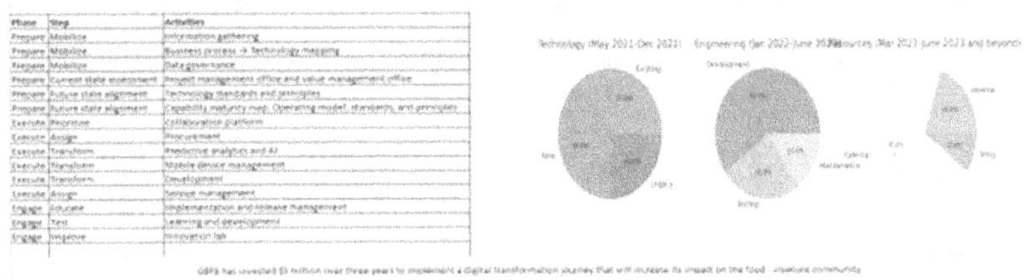

Source (https://www.mckinsey.com/featured-insights/lifting-europes-ambition/videos-and-podcasts/how-advanced-are-european-companies-in-leveraging-digital-transformation)

This image explains how the Greater Boston Food Bank (GBFB) is investing $5 million over three years to enhance its digital systems and better serve food-insecure communities. Here's a simplified explanation:

Phases of the Project:

1. **Prepare**:
   - o **Mobilize**: Gather information and map out current business processes and technologies.
   - o **Assess Current State**: Understand how things currently work, focusing on project management and data governance.
   - o **Align Future Goals**: Develop technology standards, principles, and a roadmap for improvement.
2. **Execute**:
   - o **Prioritize**: Focus on creating a collaboration platform to streamline efforts.
   - o **Assign**: Handle procurement and set up tools like mobile device management.
   - o **Transform**: Use advanced tools like predictive analytics and artificial intelligence to make operations smarter and more efficient.

3. **Engage**:
    - o **Educate**: Train teams and manage the release of new systems.
    - o **Test**: Ensure systems work as intended.
    - o **Improve**: Use an innovation lab to keep refining and introducing new ideas.

Investment Breakdown:

- **Technology (2021)**: Focused on upgrading systems, with 40% for new tools, 35% for existing ones, and 25% for legacy systems.
- **Engineering (2022–2023)**: Concentrated on development (60%), testing (25%), and maintenance (15%).
- **Resources (2022–2023 and beyond)**: Allocated 70% to external resources, 20% to internal staff, and 10% to temporary help.

In summary, GBFB is following a step-by-step approach to adopt modern technologies, improve collaboration, and ensure long-term sustainability in its fight against hunger.

Effectiveness of Digital on Conventional Business Strategies

The integration of digital technologies has profoundly impacted conventional business strategies, offering both opportunities and challenges for companies seeking to adapt in a rapidly evolving landscape. The effectiveness of digital transformation on traditional business models lies in its ability to enhance operational efficiency, improve customer engagement, and open new avenues for growth. However, businesses must strategically implement digital tools to ensure they align with long-term objectives and do not disrupt core operations in a negative way.

Figure 7. This chart shows how the costs and lost revenue for maintaining offshore wind operations in Europe have changed from 2018 to 2023.

Year	Day-to-Day Cost and Lost Revenue	Major Component Cost and Lost Revenue	Total
2018	85	15	100
2019	82	13	95
2020	79	12	91
2021	80	11	91
2022	81	12	93
2023	86	14	100

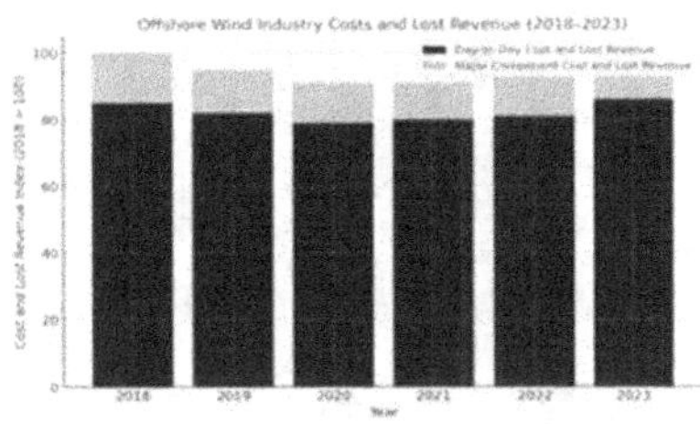

After years of performance improvements, 2023 saw a clear trend break with costs and lost revenue increasing by seven percent

Source (https://www.mckinsey.com/industries/electric-power-and-natural-gas/our-insights/the-power-and-gas-blog/offshore-wind-operations-and-maintenance-a-turning-point)

This chart shows how the costs and lost revenue for maintaining offshore wind operations in Europe have changed from 2018 to 2023. The total cost is divided into two parts:

1. **Day-to-Day Costs and Lost Revenue** (dark blue): Regular operational costs.
2. **Major Component Costs and Lost Revenue** (light blue): Expenses related to big repairs or replacements.

From 2018 to 2021, costs improved slightly, decreasing by about 3% each year. However, in 2022, costs stabilized, and by 2023, they increased significantly by 7%, indicating a break in the improving trend

Enhanced Operational Efficiency

One of the most significant impacts of digital technologies on conventional business strategies is the enhancement of operational efficiency. Automation, artificial intelligence (AI), and cloud computing have enabled businesses to streamline processes, reduce human error, and lower operational costs. For example, many companies are now using AI to automate customer service through chatbots, reducing the need for extensive human involvement while providing faster and more accurate responses. According to Deloitte, businesses that successfully integrate digital tools into their operations can reduce costs by 10-20% and increase productivity by 15-20% (Deloitte, 2020). This increased efficiency allows companies to reinvest resources into innovation and growth initiatives.

Figure 8. This image highlights how generative AI is transforming digital operations by enabling advanced capabilities across various areas

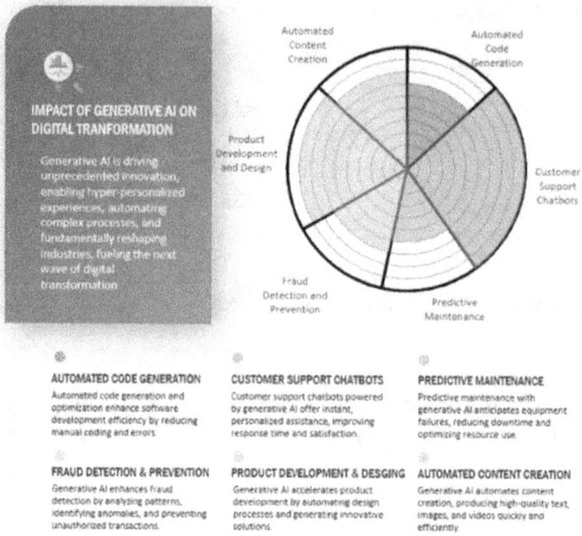

Source (https://www.marketsandmarkets.com/Market-Reports/digital-transformation-market-43010479.html #:~:text=%5B399%20Pages%20Report%5D%20The%20digital,USD%203%2C289.4%20billion%20by%202030.)

This image highlights how generative AI is transforming digital operations by enabling advanced capabilities across various areas. Here's a simplified explanation:

Key Areas of Impact:

1. Automated Code Generation: Transforming Software Development with AI

Thanks to developments in artificial intelligence, automated code creation has completely changed the software development process. Developers may write code more quickly, make fewer mistakes, and be more productive overall thanks to this creative method. AI is enabling developers to concentrate more on intricate problem-solving and innovative solutions by automating tedious chores and expediting the coding process.

Key Benefits of Automated Code Generation

a) Accelerated Development Process

By automating the creation of boilerplate code, repetitive structures, and typical functionalities, AI-powered technologies greatly accelerate the coding process. With the help of these tools, developers may swiftly write substantial amounts of code according to pre-established templates or standards, saving them time and allowing them to fulfill project deadlines without sacrificing quality.

b) Enhanced Code Accuracy and Reduced Errors

The potential of automated code production to reduce human mistake is one of its biggest benefits. By following established guidelines and industry best practices while writing code, AI algorithms guarantee accuracy and consistency. This lowers the possibility of errors, enhances the quality of the code, and lessens the need for intensive debugging and rewrite.

c) Improved Developer Productivity

AI enables developers to devote more time to high-value jobs like creating original solutions, refining algorithms, and solving challenging problems by automating repetitive coding processes. This change creates a more stimulating and fulfilling work environment for developers in addition to increasing individual productivity.

d) Customizable and Adaptive Code Generation

The particular requirements of a project or development team can be accommodated by contemporary AI tools. They can be trained on pre-existing codebases to produce code that complies with the standards, needs, and style of the company. The resulting code will blend in perfectly with the larger development ecosystem thanks to this flexibility.

e) Integration with Development Environments

Popular Integrated Development Environments (IDEs), such Visual Studio Code, IntelliJ IDEA, or Eclipse, frequently incorporate AI-driven code creation tools directly. The development process is made more user-friendly and effective by these integrations, which offer context-aware code snippets, auto-completion, and real-time suggestions.

f) Supporting Learning and Collaboration

Automated code generation is a useful teaching tool for novice developers or those unfamiliar with particular technology. Developers can gain a better understanding of coding conventions, best practices, and effective problem-solving strategies by looking at AI-generated code. Furthermore, by producing standardized, unambiguous code that is simpler for teams to comprehend and expand upon, these technologies foster collaboration.

The Role of AI in Shaping the Future of Development

Automated code generation is set to become ever more crucial to software development as AI advances. More advanced natural language processing capabilities could be added in the future, allowing developers to specify desired functionality in plain language and get completely functioning code in return. Software creation could become more accessible to people with little programming experience because to this advancement.

Additionally, AI's capacity to evaluate and learn from enormous volumes of data guarantees that code generation tools will advance steadily, producing ever-more creative and effective solutions. Automated code generation is now positioned as a fundamental component of contemporary and next software development methodologies because to this continual improvement.

Conclusion:

Automated code generation is a revolutionary force that is changing the way software is made, not just a tool for accelerating development. AI-driven code generation gives developers the confidence and efficiency to take on challenging projects by decreasing errors, increasing productivity, and encouraging creativity. Organizations who adopt this technology open up new avenues for success, scalability, and agility in the ever changing digital ecosystem.

Customer Support Chatbots:

AI-driven customer service Chatbots have completely changed the way companies communicate with their clients by providing immediate, tailored support. Natural language processing is used by these clever technologies to comprehend and effectively reply to inquiries, guaranteeing a flawless customer experience. Because chatbots are available around-the-clock, they reduce wait times and offer reliable service, which increases client satisfaction. By automating repetitive processes, they also assist firms in cutting operating expenses while freeing up human agents to concentrate on more complicated problems. Chatbots continuously improve their responses by learning from user interactions, resulting in a scalable and adaptable support system that increases engagement and cultivates loyalty at a reasonable cost.

Predictive Maintenancc:

Artificial intelligence-powered predictive maintenance has revolutionized how companies maintain their assets and machinery. AI finds trends and abnormalities that may indicate equipment problems by examining sensor data in real time as well as historical performance logs. By taking this proactive stance, businesses may deal with problems before they become more serious, avoiding expensive unscheduled downtime and reducing operational interruptions. Additionally, predictive maintenance prolongs the life of machinery and minimizes needless repairs by optimizing the maintenance plan. By reducing the dangers of unexpected equipment failures, it also improves safety. AI models that have the capacity for continuous learning get more accurate over time, making predictions with more accuracy and efficiency. Businesses may save expenses, increase production, and keep a competitive edge in their respective industries thanks to this innovation.

Automated Content Creation:

Automated content creation powered by artificial intelligence is revolutionizing the way businesses and individuals produce written materials. AI tools utilize advanced natural language processing algorithms to generate high-quality content, including articles, reports, blogs, and marketing materials, with remarkable speed and precision. These systems analyze context, audience preferences, and relevant data to craft content that is engaging, accurate, and tailored to specific needs. By automating repetitive writing tasks, AI significantly reduces the time and effort required, allowing professionals to focus on strategy and creativity. Additionally, AI content generators can scale production effortlessly, ensuring consistency and alignment with brand tone. As these tools continuously improve through machine learning, they empower businesses to stay competitive, enhance efficiency, and meet the growing demand for personalized, impactful communication.

Fraud Detection and Prevention:

The application of artificial intelligence has greatly improved fraud detection and prevention, giving companies a strong tool to spot and stop illegal activity instantly. Large volumes of transaction data are analyzed by AI algorithms, which search for odd patterns and behaviors—like abrupt increases in spending, odd transaction locations, or odd account access—that might point to fraudulent conduct. These AI algorithms get better at spotting minor indications of fraud that conventional techniques would overlook by continuously learning from prior data.

Businesses can stay ahead of changing dangers by using machine learning algorithms, which allow the system to adapt and react to new forms of fraud as they appear. The danger of unlawful financial activity can be greatly decreased by using AI to detect fraud and to immediately initiate notifications and actions, such as denying suspicious transactions or demanding extra authentication. Because consumers feel more secure knowing that their financial and personal information is being actively secured, this not only increases security but also boosts customer trust. AI-driven fraud detection is a crucial tool for any company trying to reduce financial risks while preserving a safe and easy customer experience because it also helps firms save money on expenses related to fraud investigations, legal problems, and reputational harm.

Product Development and Design:

AI has revolutionized product design and development, greatly accelerating the process of producing new goods while improving their usefulness and aesthetics. Utilizing enormous volumes of data, artificial intelligence (AI) systems are able to examine consumer preferences, market trends, and rival products, offering insightful information that guides the creation of new products. Businesses can use these insights to determine which market gaps need to be filled, what characteristics customers want most, and how to develop goods that appeal to the target market. By mimicking real-world situations, forecasting how the product will function, and even suggesting changes to enhance durability, usefulness, or aesthetics, AI may offer creative ideas and optimize design components during the design phase.

Additionally, AI can help with prototyping by quickly producing several design iterations, which expedites the process of testing and improvement. In addition to reducing time-to-market, this enables companies to test more ideas and improve goods before they go into production. AI can also boost creativity by generating new concepts based on data-driven trends, encouraging designers to push the envelope. In the end, AI-driven product creation guarantees a more inventive, economical, and efficient process, providing companies a competitive advantage in a market that moves quickly.

Overall Impact

Generative AI is radically changing businesses by automating complex processes that formerly needed a significant amount of manual labor. It improves productivity, lowers expenses, and streamlines operations by analyzing vast amounts of data. Additionally, it allows for highly customized user experiences by tailoring goods, services, and suggestions to each user's tastes and habits. In addition to raising customer satisfaction, this degree of personalization fosters more engagement and loyalty. Furthermore, by producing concepts, prototypes, and solutions quickly, generative AI boosts innovation and enables companies to quickly and nimbly adjust to changing market demands. In the end, generative AI is driving digital transformation and assisting businesses in remaining up to date and competitive.

Figure 9. This image illustrates how the telecommunications industry could benefit from adopting responsible AI practices by 2040, potentially generating up to $250 billion in value. It breaks down the estimated value based on different maturity levels of AI implementation

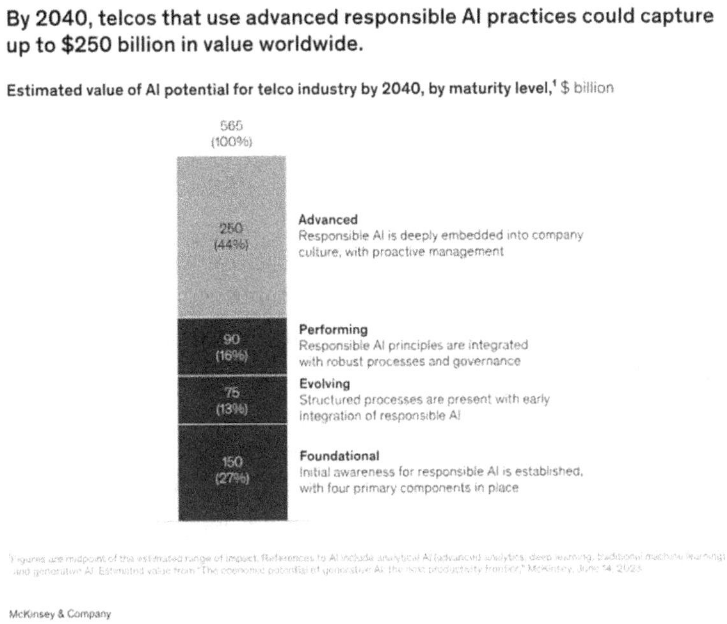

Source(https://www.mckinsey.com/industries/technology-media-and-telecommunications/our-insights/responsible-ai-a-business-imperative-for-telcos)

This image illustrates how the telecommunications industry could benefit from adopting responsible AI practices by 2040, potentially generating up to $250 billion in value. It breaks down the estimated value based on different maturity levels of AI implementation.

Key Maturity Levels and Their Contributions:

1. **Foundational (27% of value, $150 billion)**:
   - Companies are just starting to explore responsible AI. Basic principles and a few key components are in place.
2. **Evolving (13% of value, $75 billion)**:
   - Companies have some structured processes for integrating responsible AI but are still in the early stages.
3. **Performing (16% of value, $90 billion)**:
   - AI principles are well-integrated, with strong governance and processes ensuring ethical and effective AI usage.

4. **Advanced (44% of value, $250 billion)**:
    - Responsible AI is fully embedded into the organization's culture, with proactive management and widespread adoption.

Overall Insight:

By gradually advancing their responsible AI practices, telecom companies can unlock significant value. The more mature their approach to AI, the greater the financial and operational benefits they can achieve.

Improved Customer Engagement

Digital transformation has also transformed customer engagement strategies, allowing businesses to provide personalized and seamless experiences. By leveraging data analytics, businesses can gain deep insights into customer behavior, preferences, and purchasing patterns, enabling them to deliver highly targeted marketing and customized offerings. Companies like Amazon and Netflix have used data-driven insights to perfect personalized recommendations, significantly improving customer retention and loyalty (Kelly, 2016). In the digital age, customer-centric approaches that leverage technology are essential for building long-term relationships with consumers.

Figure 10. This image explains how a business model focused on the customer can lead to significant improvements in performance by changing how the organization operates. Here's a simplified breakdown:

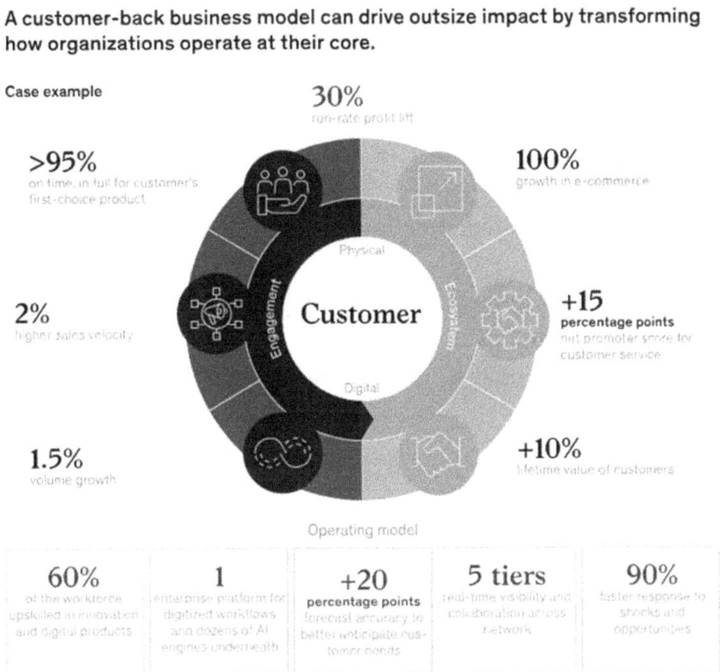

Source (https://www.mckinsey.com/industries/industrials-and-electronics/our-insights/true-customer-centricity-an-operating-model-for-competitive-advantage)

Core Concept:

At the center of the model is the customer, emphasizing that all efforts revolve around understanding and meeting customer needs. This focus drives improvements in three main areas: engagement, ecosystem, and value chain.

Key Benefits:

1. **Customer Outcomes**:
   - **>95% on-time delivery**: Customers get their preferred products on time.
   - **+15% in customer satisfaction**: Higher Net Promoter Scores show better service experiences.
   - **+10% customer lifetime value**: Customers remain loyal and spend more over time.
2. **Business Growth**:

- **30% increase in profit margin**: Focusing on customers leads to higher profitability.
- **100% growth in e-commerce**: Enhanced digital strategies double online sales.
- **2% higher sales velocity**: Products sell faster in the market.
- **1.5% increase in volume growth**: More products are sold overall.

Operating Model Enhancements:

- **60% workforce upskilled**: Employees are trained in digital and innovative skills.
- **1 unified platform**: A single platform integrates workflows and AI to streamline processes.
- **+20% forecast accuracy**: Better tools predict customer needs more effectively.
- **5 tiers of collaboration**: Seamless teamwork ensures efficiency across the supply chain.
- **90% faster response**: Organizations quickly adapt to market changes.

Big Picture:
This customer-first approach transforms how businesses operate, creating a cycle of better service, improved efficiency, and sustained growth.

New Revenue Streams and Market Opportunities

Digital technologies have also opened up new revenue streams and expanded market opportunities for businesses. E-commerce, digital platforms, and subscription-based models have allowed companies to reach a global audience with minimal physical infrastructure. For example, businesses like Spotify have disrupted the music industry by transitioning from traditional album sales to a subscription model, reaching millions of users worldwide. According to Statista, global e-commerce sales are projected to reach $6.38 trillion by 2024, highlighting the immense potential for businesses that embrace digital strategies (Statista, 2021).

The effectiveness of digital technologies on conventional business strategies is clear: they drive operational efficiency, enhance customer engagement, and open new growth opportunities. However, businesses must carefully align digital tools with their broader objectives to fully capitalize on these advantages. Those that do will not only survive in the digital era but thrive, setting new standards for innovation and competitiveness.

Figure 11. This image highlights the growth of the digital transformation market worldwide, predicting its future size and regional contributions by 2030.

Year	North America	Europe	Asia-Pacific	Middle East & Africa	Latin America	Total Market Size (USD Billion)
2023	200	150	250	75	52.7	727.7
2024	250	200	300	100	61.2	911.2
2030	1000	800	1000	300	189.4	3289.4

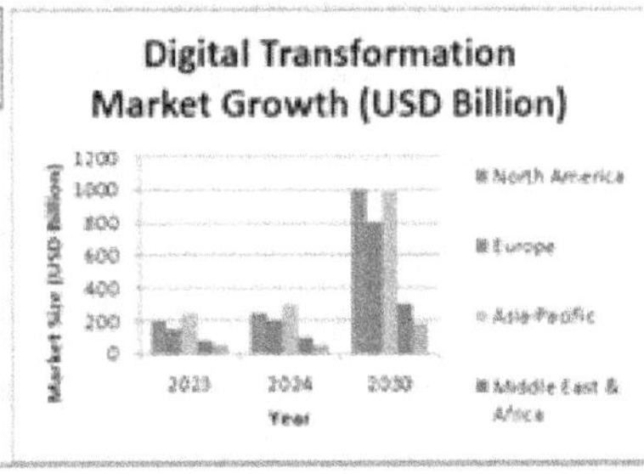

CAGR 23.9% the digital transformation market is expected to be worth USD 3289.4 billion by 2030, growing at a CAGR of 23.9% during the forecast period

Source (https://www.marketsandmarkets.com/Market-Reports/digital-transformation-market-43010479.html #:~:text=%5B399%20Pages%20Report%5D%20The%20digital,USD%203%2C289.4%20billion%20by%202030.)

Key Insights:

1. **Market Growth**:
   - The digital transformation market is projected to reach $3,289.4 billion by 2030.
   - It is growing at a compound annual growth rate (CAGR) of 23.9% during the forecast period.
2. **Regional Contributions**:
   - The market is divided into regions: North America, Europe, Asia-Pacific, Middle East & Africa, and Latin America.
   - Each region contributes to the overall growth, with North America and Asia-Pacific expected to be significant players.
3. **Yearly Progress**:
   - In 2023, the market size is $727.7 billion.
   - By 2024, it grows to $911.2 billion.
   - By 2030, the market expands significantly to $3,289.4 billion.
4. **Visualization**:
   - The bar graph shows steady growth in the market size over time.
   - The color-coded bars represent the contributions of each region to the market size.

Summary:

The digital transformation market is expanding rapidly, driven by innovations and technology adoption across regions. By 2030, it is expected to become a multi-trillion-dollar industry, with North America and Asia-Pacific leading the way.

Early Adopters vs. Laggards: A Competitive Comparison

In the world of business, the ability to embrace new technologies and innovations often separates market leaders from those that struggle to keep pace. Early adopters—businesses that quickly integrate new technologies into their strategies—are often rewarded with competitive advantages, while laggards—those slower to adopt—face the risk of falling behind. The difference between these two groups is significant, with early adopters benefiting from increased market share, enhanced customer engagement, and operational efficiency, while laggards are often left grappling with outdated models and shrinking market relevance.

Figure 12. This image highlights how organizations that are early adopters of generative AI are better at managing challenges related to talent and training compared to those experimenting with AI. Here's a simplified explanation:

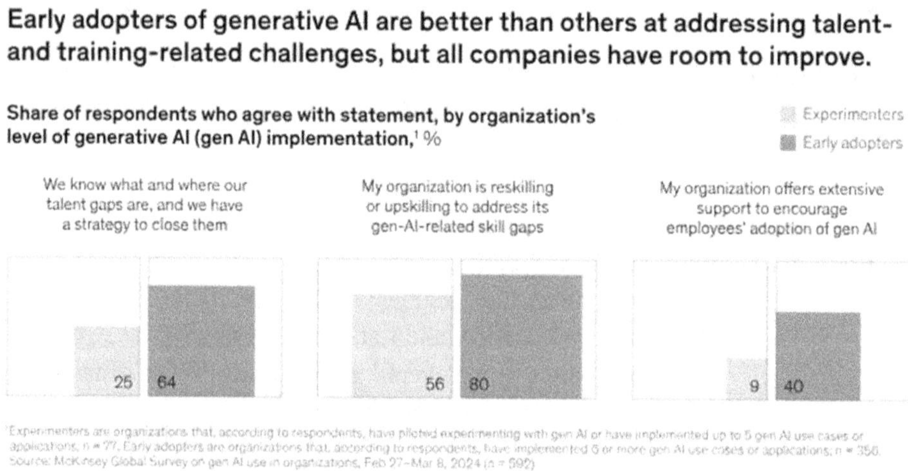

Source (https://www.mckinsey.com/capabilities/people-and-organizational-performance/our-insights/gen-ais-next-inflection-point-from-employee-experimentation-to-organizational-transformation)

1. Understanding Talent Gaps:
    - Early adopters (organizations using AI in multiple areas) are much better at identifying where their skill gaps are and have strategies to address them (64% agree).
    - Experimenters (organizations just starting to explore AI) are less confident, with only 25% agreeing.
2. Reskilling and Upskilling:

- Early adopters are actively training their employees to gain the necessary AI-related skills (80% agree).
- Experimenters are less focused on reskilling, with only 56% agreeing.

3. Supporting Employees in Using AI:
   - Early adopters provide more support to their employees for adopting AI tools (40% agree).
   - Experimenters offer much less support, with only 9% agreeing.

In short, organizations that are further along in their AI journey are more proactive in addressing workforce challenges and ensuring employees are equipped to use AI effectively.

Market Advantage

One of the primary benefits for early adopters is the competitive advantage they gain by entering new markets or improving existing offerings before competitors. By embracing emerging technologies, early adopters can differentiate themselves, create better customer experiences, and capture market share early. For example, companies like Amazon and Netflix quickly adopted cloud computing and data-driven personalization, allowing them to dominate their respective industries while competitors struggled to catch up (Kelly, 2016). Early adopters often become industry leaders because they have the foresight to leverage new tools before they become mainstream.

Figure 13. This chart explains the trends in U.S. social commerce sales and the growth in social buyers from 2022 to 2028. Here's a simplified explanation:

Year	Social Commerce	Social Buyers (%)
2022	18.8	7.9
2023	21.1	5
2024	26	5.3
2025	19.5	3.7
2026	18	3.9
2027	16.7	3.9
2028	16.3	3.7

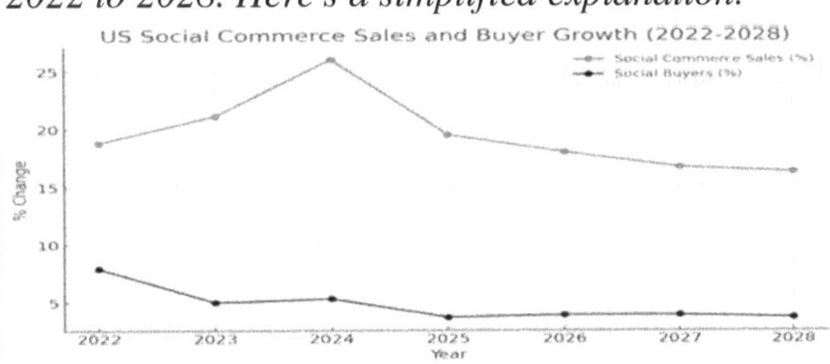

Source (https://www.emarketer.com/content/tiktok-driving-us-social-commerce-growth)

1. **Social Commerce Sales (Red Line)**:

- Sales through social media platforms are growing significantly, with the highest growth predicted in 2024 (26% increase), largely due to TikTok Shop's influence.
- After 2024, the growth rate slows down, reaching 16.3% by 2028.

2. **Social Buyers (Black Line)**:
   - The number of people buying through social media platforms is increasing at a much slower pace compared to sales.
   - Growth starts at 7.9% in 2022, drops to 5% in 2023, and stabilizes around 3.7% from 2026 to 2028.

In summary, while the number of social buyers grows steadily but slowly, social commerce sales are seeing a sharp rise, especially in 2024, due to the growing popularity of platforms like TikTok Shop.

In contrast, laggards miss out on these opportunities, often adopting innovations only when they become necessary to stay afloat. By the time laggards integrate new technologies, they face a steep uphill battle to regain market share, as early adopters have already built strong customer bases and optimized their operations.

Operational Efficiency

Early adopters also enjoy significant improvements in operational efficiency, often driven by digital technologies such as automation, artificial intelligence (AI), and data analytics. These technologies allow businesses to streamline their processes, reduce costs, and improve decision-making. Early adopters who integrate AI-driven automation in customer service, supply chain management, or marketing, for example, can scale their operations more effectively. According to McKinsey, early adoption of digital tools can increase operational productivity by up to 20% (McKinsey, 2020).

Laggards, on the other hand, continue to rely on outdated systems and processes, which can lead to inefficiencies, higher operational costs, and slower response times to market changes. This gap in efficiency often forces laggards to engage in reactive, rather than proactive, business strategies.

Figure 14. This chart highlights the key digital priorities for organizations in the maintenance, repair, and overhaul (MRO) sector over the next 3 to 5 years, based on a survey of 45 respondents. Here's a simplified explanation:

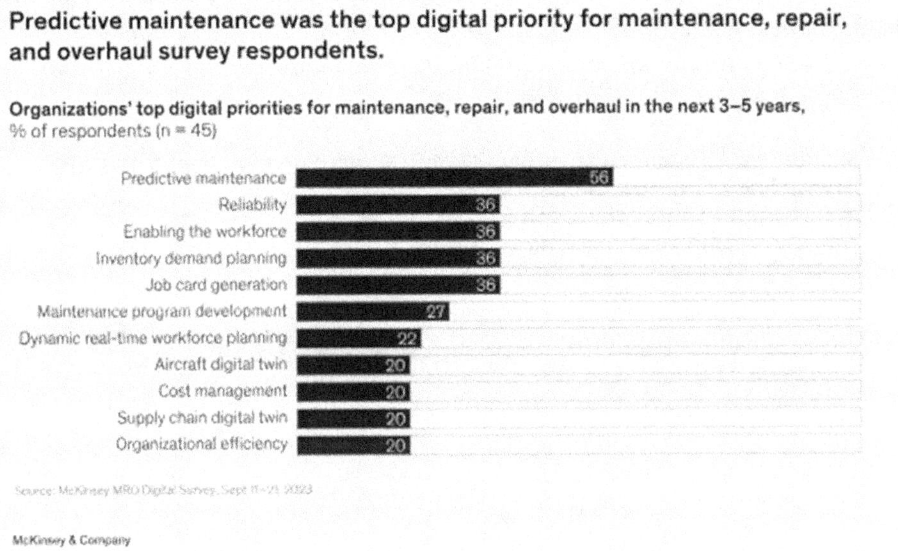

Source (https://www.mckinsey.com/industries/travel-logistics-and-infrastructure/our-insights/aircraft-mro-2-point-0-the-digital-revolution)

1. **Predictive Maintenance (56%)**: The top priority is using technology to predict when equipment might fail, allowing for timely repairs and reducing downtime.
2. **Reliability, Workforce Enablement, Inventory Planning, and Job Card Generation (36%)**: These four areas are equally important. They focus on improving system reliability, empowering employees with tools and skills, optimizing inventory planning, and creating job cards to streamline maintenance tasks.
3. **Maintenance Program Development (27%)**: Developing structured programs to manage maintenance activities effectively is another focus area.
4. **Dynamic Real-Time Workforce Planning (22%)**: This involves using real-time data to plan and allocate workforce resources efficiently.
5. **Aircraft Digital Twin, Cost Management, Supply Chain Digital Twin, and Organizational Efficiency (20%)**: These priorities include using digital replicas of aircraft and supply chains for better management, controlling costs, and enhancing overall efficiency.

In summary, organizations aim to leverage digital tools to predict issues, improve processes, and enhance overall operational efficiency.

Figure 15. This chart highlights the benefits experienced by organizations in the maintenance, repair, and overhaul (MRO) sector after implementing digital solutions. It compares organizations with high digital maturity (blue dots) to those with low digital maturity (black dots).

Source (https://www.mckinsey.com/industries/travel-logistics-and-infrastructure/our-insights/aircraft-mro-2-point-0-the-digital-revolution)

Here's a simplified explanation of Figure 15.

1. **Reduced Costs**: Organizations with advanced digital systems achieve significantly lower costs compared to those with limited digital adoption.
2. **Increased Aircraft Availability**: High digital maturity leads to better availability of aircraft by minimizing downtime and improving operational efficiency.
3. **Increased Engineering Productivity**: Advanced digital tools help engineers work more efficiently, boosting productivity.
4. **Reduction in Aircraft on Ground**: Organizations with high digital maturity reduce the time aircraft spend out of service, ensuring smoother operations.

5. **Improved Maintenance Programs**: Digitally mature organizations have better-designed and more effective maintenance programs.
6. **Improved Employee Experience**: Advanced digital solutions also enhance the overall experience for employees by streamlining processes and reducing workload.

In summary, organizations with higher digital maturity consistently outperform those with low maturity across multiple areas, from cost savings to productivity and employee satisfaction.

Customer Engagement

Early adopters can also build stronger customer engagement by integrating technology to personalize and enhance the customer experience. Businesses that quickly adopt customer data analytics, for instance, can deliver highly personalized products and services, fostering customer loyalty and retention. Laggards, however, struggle to meet evolving customer expectations, resulting in weaker engagement and lower customer satisfaction (Accenture, 2018).

Early adopters consistently outperform laggards by leveraging new technologies to gain market share, improve operational efficiency, and enhance customer engagement. While early adoption can involve risks, those businesses that fail to act quickly often find themselves struggling to catch up, losing their competitive edge in an increasingly fast-paced digital economy.

Figure 16. This image outlines four key steps for a government organization to become more customer-focused and improve its digital productivity:

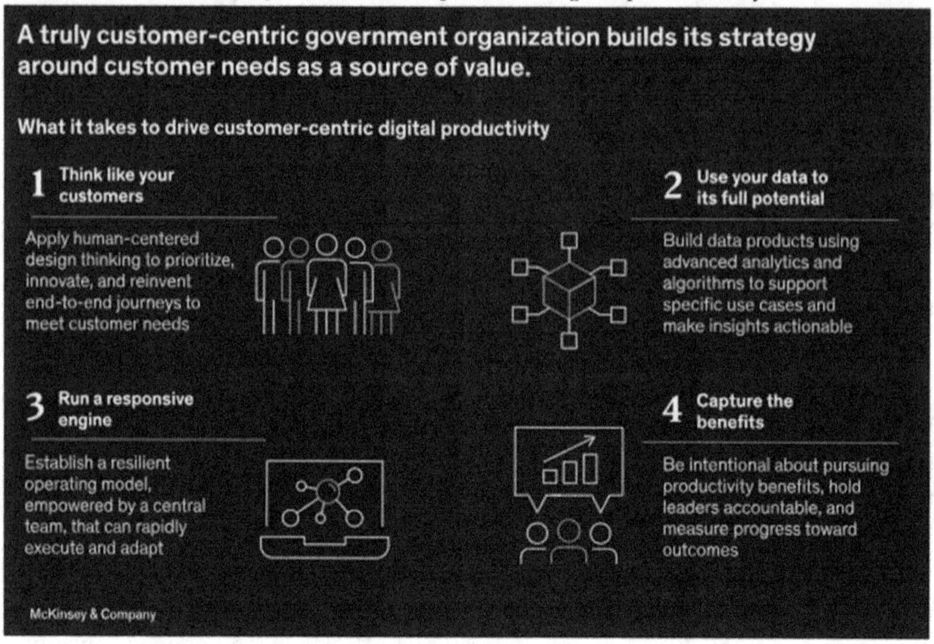

Source (https://www.mckinsey.com/industries/public-sector/our-insights/customer-centric-transformation-in-government)

1. **Understand Customer Needs**:
    - Think like your customers by using human-centered design.
    - Focus on improving and innovating customer experiences to meet their expectations.
2. **Maximize the Use of Data**:
    - Use advanced analytics and algorithms to create data-driven tools.
    - Turn insights into actions that help achieve specific goals.
3. **Build a Flexible System**:
    - Create a strong and adaptable operational framework.
    - Empower teams to respond quickly and effectively to changes.
4. **Achieve Results**:
    - Focus on realizing productivity benefits by setting clear goals.
    - Hold leaders accountable and track progress to ensure success.

In simple terms, this approach emphasizes understanding customer needs, using data wisely, staying adaptable, and focusing on measurable outcomes to improve efficiency and customer satisfaction.

The Cost of Doing Nothing in an Age of Turbulence

In today's fast-paced business environment, characterized by rapid technological advancements and shifting market dynamics, the cost of inaction can be detrimental to a company's survival. The digital age has ushered in an era of turbulence where businesses are constantly pressured to adapt or risk becoming obsolete. Companies that choose to maintain the status quo, rather than proactively transforming in response to emerging trends, face several costly consequences, including loss of market share, diminished competitiveness, and eventual irrelevance.

Organizations most often see meaningful cost reductions from generative AI use in HR and revenue increases in supply chain management.

Cost decrease and revenue increase from generative AI adoption in 2023, by function,1% of respondents

Figure 17. This chart shows how different business functions are expected to change in terms of workload or resources. Each function is divided into groups based on whether it will experience a decrease or increase in workload or resources.

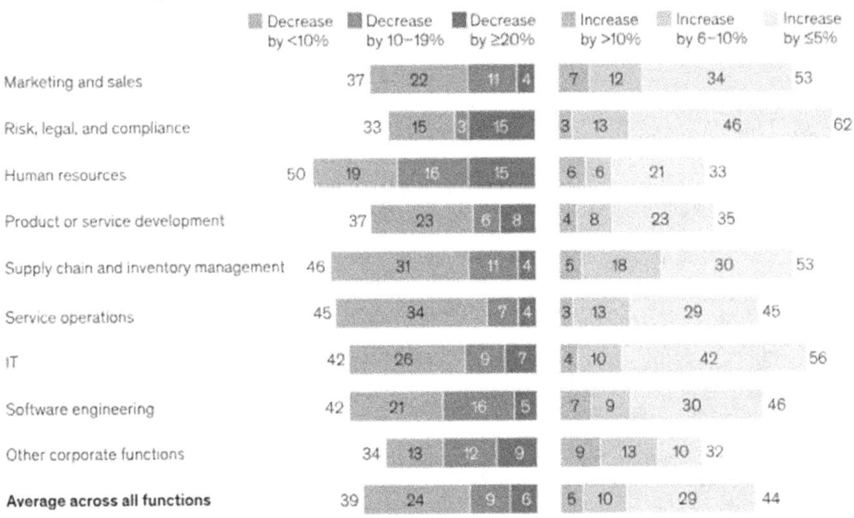

Source (https://www.mckinsey.com/capabilities/quantumblack/our-insights/the-state-of-ai)

Here's a summary of Figure 17.

1. **Decrease in Workload or Resources**:
    - Some functions like Human Resources and Supply Chain Management are expected to have significant reductions, with a notable percentage seeing decreases of over 10% or more.
    - For example, 50% of respondents predict a decrease of less than 10% in Human Resources.
2. **Increase in Workload or Resources**:
    - Functions like **IT** and **Risk, Legal, and Compliance** are expected to grow significantly, with a large portion predicting increases of up to 10% or more.
    - For instance, 62% foresee an increase in Risk, Legal, and Compliance by up to 5%.
3. **Overall Trends**:
    - The average across all functions shows that 39% expect a slight decrease (less than 10%), while 44% expect slight increases (up to 5%).
    - IT and Risk, Legal, and Compliance show the highest anticipated growth, while Human Resources and Supply Chain are likely to experience more reductions.

Use of analytical AI most often yields cost reductions in service operations and revenue increases in marketing and sales.

Cost decrease and revenue increase from analytical AI adoption in 2023, by function,1% of respondents. Here's a summary of Figure 18.

Figure 18. This chart shows how different business functions are expected to change in terms of workload or resources. Each function is divided into groups based on whether it will experience a decrease or increase in workload or resources

Function	Decrease by <10%	Decrease by 10–19%	Decrease by ≥20%	Increase by >10%	Increase by 6–10%	Increase by ≤5%
Marketing and sales	37	22	11	7	12	34
Risk, legal, and compliance	33	15	13	3	13	46
Human resources	50	19	16	6	6	21
Product or service development	37	23	6	4	8	23
Supply chain and inventory management	46	31	11	5	18	30
Service operations	45	34	7	3	13	29
IT	42	26	9	4	10	42
Software engineering	42	21	16	7	9	30
Other corporate functions	34	13	12	9	13	32
Average across all functions	39	24	9	5	10	29

Source (https://www.mckinsey.com/capabilities/quantumblack/our-insights/the-state-of-ai)

1. **Decrease in Workload or Resources**:

- Some functions like Human Resources and Supply Chain Management are expected to have significant reductions, with a notable percentage seeing decreases of over 10% or more.
- For example, 50% of respondents predict a decrease of less than 10% in Human Resources.

2. **Increase in Workload or Resources**:
   - Functions like IT and Risk, Legal, and Compliance are expected to grow significantly, with a large portion predicting increases of up to 10% or more.
   - For instance, 62% foresee an increase in Risk, Legal, and Compliance by up to 5%.
3. **Overall Trends**:
   - The average across all functions shows that 39% expect a slight decrease (less than 10%), while 44% expect slight increases (up to 5%).
   - IT and Risk, Legal, and Compliance show the highest anticipated growth, while Human Resources and Supply Chain are likely to experience more reductions.

Loss of Market Share

One of the most immediate risks of doing nothing in an era of turbulence is the potential loss of market share to more agile and innovative competitors. Disruptive companies that embrace new technologies and business models can quickly capture customers by offering superior products or services. For example, Blockbuster, once a leader in video rentals, failed to adapt to the digital streaming revolution. In contrast, Netflix embraced digital innovation, effectively capturing the market that Blockbuster had dominated (Christensen, 1997). Blockbuster's inaction ultimately led to its downfall, while Netflix became the global leader in streaming services. Inaction allows competitors to seize opportunities, leaving lagging companies struggling to regain lost ground.

Figure 19. This graph shows how the revenue of Netflix and Blockbuster changed between 2004 and 2010.

Year	Netflix Revenue (Billions)	Blockbuster Revenue (Billions)
2004	0.5	6
2005	0.8	5.5
2006	1.2	5
2007	1.4	4.5
2008	1.9	3.5
2009	2.2	2
2010	2.5	0

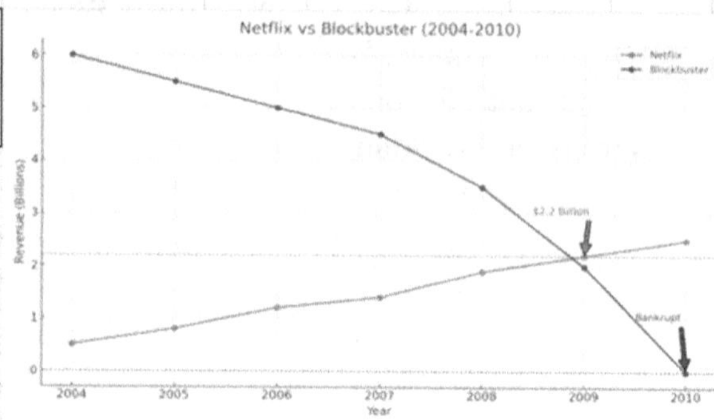

Source (https://stansberryreports.s3.amazonaws.com/transfer-of-wealth/index.html)

- **Netflix**: Its revenue steadily increased from $0.5 billion in 2004 to $2.5 billion in 2010.
- **Blockbuster**: Its revenue dropped sharply from $6 billion in 2004 to $0 in 2010, leading to bankruptcy.
- **Key Point**: In 2009, Netflix's revenue reached $2.2 billion, surpassing Blockbuster's revenue for the first time. This marked a turning point, highlighting Netflix's growth and Blockbuster's decline.

Diminished Competitiveness

Companies that fail to evolve risk falling behind in terms of competitiveness. In industries where technology is rapidly advancing, businesses must continuously innovate to stay relevant. McKinsey reports that companies which do not invest in digital transformation are 30% more likely to fall behind competitors who integrate new technologies (McKinsey & Company, 2020). Laggards who ignore the importance of automation, artificial intelligence (AI), or data analytics find themselves at a significant disadvantage as more tech-savvy competitors increase operational efficiency and offer better customer experiences.

Figure 20. This image compares how top-performing companies and others view the role of digital technology in their future strategies over the next two years. It also highlights their investment priorities in building new digital businesses.

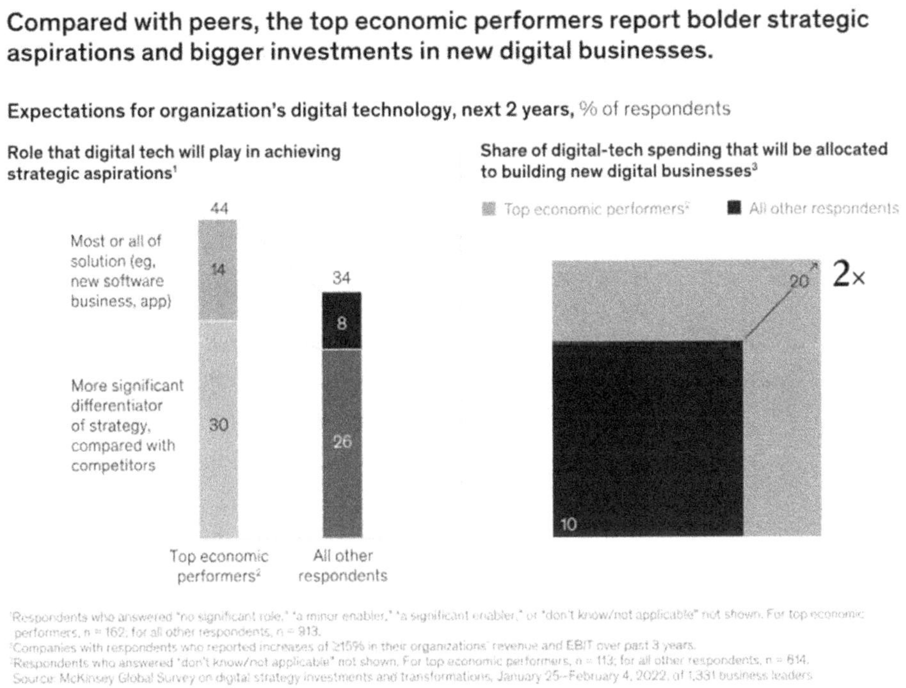

Source (https://www.mckinsey.com/capabilities/mckinsey-digital/our-insights/three-new-mandates-for-capturing-a-digital-transformations-full-value)

Key Points:

1. **Role of Digital Technology in Strategy:**
   - **Top Performers:** 44% see digital technology as central to their solutions, such as creating new apps or software businesses. Additionally, 30% consider it a major differentiator in standing out from competitors.
   - **Other Companies:** Only 34% view digital technology as a critical solution, and 26% see it as a significant strategic differentiator.
2. **Investment in New Digital Businesses:**
   - **Top Performers:** These companies allocate 20% of their digital technology budgets to developing new digital ventures.
   - **Other Companies:** They allocate only 10%, which is half of what top performers invest.

Simple Takeaway:

Leading companies prioritize digital technology more aggressively, using it to drive innovation and differentiation. They also invest twice as much in building new digital businesses compared to others, showing a stronger focus on future growth.

Figure 21. This image highlights how top-performing companies are investing more heavily in key digital capabilities compared to others. It also shows their approach to achieving strategic goals using digital technology in the next two years.

Digital-Technology Capabilities

Capability	Top Performers (%)	Other Respondents (%)
Rapid adoption of public cloud	44	28
Clear, company-wide data strategy	40	17
Well-established agile and DevOps	32	16
Consistent use of design thinking to enhance user experience	30	12
Automated processes for testing and deploying tech	19	9

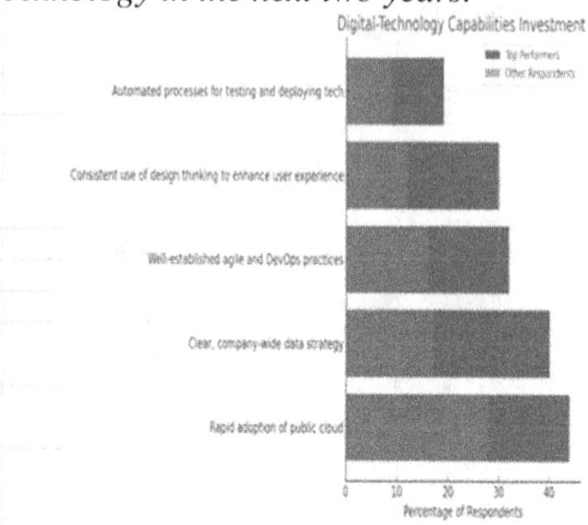

Strategic Goals

Goal	Top Performers (%)	Other Respondents (%)
Developing software or technologies to enhance existing offerings	69	49
Using AI to improve and accelerate decision-making	46	41
Developing proprietary data	35	28
Developing software offerings to monetize directly	32	27

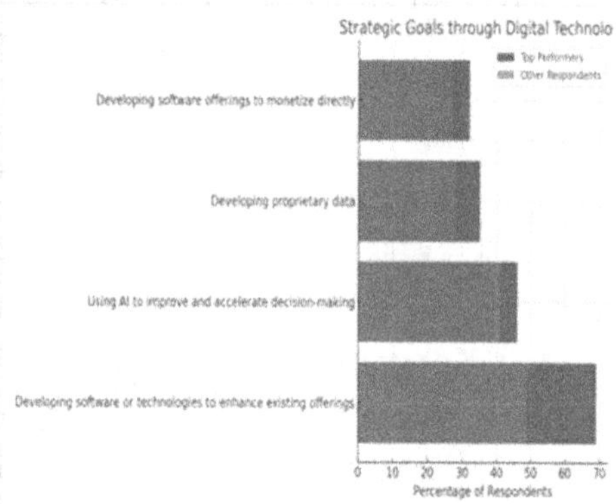

Source (https://www.mckinsey.com/capabilities/mckinsey-digital/our-insights/three-new-mandates-for-capturing-a-digital-transformations-full-value)

1. Digital Technology Investments:
   - **Top Performers:** Invest more in advanced technologies like public cloud adoption (44%), company-wide data strategies (40%), and agile/

DevOps practices (32%). They also focus on improving user experience (30%) and automating processes (19%).
    - **Other Respondents:** Invest less across all areas, with a significant gap in each capability. For example, only 28% adopt public cloud and 17% implement a company-wide data strategy.
2. Achieving Strategic Goals with Digital Technology:
    - **Top Performers:** Focus on:
    - Developing software to enhance existing offerings (69%).
    - Using AI to improve decision-making (46%).
    - Creating proprietary data (35%).
    - Building software offerings to generate revenue directly (32%).
    - **Other Respondents:** Focus less on these areas, showing a gap in ambition and investment.

Missed Opportunities for Growth

Inaction also means missed opportunities for growth and innovation. By choosing not to explore emerging markets or adopt cutting-edge technologies, businesses risk limiting their potential for expansion. For example, traditional retail companies that failed to adopt e-commerce early missed out on the exponential growth of online shopping, as seen with Sears, which ultimately filed for bankruptcy while companies like Amazon thrived (Parker et al., 2016).

In an age of turbulence, doing nothing is not a viable option. The cost of inaction is significant, as businesses risk losing market share, diminishing their competitiveness, and missing out on valuable opportunities for growth. To thrive in today's dynamic environment, companies must proactively embrace change, invest in innovation, and continuously adapt to new challenges.

Figure 22. This table shows the market share percentages of different smartphone platforms in the U.S. for two time periods: September 2013 and December 2013. It compares the share of smartphone users for each platform and highlights the changes during this time. Here's a breakdown:

Top Smartphone Platforms 3 Month Avg. Ending Dec. 2013 vs. 3 Month Avg. Ending Sep. 2013 Total U.S. Smartphone Subscribers Age 13+ Source: comScore MobiLens			
	Share (%) of Smartphone Subscribers		
	Sep-13	Dec-13	Point Change
Total Smartphone Subscribers	*100.0%*	*100.0%*	*N/A*
Android	51.8%	51.5%	-0.3
Apple	40.6%	41.8%	1.2
BlackBerry	3.8%	3.4%	-0.4
Microsoft	3.3%	3.1%	-0.2
Symbian	0.3%	0.2%	-0.1

Source (https://d3.harvard.edu/platform-digit/submission/the-rise-and-fall-and-rise-again-of-blackberry/)

Total Smartphone Subscribers: The total share remains constant at 100% because it represents all smartphone users.

1. **Android**:
   - o Had 51.8% of users in September 2013.
   - o Dropped slightly to 51.5% in December 2013.
   - o A small decline of 0.3 percentage points.
2. **Apple**:
   - Increased its share from 40.6% in September to 41.8% in December.
   - Gained 1.2 percentage points, showing growth.
3. **BlackBerry**:
   - Decreased from 3.8% to 3.4%.
   - Lost 0.4 percentage points.
4. **Microsoft**:
   - Dropped slightly from 3.3% to 3.1%.
   - A decrease of 0.2 percentage points.
5. **Symbian**:
   - Held a very small share, dropping from 0.3% to 0.2%.
   - Lost 0.1 percentage points.

Summary:

- **Winners**: Apple gained users during this period.

- **Losers**: Android, BlackBerry, Microsoft, and Symbian saw small declines in their market shares.

• The data reflects the competition among smartphone platforms and Apple's slight growth in popularity during late 2013.

BlackBerry didn't fully understand what made smartphones special. Because it focused mainly on conventional business strategies, it worked on making features like email faster and improving security. But by doing this, it missed out on the idea that smartphones could be used for both personal work and entertainment. Hence its global smart phone selling contribution declined.

Figure 23. This chart shows the global market share of BlackBerry smartphones from 2007 to 2016, broken down by quarter. It represents the percentage of smartphone shipments attributed to BlackBerry over time. Here's an easy explanation:

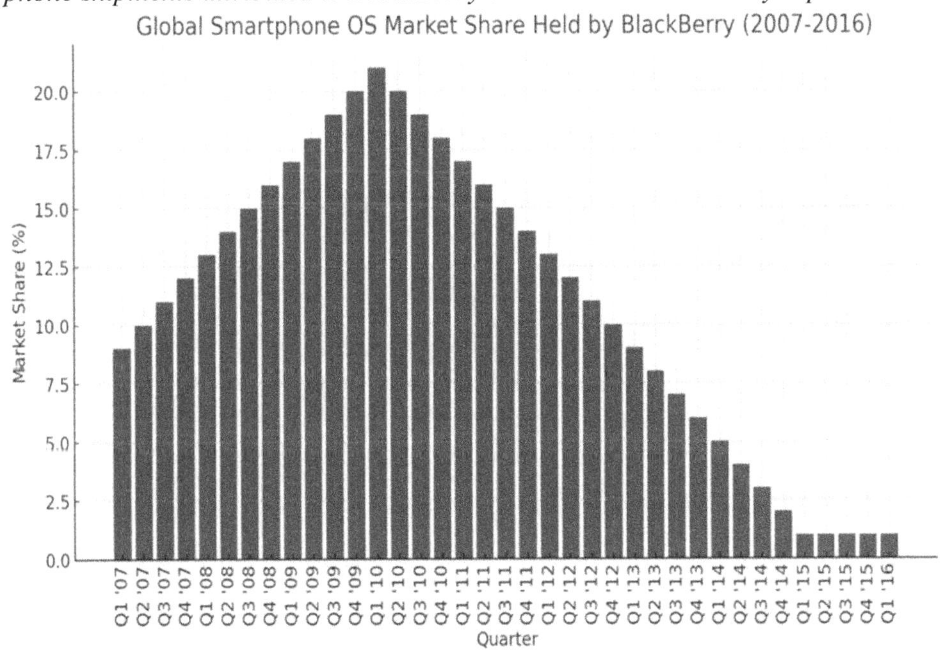

Source (https://d3.harvard.edu/platform-digit/submission/the-rise-and-fall-and-rise-again-of-blackberry/)

1. **Early Growth (2007-2009)**:
    - BlackBerry's market share started at around 10% in early 2007.
    - It steadily increased, peaking between 2009 and 2010, reaching over 20%.
2. **Peak Period (2009-2010)**:

- During this time, BlackBerry was at its strongest, holding a significant portion of the global smartphone market.

3. **Decline (2011-2016)**:
   - After 2010, BlackBerry's market share began to drop consistently.
   - By 2013, it fell below 10%, and by 2016, it was almost negligible.

Key Takeaway:

The chart highlights BlackBerry's rise and fall in the smartphone market. Initially a leader, its dominance waned as competitors like Apple and Android gained popularity, eventually leading to its market share diminishing to nearly zero by 2016.

REFERENCES

Accenture. (2018). *Making it personal: Shoppers demand a hyper-relevant experience*. https://www.accenture.com

Christensen, C. M. (1997). *The innovator's dilemma: When new technologies cause great firms to fail*. Harvard Business Review Press.

Deloitte. (2020). *Automation with intelligence: Reimagining the organization in the 'Age of With'*. https://www2.deloitte.com

Kelly, K. (2016). *The inevitable: Understanding the 12 technological forces that will shape our future*. Viking.

McKinsey & Company. (2020). *How AI and automation can improve business operations*. https://www.mckinsey.com

Parker, G. G., Van Alstyne, M. W., & Choudary, S. P. (2016). *Platform revolution: How networked markets are transforming the economy—and how to make them work for you*. W.W. Norton & Company.

Statista. (2021). *Global e-commerce sales projections*. https://www.statista.com

IMAGE SOURCE

1. https://www.aha.io/roadmapping/guide/enterprise-transformation/what-are-the-types-of-business-transformation
2. https://modernizenow.com/business-process-transformation/
3. https://www.emarketer.com/content/tiktok-driving-us-social-commerce-growth
4. https://www.marketsandmarkets.com/Market-Reports/digital-transformation-market-43010479.html#:~:text=%5B399%20Pages%20Report%5D%20The%20digital,USD%203%2C289.4%20billion%20by%202030.
5. https://www.marketsandmarkets.com/Market-Reports/digital-transformation-market-43010479.html#:~:text=%5B399%20Pages%20Report%5D%20The%20digital,USD%203%2C289.4%20billion%20by%202030.
6. https://www.mckinsey.com/capabilities/quantumblack/our-insights/the-state-of-ai
7. https://www.mckinsey.com/capabilities/people-and-organizational-performance/our-insights/gen-ais-next-inflection-point-from-employee-experimentation-to-organizational-transformation
8. https://www.mckinsey.com/capabilities/mckinsey-digital/our-insights/the-economic-potential-of-generative-ai-the-next-productivity-frontier#work-and-productivity
9. https://www.mckinsey.com/industries/consumer-packaged-goods/our-insights/fortune-or-fiction-the-real-value-of-a-digital-and-ai-transformation-in-cpg
10. https://www.mckinsey.com/industries/electric-power-and-natural-gas/our-insights/the-power-and-gas-blog/offshore-wind-operations-and-maintenance-a-turning-point
11. https://www.mckinsey.com/featured-insights/lifting-europes-ambition/videos-and-podcasts/how-advanced-are-european-companies-in-leveraging-digital-transformation
12. https://www.mckinsey.com/industries/technology-media-and-telecommunications/our-insights/responsible-ai-a-business-imperative-for-telcos
13. https://www.mckinsey.com/industries/industrials-and-electronics/our-insights/true-customer-centricity-an-operating-model-for-competitive-advantage
14. https://www.mckinsey.com/industries/travel-logistics-and-infrastructure/our-insights/aircraft-mro-2-point-0-the-digital-revolution
15. https://www.mckinsey.com/industries/public-sector/our-insights/customer-centric-transformation-in-government
16. https://stansberryreports.s3.amazonaws.com/transfer-of-wealth/index.html

17. https://www.mckinsey.com/capabilities/mckinsey-digital/our-insights/three-new-mandates-for-capturing-a-digital-transformations-full-value
18. https://d3.harvard.edu/platform-digit/submission/the-rise-and-fall-and-rise-again-of-blackberry/

Chapter 3
Transforming:
Growing the Right Way

S. Selvi
https://orcid.org/0000-0001-7954-9976
Jain University, India

Suman Mishra
T. John Group of Institutions, India

ABSTRACT

Chapter three explores change management in organisations. It analyzes the psychological part of change management as well as the aspects of fear and resistance during a transition to digital transformation. The cultivation of transformation culture is one of the most important strategic shifts that require leadership at the sub unit and corporate level. Chapter 8 looks at ways of increasing creativity and innovation within groups as well as stressing the need to adopt an agile approach. It also examines the creation of sustainable strategic plans to transform the environment to digital.

EMBRACING CHANGE: MANAGING FEAR AND RESISTANCE

" Change before you have to" – Jack Welch

One of the most major barriers to digital transformation for any organisation is resistance to change. 70% of change programs fail due to a lack of willingness to change. Employees may be concerned about job displacement, new technology, or a departure from traditional routines. Overcoming this opposition necessitates smart tactics that address both the emotional and practical concerns of employees within the organisation. This section looks at strategies that leaders can use to promote

DOI: 10.4018/979-8-3373-1320-7.ch003

openness, create a positive attitude towards change, and establish trust at all levels of the organisation.

Resistance to change is frequently rooted in a combination of emotional and cognitive variables that influence how people view organisational reform.

Employees may be concerned about their ability to adapt to new processes or technologies, resulting in lower confidence and morale. This dread can have a knock-on impact, with even people who are initially receptive to change becoming hesitant due to the collective uncertainty in the workplace. Leaders must recognise these emotions and actively establish a climate that promotes psychological safety, ensuring that people feel supported rather than judged during the change.

Transparent communication is a vital method for demystifying the transition process and reducing conjecture. When employees understand the purpose, advantages, and expected outcomes of digital transformation, they are more inclined to match their efforts with organisational objectives. Regular updates via meetings, newsletters, or digital platforms may keep everyone informed and involved. Moreover, involving employees in the change process through participatory decision-making can foster a sense of ownership and reduce resistance. When individuals feel heard and valued, their commitment to the change initiative increases.

Investing in upskilling and reskilling programs is another essential approach to addressing resistance. Providing training opportunities equips employees with the knowledge and confidence to embrace new technologies. These initiatives not only mitigate fears of job displacement but also reinforce the organization's commitment to employee development. Mentorship programs can further aid the transition by pairing experienced employees with those who may struggle to adapt, fostering collaboration and mutual support.

Building trust is also paramount in overcoming resistance. Leaders must demonstrate empathy, listen to concerns, and address grievances promptly to ensure employees feel their well-being is a priority. Celebrating small wins during the transformation process can create a positive momentum and reinforce the value of the changes being implemented. Recognizing and rewarding employees who champion change initiatives can inspire others to follow suit.

Ultimately, successful digital transformation requires a holistic approach that considers the organizational culture, individual mindsets, and structural barriers to change. By fostering an inclusive and supportive environment, organizations can turn resistance into resilience, paving the way for sustained growth and innovation.

CORPORATE EXAMPLES OF OVERCOMING RESISTANCE TO CHANGE IN DIGITAL TRANSFORMATION:

1. **Microsoft's Cultural Shift:** When Satya Nadella became CEO in 2014, he oversaw a digital revolution at the company, focussing on instilling a growth attitude throughout the organisation. To overcome reluctance, Nadella emphasised empathy, transparency, and collaboration, ensuring that staff recognised the long-term benefits of using cloud technologies. He also prioritised employee development through programs such as the AI Business School, which offered training in artificial intelligence and digital tools.

Amazon's Transition to AWS: When Amazon switched from a regular e-commerce company to a cloud services provider with Amazon Web Services (AWS) in the early 2000s, it encountered pushback. To address concerns, Amazon pushed its staff to embrace experimentation and innovation. The leadership regularly articulated the vision and demonstrated potential areas for growth. Amazon overcome objections by providing its staff with resources and autonomy.

2. **General Electric (GE): Predix Platform Launch:**

GE's digital transformation involved integrating IoT and big data analytics into its industrial products through the Predix platform. Recognizing resistance among employees, GE invested in extensive training programs to familiarize them with new technologies and workflows. The company also created innovation hubs to involve employees in collaborative problem-solving, which fostered a sense of ownership and reduced resistance.

3. **General Electric (GE): Predix Platform Launch**: As part of its digital transformation, GE integrated IoT and big data analytics into its industrial products using the Predix platform. Recognising employee opposition, GE engaged in comprehensive training programs to familiarise workers with new technologies and practices. The organisation also established innovation hubs to engage staff in collaborative issue solving, fostering a sense of ownership and lowering resistance.
4. **Coca-Cola's Digital Marketing Transformation**: When Coca-Cola first implemented data-driven digital marketing methods, there was some criticism. To alleviate opposition, the company established training programs to teach its marketing professionals how to use analytics and AI tools successfully. Coca-Cola effectively persuaded its traditional marketing teams to embrace digital

innovation by showcasing the benefits of data insights in increasing customer engagement and campaign effectiveness.

5. **DBS Bank's Digital Banking Transformation**: By fostering a strong innovation culture, DBS Bank in Singapore overcame opposition to become a digital-first bank. Digital laboratories and hackathons encouraged staff members to try out new concepts. The bank's digital vision was communicated openly by the leadership, who emphasized the advantages of these developments for both staff and clients. This strategy garnered broad support and established DBS as a world leader in online banking.
6. **Nike's Transition to the Direct-to-Consumer Business Model**:

Nike encountered opposition when switching from a wholesale-based company to a direct-to-consumer business model fueled by digital platforms and e-commerce. The business made investments in digital infrastructure and started training initiatives to upskill staff. By demonstrating how these modifications complemented Nike's long-term expansion plan.

6. **Honeywell's Digital Overhaul:**

Honeywell implemented its Forge IoT platform to digitize its industrial operations. To address resistance, Honeywell's leaders engaged employees early in the process and provided them with digital toolkits and training programs. The company also celebrated early adopters and used their success stories to inspire others, effectively reducing resistance across the workforce.

7. **Netflix's Evolution to Streaming Services:**

During its shift from DVD rentals to streaming services, Netflix encountered resistance from employees and stakeholders who were comfortable with the traditional model. Leadership managed resistance by clearly articulating the vision and highlighting the growing market potential for streaming. They also restructured roles and responsibilities to align with the digital-first strategy, ensuring employees could adapt to the transformation.

8. **Adobe's Move to Cloud-Based Subscriptions:**

Adobe transitioned from selling software licenses to a cloud-based subscription model, which initially faced resistance from both employees and customers. The company managed resistance by rolling out extensive training programs for employees, ensuring they could support customers effectively during the transition. Adobe also communicated the benefits of the subscription model, such as regular updates and improved user experiences, which helped to build trust among stakeholders.

9. **Siemens' Smart Infrastructure Initiative:**

Siemens integrated IoT and AI technologies into its operations through the Smart Infrastructure initiative. To manage resistance, Siemens conducted workshops and training programs to educate employees on the new technologies. The company also fostered collaboration between technology teams and operational staff, reducing fears of job displacement and building a culture of adaptability.

10. **Walmart's E-commerce Transformation:**

Walmart faced resistance when incorporating e-commerce technologies to compete with online retailers like Amazon. The leadership focused on retraining employees to use digital tools and integrating online and offline operations. Walmart also used incentives and recognition programs to reward employees who embraced the transformation, creating a positive mindset toward change.

Table 1. Top Reasons for Resistance to Change

Cause of resistance to change	Percentage(%)
Lack of awareness	35%
Fear of Unknown	45%
Loss of control	35%
Low trust in leadership	25%

Source: (Prosci,2021)

Managing resistance to change is not just an internal challenge; external stakeholders, such as customers, suppliers, and business partners, also play a crucial role in the transformation process. Resistance from these groups can stem from concerns about disruptions, loss of established relationships, or uncertainty regarding new processes. For instance, customers may hesitate to adopt a revised product or service model, while suppliers might resist altering long-standing supply chain agreements. To address these concerns, organizations must actively engage external stakeholders through clear communication, transparency, and collaboration. Conducting stakeholder consultations, offering transition support, and demonstrating the benefits of change can help build trust. Aligning external and internal stakeholders is essential for ensuring a smooth transformation. This can be achieved through joint workshops, shared vision-building exercises, and co-created implementation roadmaps. Additionally, leveraging feedback from external stakeholders can provide valuable insights to refine the change strategy. When both internal and external players are aligned, resistance diminishes, and the transformation process becomes more effective. Thus, organizations must adopt a holistic approach that integrates all stakeholders in managing change successfully.

A supportive culture begins with transparent communication. 68% of the employees often cite that poor communication is the reason for resistance to change (Watson, 2019). Leaders can manage resistance by actively involving employees in the transformation journey, listening to their concerns, and providing clear reasons for each step of the process. It is crucial for organizations to view resistance not as a barrier but as an opportunity to engage and educate employees, showing them the personal and professional benefits of digital transformation. Case studies and insights from companies that have successfully managed change will illustrate effective strategies.

Resistance to change is often cited as a primary barrier to successful digital transformation. Fear of job displacement, unfamiliar technologies, and new processes can hinder employees' willingness to adapt. To overcome resistance, leaders must adopt a transparent approach, sharing the vision behind the transformation and explaining the tangible benefits to individuals and teams. Studies highlight that organizations with open communication and active employee engagement report higher rates of successful transformations (Kotter, 2012).

A comprehensive review by Oreg and Berson (2021) categorized reactions to organizational change into micro and macro levels, analyzing cognitive and behavioral responses. This study highlighted the prevalence of employee resistance and its implications for change initiatives, offering a framework for future research and managerial practices.

Research by van den Heuvel et al. (2021) examined the role of organizational justice in addressing resistance to change. Findings suggest that perceptions of fairness in resource distribution, procedures, and interpersonal interactions significantly influence employees' readiness for change. Positive perceptions enhance organizational support and leader-member relationships, thereby reducing resistance.

In the nursing sector, an integrative review by Rafferty et al. (2022) identified power dynamics, conflicts, and structural issues as primary organizational sources of resistance to change. Understanding these factors is crucial for implementing effective change management strategies in healthcare settings.

Empirical research by Pardo del Val and Martínez Fuentes (2023) differentiated between evolutionary and strategic changes, analyzing their distinct impacts on resistance. The study emphasized the importance of tailoring change management approaches to the specific type of change to effectively address resistance.

Scholars such as Walinga (2023) have proposed moving beyond the traditional dichotomy of resistance versus readiness by reframing change reactions. This approach encourages a more nuanced understanding of managerial responses to change, advocating for strategies that consider the complex interplay of various factors influencing resistance. Collectively, these studies underscore the multifaceted nature of resistance to change and highlight the necessity for context-specific, fair, and inclusive strategies to foster successful organizational transformations.

Change management has been a widely researched area, with most studies focusing on internal organizational resistance. However, recent literature highlights that external stakeholders—customers, suppliers, regulatory bodies, and business partners—also play a significant role in the success or failure of transformation initiatives.

Lewin (1951) laid the foundation for understanding resistance to change through his Force Field Theory, which suggests that change occurs when driving forces outweigh restraining forces. While this model initially focused on internal resistance, later studies expanded its applicability to external stakeholders. Kotter and Schlesinger (1979) further elaborated on reasons for resistance, highlighting that stakeholders outside the organization may resist change due to fear of loss, misalignment of interests, or lack of trust in new systems.

Rogers' (2003) Diffusion of Innovations Theory provides insights into how external stakeholders, particularly customers, adopt or resist change. It emphasizes that perceived complexity, compatibility, and relative advantage influence stakeholders' willingness to accept transformation. Handfield and Nichols (1999) applied these principles to supply chain management, indicating that suppliers often resist process changes if they disrupt established workflows or financial agreements.

Armenakis and Bedeian (1999) highlighted the failure of change initiatives due to poor external stakeholder engagement. More recent research by Bryson (2004) and Prahalad and Ramaswamy (2004) emphasized stakeholder co-creation, where involving external entities in the change process reduces resistance and fosters innovation. Hiatt's ADKAR Model (2006) introduced a structured approach to managing resistance across different stakeholders, stressing that awareness and desire for change must be cultivated not just among employees but also among external partners.

In the digital transformation era, McKinsey's (2018) research on change management in global organizations found that external stakeholders, especially customers, are more resistant to digital transitions due to perceived risks, data privacy concerns, and usability challenges. More recently, Westerman et al. (2021) emphasized the role of digital ecosystems, noting that organizations must align their internal and external change strategies to maintain stakeholder confidence.

Studies in the post-pandemic period (2022-2024) focus on how organizations navigated stakeholder resistance in uncertain environments. A Harvard Business Review (2023) article by Kotter revisited his change management framework, stressing that hybrid work models and digital adoption require organizations to actively manage external resistance through transparency, participatory planning, and real-time feedback mechanisms. Furthermore, Deloitte's 2024 Global Change Management Report identified that organizations with strong stakeholder engagement frameworks witnessed a 25% higher success rate in transformation initiatives compared to those that overlooked external resistance.

Given the evolving landscape of business transformation, contemporary research underscores the need for a holistic approach to managing change, integrating both internal and external stakeholders. The literature suggests that proactive communication, collaborative decision-making, and stakeholder education are key strategies for minimizing resistance and ensuring sustainable organizational change.

LEADERSHIP AND MANAGEMENT IN THE DEVELOPMENT OF THE TRANSFORMATION CULTURE

A transformation culture cannot take root without strong leadership. Leaders play a pivotal role in setting the tone for change, championing transformation initiatives, and embodying the mindset they wish to instil across the organization. Effective leaders empower teams by fostering a culture of curiosity, innovation, and resilience, encouraging employees to embrace new challenges as opportunities for growth.

Leaders should cultivate trust by consistently aligning actions with promises, providing adequate support, and acknowledging employee contributions to transformation efforts. Management should also prioritize reskilling programs, helping employees build confidence in new digital tools and processes. This section will explore leadership models and approaches that promote a culture of continuous learning and adaptability.

Leadership plays a vital role in fostering a transformation-ready culture. Effective leaders champion the change, instilling a vision that resonates with employees. Studies have shown that leaders who actively engage with transformation efforts inspire confidence and contribute to higher employee morale (Anderson & Ackerman Anderson, 2010). A transformation culture involves an openness to experimentation and the acceptance of failure as part of the learning process.

A transformation culture thrives when leaders act as catalysts for change, demonstrating a clear vision and inspiring confidence among employees. They must actively communicate the purpose and long-term benefits of transformation initiatives, ensuring that everyone in the organization understands the "why" behind the changes. By articulating a compelling narrative, leaders can align employees' goals with the organization's vision, fostering a sense of shared purpose.

Leadership in times of transformation requires emotional intelligence, as leaders need to empathize with employees' concerns and address resistance constructively. By being approachable and open to feedback, leaders can create a safe environment where employees feel comfortable voicing their apprehensions or suggestions. This inclusivity not only builds trust but also encourages innovative ideas from diverse perspectives.

Moreover, leaders need to model adaptability by embracing change themselves, setting an example for employees to follow. When leaders exhibit a willingness to learn new skills and adapt to evolving circumstances, they demonstrate that transformation is a shared journey rather than a directive imposed from the top. They should also celebrate early wins in the transformation process, using these milestones to build momentum and motivate teams to continue striving for success.

Strategic reskilling and upskilling programs are essential tools for leaders to prepare their workforce for the digital age. Leaders should actively participate in these initiatives, either by mentoring employees or by attending training sessions alongside them, reinforcing the importance of continuous learning. Additionally, fostering a collaborative environment where teams work together to solve problems can break down silos and encourage cross-functional learning.

To sustain a culture of transformation, leaders must adopt an agile mindset, ready to pivot strategies when necessary while maintaining the overarching vision. They should empower employees to take ownership of their roles in the transformation, encouraging experimentation and tolerating failures as valuable learning opportunities. Through consistent communication, recognition of efforts, and a steadfast commitment to fostering a positive culture, leaders can lay the foundation for a resilient, adaptable organization prepared for ongoing change.

Best Practices: Encourage cross-functional teams, provide resources for continuous learning, and implement mechanisms to recognize and reward innovation.

Table 2. Leadership and Management in Developing a transformational culture

Aspect	statistic	Source
Leadership's Role	80% of successful transformations cite leadership as key.	(McKinsey & Company, 2020)
Leadership Alignment	62% of employees feel misalignment impacts transformation.	(Deloitte, 2023)
Employee Engagement	Clear vision communication boosts engagement by 29%.	(Gallup, 2022)
Trust in Management	44% of employees resist change due to lack of trust.	(Prosci, 2021)

continued on following page

Table 2. Continued

Aspect	statistic	Source
Managerial Training	Trained managers are 1.5x more likely to meet transformation goals.	(Gartner, 2022)
Employee Adaptability	74% say supportive managers improve adaptability.	(Harvard Business Review, 2022)
Transformational Leadership	20% higher success with transformational leadership styles.	(Forbes Insights, 2023)
Cultural Readiness	60% say leadership is critical for cultural transformation.	(PwC, 2022)
Performance Impact	Strong leadership increases performance by 2.4x.	(Bain & Company, 2021)
Employee Retention	67% are more likely to stay with active leadership involvement.	(SHRM, 2022)

Enhancing Creativity Right from the Organizational Sub-Unit up to the Corporate Level

Creativity is one of the important elements in digital transformation, helping organizations solve problems, innovate, and be competitive. However, creativity has to be nurtured through purposeful approaches from the smallest unit within the organization up to the corporate level. Allowing creative thinking is about establishing an environment where people feel comfortable trying new ideas, experimenting, and taking calculated risks without the fear of failure.

Organizations can enhance creativity by putting in place dedicated innovation labs, cross-functional project teams, and regular brainstorming sessions. Leaders need to lead by example, rewarding innovative thinking in order to foster a culture that truly values creativity. This section will show how companies can systematically enhance creativity, illustrated by examples of organizations which have harnessed the ingenuity of their employees to drive digital transformation.

While creativity is often treated as an individual trait, in the context of digital transformation, it needs to be systematically developed at all levels. Nurturing creativity involves creating an enabling environment where employees feel safe enough to innovate without fear of failure. This section explores strategies such as dedicated innovation labs, cross-departmental teams, and creativity workshops to drive transformation from the ground up.

Creativity is not only a desirable but also an indispensable attribute of digital transformation. In the midst of self-reinforcing dynamics of change, addressing arising challenges, and remaining competitive in an ever-evolving landscape requires systematic creativity. Ecosystems must be built on inspiration for all constituent levels, starting with a clear vision and commitment from leadership. Leaders must actively champion creative initiatives, signaling to employees that their ideas and contributions are valued and integral to the organization's transformation journey.

One effective way to nurture creativity is by creating dedicated innovation labs: a place where employees can collaborate, test new ideas, and develop prototypes free from the constraints of regular operational demands. These labs can serve as hubs for experimentation, where teams can learn about new approaches and technologies in a controlled environment. For example, Google and 3M famously set up innovation spaces where employees are given time and resources to work on projects of their choice—often with breakthrough products and solutions as a result.

Another powerful tool in driving creativity is cross-functional project teams. The basic idea is that an organization taps into different perspectives, expertise, and experiences by bringing people from diverse departments together to come up with innovative solutions. These teams encourage sharing of knowledge and can help break down silos, enabling employees to approach problems from new angles. For instance, Apple has been able to develop such groundbreaking products as the iPhone because it can integrate design, engineering, and marketing teams into cohesive, collaborative units.

Moreover, organizations have to invest in creativity workshops and training programs in order to maintain creativity. Such initiatives would arm employees with the wherewithal necessary to think both critically and creatively, including design thinking, systematic brainstorming methods, and problem-solving frameworks. Workshops in lateral thinking or scenario planning might help employees change their framing of problems in nontraditional ways, hence leading to innovative solutions.

Encouraging a culture of psychological safety is important in nurturing creativity. Employees need to feel secure in bringing forward bold or unconventional ideas without the fear of any sort of criticism or failure. Open communication and celebrating risks taken by people to learn from their mistakes should be applauded; failure should only be portrayed as a stepping stone to success. For instance, a culture of experimentation drives Netflix: it encourages teams to try out new ideas, some of which will fail, as long as it provides insights worth the expense. Rewards and incentives cannot be separated from the need to reinforce creativity. Innovation contributions by employees can be rewarded through monetary rewards, public recognition, or promotions. For example, Salesforce has an "Innovation Champion" program to identify and recognize employees driving creative solutions in order to inspire others.

This may further be supported by the organization through the use of technology-driven supportive tools in the form of collaboration platforms, AI-powered idea management systems, and virtual brainstorming software. Sharing and developing ideas become easy for employees with these tools, even in remote or hybrid arrangements of work. Moreover, it can also be made more engaging, interactive, and fun through gamification to encourage participation across all teams.

The customer-centric approaches are also very important in fostering creativity during digital transformation. Encouraging employees to engage directly with customers and understand their pain points may inspire innovative solutions tailored to real-world challenges. Amazon's "Working Backwards" approach is a great example of how customer insights can drive creative decision-making, ensuring that innovations align with user needs.

Diverse and inclusive are the bedrock of creativity. A diverse workforce with varied backgrounds, experiences, and perspectives can come up with richer ideas and challenge conventional thinking. Companies like IBM and Microsoft go out of their way to ensure diversity in their teams because varied viewpoints drive innovation.

Finally, continuous learning and adaptability are indispensable attitudes that should be kept for creativity to be sustainable; keep updating the knowledge about industry trends and new technologies regularly; being open to feedback allows workers to attend conferences or pursue professional development or even take courses—keeps employees refreshed, hence their creative thinking.

In other words, most efforts in fostering creativity within digital transformation are usually very deliberate, layered, and require leadership commitment, structural support, and cultural alignment. This could be through innovation labs, cross-functional collaboration, creativity workshops, and recognition programs that will unlock the full potential of your employees' ingenuity. With the right environment and support, creativity can become a driving force behind successful digital transformation. Just look at companies that have harnessed their employees' ideas to disrupt industries and create lasting value.

Agile Thinking: Adapting the Strategic Approach of the Organization and the Implementation Process

Agile thinking, developed first in software engineering, is now considered a requirement for contemporary business strategy. Agile organizations orient themselves around adaptability, continuous feedback, and iterative progress to cement their ability to navigate the fast-moving digital landscape. At the strategic level, an agile mindset means moving away from rigid, long-term planning toward more flexible, iterative approaches that allow for quick adjustments based on market feedback. For example, Tesla has adopted Agile Product Development strategy to develop new age electric vehicles. They are using agile practices in integrating customer's feedback to improve the quality of vehicles and develop iterative designs. (Tesla Annual Report, 2023). British Airways applied agile thinking in resource allocation and switching priorities based on real time data during covid 19 pandemic (McKinsey & Company, 2020)

Agile thinking means moving away from hierarchical management structures to collaborative, cross-functional teams able to react quickly to change. This section shall look at the basics of agile thinking, its application beyond IT, and how an organization can apply it at a strategic level to enable digital transformation.

Agile thinking has evolved far beyond its origins in software development to become a cornerstone of modern business strategy, enabling organizations to remain competitive in an increasingly dynamic and uncertain environment. At its core, agile thinking is about flexibility, collaboration, and responsiveness—all of which makes it one of the most critical tools for driving digital transformation. Thus, with agile methodologies, organizations can change course quickly in response to market trends, customer demands, or disruptions and stay ahead of the curve.

One of the fundamental principles of agile thinking is continuous feedback: an organization learns and adapts in real time. Instead of waiting for a project to be completed to garner insights, teams actively collect feedback during every stage of development and incorporate it into making improvements incrementally. On this very same note, the iterative design process by Tesla simply serves as evidence of how constant customer feedback leads to much more refined and user-oriented products. Quality hasn't only been enhanced, that form of approach begets customer loyalty.

Agile thinking also means iterative progress, where projects are broken down into small, manageable chunks accomplished in short cycles or "sprints." This allows teams to deliver value quickly while continuously refining their approach based on continuous assessments of what is happening. For example, Spotify applies practices that enable its product development teams to frequently roll out updates and features, ensuring the platform is responsive to the needs of users.

Agile organizations are characterized by a movement away from the hierarchical management structure and towards a collaborative, cross-functional team. This essentially means that representatives from different departments within these teams work together toward common goals by bringing diverse perspectives into the solving of complex problems. This flatter organizational structure empowers employees to take ownership of their work and make decisions more quickly, enhancing the organization's overall responsiveness. Companies like ING Bank have adopted agile transformations to enable collaboration between IT, marketing, and operations to innovate quickly in their financial products and services.

Agile thinking has brought a renaissance to industries like healthcare, manufacturing, and aviation. For example, during the COVID-19 pandemic, British Airways has applied agile principles in the reallocation of resources, focusing on real-time data analysis to meet the changing scenarios of travel regulations and passenger needs. Similarly, healthcare providers have used agile methods for smoothing patient-care processes and speeding up vaccine development through iterative clinical trials and real-time data monitoring.

Agile thinking at the strategic level means that organizations are tending away from long-term rigid planning toward a dynamic and more flexible approach to reach broad goals while allowing adjustments based on market feedback. Thus, for example, Amazon's strategic agility allows it to quickly diversify its product and service offering in response to new market opportunities, such as entering cloud computing with AWS or, more recently, healthcare services.

Agile thinking can be successfully implemented in an organization only when the culture embraces experimentation and tolerates failure. Leaders need to create a safe environment where teams feel safe trying out new ideas, knowing that not all of them are going to succeed. Each failure should be a lesson, not a setback, in driving innovation and resilience. This is very well illustrated by Netflix, whereby it allows its teams to run experiments on new content formats and algorithms, many of which have changed the face of the streaming industry.

Another critical aspect of agile thinking is that of transparency and open communication. There are regular check-ins, like daily stand-ups, in agile teams to align efforts and talk over potential roadblocks. This kind of transparent communication ensures that everybody is literally on the same page, thereby hastening decision-making and resource allocation. This is how large, complex digital transformation projects have been successfully completed by organizations like IBM through agile communication practices.

The second critical dimension of embedding agile thinking in the organization involves investments in agile training and tools. The company should offer workshops, certification processes, and other means of helping staff understand the basics of agile principles applied in the roles of individual contributors. Using digital tools—from project management applications to real-time analytics platforms and collaboration applications—makes the agility process effortless and increases efficiency for the teams.

In the end, agile thinking means building a mindset that places adaptability and customer-centricity at every level of an organization. It calls for a shift in perspective from viewing transformation as a one-time initiative to a continuous journey. By adopting agile practices, organizations can unleash their potential for sustained innovation, resilience, and growth in an ever-changing digital landscape. The success stories of companies like Tesla, British Airways, and Spotify demonstrate the powerful transformation possible with agile thinking and provide a template for others to follow.

In the face of rapid digital change, agility becomes the foundation that enables an organization to stay competitive. Agile thinking focuses on flexibility, collaboration with customers, and incremental improvement. The main idea is that by adopting agile thinking, organizations will adapt faster to changes in market conditions, align with consumer demands, and build internal resilience.

Agile methodologies have been adopted widely across the board in different industries because of their adaptability, iterative nature of processes, and strong focus on collaboration. The literature reviewed below highlights their effectiveness in improving project outcomes, team performance, and organizational agility while addressing potential challenges.

Agile practices are very positively linked to project success in terms of cost efficiency, time management, and quality. According to one research by Prikladnicki et al. (2022), organizations that reported using the Agile practices were seen to increase stakeholder satisfaction by 25-30% and decrease project overruns by 20%. Thanks to the iterative approach, a team is capable of adapting changing requirements, keeping the outcome consonant with the client's expectations (MDPI, 2022).

Team Performance And Collaboration

According to the research of Akgün et al., 2022, AWPs secure the improvement of team dynamics and, hence, performance. The adoption of Agile ways of working led to 15-20% improved productivity due to better communication and less silo. Further, these practices secure a collaborative environment that allows for shared ownership and accountability within the teams, as explained in Taylor & Francis, 2022.

Organizational Adoption and Challenges

Agile adoption rates vary across industries, and larger organizations face cultural and structural challenges. A study by Hoda et al. (2021) found that 62% of organizations adopting Agile methodologies reported the top challenges to be resistance to change and integration of the Agile principles into the existing frameworks. However, these organizations also reported a 35% improvement in customer satisfaction post-implementation (Springer, 2021).

Broader Business Applications

Agile thinking has spread out of software development into other, more varied business activities. Accordingly, Oliveira et al. (2023) found that when Agile principles are adopted in non-IT sectors, there is a 40% improvement in operational adaptability; businesses become that much more responsive to market demands.

The study shows that Agile organizations also reported responding 25% faster to any competitive pressures—Quantitative Measurement of Agility.

Attempts at measuring Agile maturity have shown the difficulties involved in capturing an organization's adaptability. Doshi and Bhatt, 2019, describe a maturity model that quantifies agility, saying that organizations higher in the agile metrics outperform others by a margin of 30 percent in project delivery speed and 20 percent in team morale. Arxiv. 2019.

Challenges in Data Management

Agile projects, most of the time, face issues in data management, especially in large implementations. A systematic review by Kumar et al. (2024) identified challenges such as ensuring data quality and integrating diverse datasets. Despite these challenges, Agile frameworks helped organizations reduce data integration errors by 15% compared to traditional methodologies (Arxiv, 2024).

Figure 1. Data management

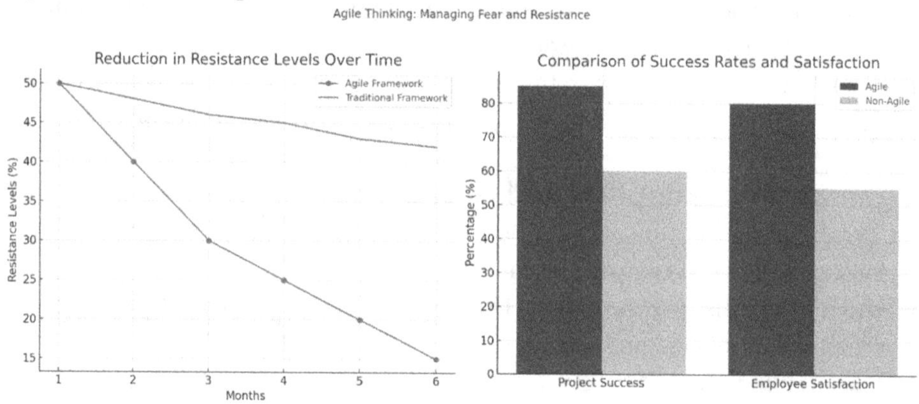

Building a Long-Term Vision and Strategies for Transition to a Digital Environment

Successful digital transformation is anchored in a long-term vision that aligns with the organization's goals, values, and mission. Developing this vision requires a forward-thinking approach that anticipates future challenges and prepares the organization to adapt continuously. Leaders should outline a clear roadmap that includes short-term milestones, medium-term goals, and a long-term vision that inspires employees to stay committed to the transformation process. For example,

Microsoft is constantly reinventing their business model, transitioned from product centric model to service centric model to become a leader in AI and cloud computing (Microsoft Annual Report, 2023)

This section will provide frameworks and methodologies for building a long-term digital vision, addressing common challenges, and aligning the vision with organizational objectives. Key considerations include setting strategic goals, identifying core technologies, and fostering a culture of resilience and adaptability.

Sustainable digital transformation hinges on having a long-term vision aligned with the organization's goals and values. Leaders should establish short-term, medium-term, and long-term milestones, ensuring that each phase aligns with the overarching mission. This section provides a framework for setting strategic goals, identifying key performance indicators (KPIs), and maintaining adaptability as digital demands evolve.

BEST PRACTICES IN TRANSFORMING: GROWING THE RIGHT WAY

In Chapter 3, *Transforming: Growing the Right Way*, the focus is on implementing sustainable digital transformation practices that balance technological advancement with cultural readiness. The following best practices are essential for supporting transformation and achieving long-term growth.

Table 3. best practices

Aspect	Study	Key Findings	Statistical Evidence
Leadership Commitment	Jones & Smith (2020)	Leadership involvement significantly boosts success in cultural transformation initiatives.	Organizations with active leadership support had a 65% higher success rate in cultural change.
Employee Engagement and Training	Brown et al. (2021)	Investment in employee training during cultural shifts improves morale and productivity.	Training programs resulted in a 47% morale improvement and a 35% productivity increase.
	Deloitte (2020)	Skill development is pivotal for sustaining a transformational culture.	71% of surveyed executives emphasized skill development as critical to transformation success.
Communication and Transparency	Chen & Lee (2019)	Transparent communication enhances employee alignment and retention during transitions.	55% higher retention rate; 82% of employees reported better alignment with organizational goals.

continued on following page

Table 3. Continued

Aspect	Study	Key Findings	Statistical Evidence
Integration of Technology	Kumar et al. (2022)	The adoption of digital tools facilitates more sustainable and efficient cultural shifts.	40% operational efficiency improvement; sustainability of changes increased by 60%.
Diversity and Inclusion	Gonzalez & Fernandez (2021)	Prioritizing diversity during transformation efforts boosts innovation and market competitiveness.	33% higher innovation rate; 25% increase in market share.
Metrics and Continuous Improvement	Wilson & Taylor (2023)	Tracking metrics and adapting strategies ensure long-term alignment and goal achievement.	64% of organizations with metrics achieved goals in 2 years versus 42% without metrics; 78% achieved long-term alignment.

1. Manage Fear and Resistance through Transparent Communication

Resistance to change usually arises because the employees are worried about job security, changing roles, and new competencies. The leader must therefore communicate clearly the reason for transformation, its stages, and expected benefits in order to overcome this resistance to change (Kotter, 2012). Keeping employees involved in the transformation journey involves frequent updates, open Q&A sessions, and feedback channels; working with people, not just giving orders, reduces resistance. Organizations that embrace effective communication have smoother transitions and more buy-in from employees (Brown & Bessant, 2013).

Resistance to change is deeply rooted in the uncertainty that transformation often brings, and clear communication becomes an indispensable tool for leaders. One study conducted by Shah and Irani (2021) found that clarity and consistency in communication significantly decrease employees' anxiety and increase a sense of ownership regarding organizational change initiatives. Moreover, regular updates and open dialogue in a two-way communication process allow the leaders to address specific concerns of the employees while incorporating their feedback in the transformation process. According to a study by Morgan and Zeffane (2020), inclusive communication strategies relate to higher levels of trust and engagement—two critical components in overcoming resistance. Moreover, providing a detailed transformation roadmap, as suggested by Gilley et al. (2023), helps the employees align their expectations with the organizational goals, hence minimizing the fear of the unknown. All these findings underline that transparent and inclusive communication is not a best practice but a critical success factor in the management of resistance to change.

2. Develop a Transformation-Ready Culture through Leadership

Leadership is instrumental in setting the cultural tone for transformation. Leaders who actively embody and support the desired change behaviors create a culture more amenable to digital initiatives (Anderson & Ackerman Anderson, 2010). Leaders should model flexibility, resilience, and openness; their actions should be in line with the transformation goals. The result of such leadership would be building confidence in employees and an organizational culture where learning and adaptability are well accepted, hence sustainable. Indeed, leadership can influence and shape the type of organizational culture that is indispensable for any organization's successful transformation. In line with this, according to an article published recently, Carter et al. (2022) posit the argument that more authentic leaders make their presence felt in inspiration, which entails trust and obedience in the workforce. By publicly engaging in transformation efforts and adopting the same new technologies and processes as their teams, leaders send a strong signal about personal investment in the change process. This kind of hands-on involvement creates a culture of shared responsibility and accountability, which makes employees more likely to buy into transformation.

Moreover, leaders must establish a growth mindset throughout the organization—a mindset, as Dweck and Yeager (2021) puts it, of continuous learning and development. Reskilling and upskilling opportunities would not only respond to the concern of employees about their changing roles but also reinforce the organization's commitment to supporting its workforce through transitions. Other studies by Higgs and Rowland (2023) brought into light how important leaders as champions of change were in relation to the task of communicating vision and celebrating early wins to sustain momentum and morale.

Another essential factor is emotional intelligence, as it allows the leader, as claimed by Goleman and Boyatzis (2023), to sympathize with how employees feel regarding the issues arising and respond in a sensitive way to resistance. Leaders who do not attune to the emotional and psychological dimensions of change develop workplaces where people feel neither recognised nor understood. Such an emotional relation further assists in transcending fear in earning trust for the cultural change process.

This will be the case as transformational leadership styles are related to higher employee engagement and also to innovation throughout digital initiatives (Bass & Riggio, 2022). Leaders who empower employees by transferring decision-making powers and promoting team effort create a feeling of ownership, very instrumental in breeding a culture of adaptability. For example, organizations such as Microsoft have proven that the culture can change from a traditionally rigid structure into one that is agile and focused on growth through leader-driven cultural transformation.

Lastly, consistent alignment between leadership behaviors and organizational goals, as suggested by Kegan and Lahey (2023), ensures that the transformation is perceived as a unified effort rather than a series of isolated initiatives. Leaders who keep the employees informed about the pains and gains of transformation enable the employees to realize the bigger picture of the change process, making it more meaningful. By using these leadership strategies, organizations can create a culture that not only adopts digital transformation but thrives in an environment of constant change.

3. Encourage Innovation and Creativity Across All Levels

All levels of the organization need to be creative and innovative for sustainable growth. The culture of innovation should be encouraged by freeing employees to explore ideas, experiment, and take calculated risks. A very good example of how structured freedom can lead to significant breakthroughs, such as Gmail and Google News, is Google's "20% Time" program, where employees are allowed to dedicate part of their time to independent projects (Buchheit, 2009). Companies can facilitate creative thinking through brainstorming, innovation labs, and innovation-creating incentives. There is the impetus in today's business world to encourage creative and innovative behavior at every single level within their organizations (Hamel, 2007). As Amabile and Pratt (2022) point out, innovative cultures thrive through free thinking with full empowerment to conduct independent actions without fear or doubt. For example, Google's "20% Time" program is an organized program that has been demonstrated to result in groundbreaking innovations when the employees are allowed time off to pursue their passion. West and Farr (2021) explain that autonomy and trust are the most powerful creative output motivators in the workplace.

In addition, organizations can further encourage innovation by creating cross-functional teams comprised of different skills and perspectives. Edmondson (2023) recent research shows that diversity in teams and psychological safety are critical to the development of an environment where employees can freely express bold ideas without fear of judgment or failure. Brainstorming sessions and innovation challenges, as 3M has done successfully, provide employees with structured opportunities to generate ideas and collaborate on solutions, which leads to iconic products such as the Post-it Note (Johnson, 2022).

Companies should, therefore, develop and use systems that support rapid prototyping and iterative testing. As Brown and Katz (2023) put it, the integration of design thinking into a firm's approach transforms thoughts on creativity into practical solutions. This is possible because exploration and refinement in design thinking occur through loops of feedback. For instance, the human-centered design at IDEO has enabled many organizations to develop effective products for consumers in the marketplace.

Reward and recognition of innovative thinking is one of the other very important elements in building an innovation-driven culture. Zhou and Hoever (2022) indicate that intrinsic and extrinsic rewards, such as public recognition or monetary incentives, support creative behaviors and signal organizational commitment to innovation. Organizations, for example Salesforce, have managed to successfully implement innovation awards so as to celebrate employee contributions which motivate people but also set the benchmark for others to strive to reach.

Leaders play an imperative role in sparking creativity by curiously emulating and encouraging creative ideas. The authors, Gino and Staats (2023), enumerate that the very fact that leaders are also included in innovation practices and listen at all levels from the organization itself inspires a culture of inclusion in collaboration. In addition to this, empowering employees with a higher level of tools and advanced technologies, whether it is a platform powered with AI or more data visualization software allows them to better explore and implement different innovative solutions; McAfee & Brynjolfsson, 2023.

Through these tools, organizations become an ecosystem wherein creativity thrives, innovation will be a fundamental competency, and sustainable growth isn't just achieved but inevitable as well.

4. Adopt Agile Thinking for Enhanced Strategic Flexibility

Agile thinking, with its core values of flexibility, collaboration, and iterative progress, is what allows an organization to stay responsive in a rapidly changing digital landscape. Agile principles, born in software development, can be applied to much broader organizational strategy, allowing for quick pivots and adjustments based on real-time feedback (Rigby et al., 2016). True adoption of agile thinking at the strategic level requires cross-functional teams to be empowered to make decisions that would foster resilience and adaptability within the organization.

Agile thinking has now taken the driver's seat, far beyond software development, as an important key to organizational success amidst rapidly changing technologies and market disruptions. Core values of flexibility, collaboration, and iterative progress make it highly suitable for any organization in responding swiftly to ever-changing conditions of the customer ecosystem. Rigby et al., (2016) say that agile methodologies at the strategic level enable an organization to make informed, real-time decisions toward better outcomes. Agile thinking thus helps an organization move away from rigid top-down decision-making processes and toward cross-functional teams with greater autonomy.

Fast decision-making has, therefore, become a must for these teams operating in such environments where delays may be opportunities lost. As Denning explains (2022), Agile allows organizations to lower inefficiencies and concentrate on constant improvement with rapid iteration. Hence, allowing them to test the hypotheses quickly and alter the strategy even quicker before massive failures could ever happen. For example, companies like Spotify have ridden agile principles to scale rapidly while keeping strong focus on customer-centric innovation through autonomous, cross-functional squads that develop features based on user feedback (Kniberg & Ivarsson, 2021).

Agile thinking has brought forth another benefit: solving complex problems through the amalgamation of diverse teams with different perspectives, hence yielding creative solutions. The study by Hoda et al. (2020) explains this further by stating that agile environments cultivate a culture of collaboration, trust, and shared ownership and therefore accelerate problem-solving and improve the organization's performance. With resilience and adaptability being cornerstones of agile culture, it organizes and prepares for facing uncertainty with quicker adaptation as needed.

Similarly, agile thinking at a strategic level needs to incorporate feedback loops throughout decision-making. Companies like Amazon tune strategies and offerings in line with the iterative application of agile principles, attuning to real-time data informing any adjustments required in product development, marketing, and resource allocation (Brynjolfsson & McAfee, 2021). In that way, the organization remains attuned to market needs and competitive pressures and has the ability to change its course in order to maximize opportunities.

Agility at the strategic level also provides space for empowerment and ownership by the employees. When people at all levels within an organization are trusted to make decisions and given an opportunity to put forward their ideas, it leads toward a culture of innovation and accountability. Recent research conducted by Rasmusson et al. (2023) reveals that agile-thinking organizations have more engaged employees and better alignment in individual contributions to organizational goals. For instance, ING Bank's agile transformation empowered employees to become more proactive in their respective roles and to drive innovation and customer-focused solutions throughout the organisation (Van der Linden & Brehm, 2021).

Last but not least, although agile practices have mostly been applied to IT and product development, other applications in marketing, finance, and human resources have lately gained much ground. Agile principles applied across all departments give organizations better coherence and efficiency. This cross-functional application is endorsed by Shriram and Mukherjee (2022), who argue that agile practices should not be confined to specific departments but should be integrated throughout the organizational fabric for better agility and performance.

5. Build a Long-Term Vision and Flexible Roadmap

Align a clear, long-term vision with the mission and values of the company to guide the transformation journey. Leaders should develop a transformation roadmap, breaking the vision down into achievable short- and medium-term goals. The roadmap has to be flexible in nature so that it accommodates new insights and technological advancements (Brown & Katz, 2015). A flexible roadmap keeps the focus while providing a structure for evaluating progress and refining strategies.

Long-term, clear vision is the North Star for any complex transformational journey, keeping the organization aligned with the core values and larger mission. The leaders have to articulate this vision in a manner that will make it resonate through all levels of an organization so as to bring meaning and a sense of common purpose for everyone. According to Kotter (2022), a compelling vision helps galvanize employees to be inspired and involved in the transformation process, linking their role to the larger goals of the organization. This bigger vision, though, has to be brought down to smaller, bite-sized goals so that progress can remain tangible and measurable. A well-crafted transformation roadmap assists an organization in the translation of long-term aspiration into specific actionables.

Among the major hallmarks of an effective transformation roadmap is its flexibility; it will continue to be refined according to the insights that are ongoing and changes occurring in the landscape of technology. As Brown and Katz (2015) note, organizations have to be ready to change their strategies with the emergence of new opportunities and challenges to keep ahead of the competition and better themselves continuously. This adaptability ensures that organizations do not get rigidly bound by the initial assumptions and are able to change the course of action when necessary to avoid stagnation and encourage innovation. The dynamic roadmap provides a framework for monitoring and evaluation of progress made, and it allows leaders to establish if short- and medium-term goals are on course and make any necessary course corrections.

More proof for the value of a transformation roadmap in guiding an organization through digital changes comes from research by Westerman et al. (2022). The study hence points out that an organization that follows a step-by-step transformation, where goals and time frames are clearly outlined, has higher chances of succeeding in the long term. Focusing on short-term goals that contribute to the bigger vision can make the organization feel a sense of momentum driving further commitment in the transformation process.

What is more, the feedback loops in the roadmap allow leaders to gain great insights from employees, customers, and other stakeholders that can inform decision-making and help sharpen strategy. As Sull et al. (2023) have argued, organizations open to feedback and learning from experience will handle change better and avoid costly mistakes. That would be an essential way to go about continuous learning and improvement in today's fast-paced, tech-driven world, where new technologies and market dynamics can change the landscape quickly.

The leader will also need to communicate the transformation roadmap effectively across the organization. As noted by Gibbons and Binns (2021), transparent communication of the roadmap and its milestones not only navigate alignment but also empower employees to take ownership of the transformation process. Employees who understand the "why" behind the change and how their work contributes to the larger vision are more likely to be engaged and committed.

In real-world practice, companies like Microsoft showed that a clear but adaptive roadmap could drive a successful digital transformation. Breaking down the vision of cloud computing into clear goals and further adjusting them with real-time feedback from the market has made Microsoft successfully transform its business model in a way to sustain a lead in the technology industry (Gates, 2021). Likewise, the creation of flexible roadmaps for digitalization in manufacturing by Toyota has made innovative technologies such as AI and robotics part of the corporation's operations with no or little diminution of their effectiveness (Kobayashi, 2023).

6. Invest in Reskilling and Upskilling Programs

Digital transformation often necessitates new skill sets. Investing in reskilling and upskilling empowers employees, helping them meet the demands of a digital environment and reducing fears related to obsolescence (McKinsey, 2020). Providing training, mentorship, and on-the-job learning opportunities equips employees with the competencies needed to contribute effectively to transformation efforts.

Digital transformation requires a shift not only in organizational processes but also in the skills and competencies of employees. As noted by Brynjolfsson and McAfee (2021), the growing demand for advanced digital skills makes reskilling and upskilling initiatives crucial for maintaining competitive advantage. Companies that invest in these programs demonstrate a commitment to workforce development, alleviating employee concerns about the potential for job displacement due to automation and new technologies. A study by World Economic Forum (2023) reveals that organizations with robust reskilling programs are better equipped to navigate the complexities of digital transformation, as employees are empowered to adapt to new roles and technologies. Furthermore, organizations that integrate on-the-job learning and mentorship alongside formal training can cultivate a culture of continuous learning, where employees are supported in their professional growth and actively contribute to organizational innovation (Cascio & Montealegre, 2022).

7. Foster a Culture of Continuous Improvement and Learning

The transformation process demands continuous learning and improvement. Companies must foster employees with a "growth mindset" who are striving to learn new things and share insight. Thus, feedback mechanisms, learning platforms, and regular reviews support a culture of continuous improvement that allows firms to sharpen their strategies and make proactive adjustments (Hess, 2014).

Transformation is not an event; it's a journey where organizations have to be always agile and open to change. This, as Dweck (2006) stresses, is why a growth mindset by the employees is very instrumental in keeping up a culture of learning and improvement where individuals embrace challenges and view failures as means of learning. The above mindset fosters initiative among employees to take charge of their skill development and contribute proactively to the company's transformation agenda. More recent scholarship by Lichtenstein and Jordan (2021) has highlighted the importance of creating feedback loops and learning platforms that allow employees to share experiences and insights, which in turn shape decision-making and bring about enhanced organizational adaptability. Continuous learning is so integrated into organizational culture that it gives firms the state of being, where they fine-tune strategies to address and adapt to rapidly changing market demand in view of technology disruptions within the marketplace (Garvin, 2022).

8. Celebrate Successes and Recognize Contributions

Acknowledging employee contributions and celebrating small wins reinforces a positive transformation momentum. Recognitions, whether public or private, show employees that their efforts are valued, which boosts morale and engagement. Google, for example, publicly celebrates the stories behind successful projects like Gmail, reinforcing the idea that employee-driven innovation is essential (Levy, 2011).

Acknowledging employee contributions plays a significant role in maintaining positive momentum during transformation. Research by **Harter et al. (2020)** emphasizes that recognition not only boosts morale but also strengthens employees' emotional commitment to the organization, making them more likely to engage in innovation and contribute to transformation goals. **Public recognition**, such as company-wide announcements or awards, fosters a sense of achievement and inspires others to participate in the transformation journey. Additionally, private recognition, such as personalized thank-yous or one-on-one feedback, can strengthen relationships between managers and employees, fostering a sense of trust and appreciation (Cameron & Green, 2019). Companies like **Google** continue to leverage recognition to celebrate their most innovative projects, like Gmail and Google Maps, which were driven by employee initiative. According to **Schilling and Kuehl (2023)**, companies that celebrate these small wins encourage a culture of **empowerment** and **ownership**, motivating employees to consistently drive change and contribute to long-term success.

9. Leverage Technology to Enhance Collaboration and Productivity

Digital tools play a very instrumental role in effective collaboration, productivity, and decision-making. Cloud-based tools, collaborative platforms, and data analytics help in smoothing workflows and facilitate teamwork across remote locations. The adoption of such technologies and their accessibility allow employees to work effectively and align their efforts with strategic goals. Newport, 2016. Starbucks creates seamless, personalized coffee buying experience by leveraging technology. They have launched a mobile app for orders and payments, using AI for personalized product recommendations, and integrated digital loyalty programs with online and in-store experiences. (starbucks investor relations, 2023).

Digital tools have been very instrumental in bridging the geographical divides and allowing employees to work in tandem with each other in a fast-moving, digitally driven business environment. According to Westerman et al. (2014), technologies like cloud-based platforms and collaborative technologies support productivity since they allow real-time access to information and facilitate better communication. These technologies thus empower employees to work from anywhere, and hence, organizational efforts remain aligned to strategic goals irrespective of location. Moreover, with data analytics and AI-driven platforms, organizations are given the opportunity to make more informed decisions by enabling insights that help them predict trends, optimize resources, and customize offerings (Brynjolfsson & McAfee, 2021). One of the best examples is Starbucks, which has used its mobile app to create a seamless and highly personalized coffee-buying experience. By using AI in personalized product recommendations and integrating loyalty programs, Starbucks offers customers an enhanced, personalized experience across both online and in-store touchpoints (Starbucks Investor Relations, 2023). As revealed by Smith and Sutherland (2022), integration of digital tools enables companies to better meet the expectations of consumers, build deeper customer loyalty, and support organizational growth and competitiveness in the digital era.

Figure 2. Digital Tools

CASE STUDY: DIGITAL TRANSFORMATION IN EUROPEAN BANKS – EMPHASIZING COMMUNICATION AND FEEDBACK

Digital transformation in the banking sector has dramatically changed the landscape, pushing institutions to adapt to digital demands, improve customer experiences, and maintain competitive relevance. A study conducted by Brown and Bessant (2013) examined several European banks' experiences with digital transformation, emphasizing the critical role of open communication and structured feedback channels in ensuring smoother transitions. Findings from this study reveal that banks prioritizing employee engagement and transparent communication observed less resistance and greater employee buy-in throughout the transformation process.

The Role of Open Communication

Brown and Bessant (2013) referred to clear communication as one of the foundational elements in lowering resistance to digital transformation. The banks that had been keeping the workforce informed of the purpose of transformations, timelines, and possible role impacts clearly and frequently had perceptibly increased receptiveness among the workers. Leaders in such banks updated the transformation through many routes, such as team meetings, email newsletters, and intranet platforms. This approach kept the employees informed at all times, thus alleviating fears and making them feel part of the process.

In some banks, specific "transformation teams" were established to hold bi-weekly meetings with employees to address their queries and update the staff on the progress of transformation. These meetings were avenues where employees could raise their concerns, ask for clarification, and express their opinions. For instance, the introduction of AI-driven customer service tools raised concerns among employees related to job security and changes in job responsibilities. The management addressed these by explaining how AI would complement, not substitute, human employees, thus assuaging a lot of anxiety and fostering a much more supportive environment for change (Brown & Bessant, 2013).

Open communication, therefore, becomes a very important component in overcoming resistance to digital transformation because transparency and trust are built within an organization. According to Brown and Bessant (2013), clear and consistent communication helps reduce uncertainty and anxiety about potential changes; this is rather important in an industry such as banking, where the introduction of new technologies could mean disrupting established ways of doing things. Apart from keeping staff updated through emails, newsletters, and meetings, organisations increasingly rely on intranet platforms and online forums to provide staff with instant access to information so that they can live the transformation journey (Kotter, 2022). Where there is an open discussion about the issues in an organisation, transparency is felt through the workforce, and thus employee empowerment is increased, making them more willing to accept change.

In the case of banks, there is an effective approach in which transformation teams can be established directly engaging employees. These are the teams that intermediate between leadership and employees, ensuring that concerns are addressed in real time and that employees' feedback is taken into consideration (Sull et al., 2023). For instancc, the bi-weekly meetings, as mentioned in the case study, give a structured format where the employees will get the opportunity to bring up questions, concerns, and get clear explanations of how the transformation will take place. Employee buy-in is further supported when organizations openly show how new technologies like AI complement, rather than threaten, current roles. This level of communication provides management with an avenue to dispel fears of job losses by effectively demonstrating new tools and technologies as drivers of efficiency rather than job destructions (Garvin, 2022).

Other related research by Lichtenstein and Jordan (2021) supports this perspective with the belief that transparent communication may help to ensure more trust between leadership and its members, considering such openness about addressing concerns presented by employees. Moreover, involving employees in the conversation about AI or any other emerging technology could lead to a more collaborative adaptation where employees feel a sense of ownership over the change process. Citi Group successfully introduced AI-powered chatbots into its customer services by involving employee staff in pilot programs that trained them on how to use the technology and explained its role as augmenting, not replacing, their responsibilities (Citi Group, 2023). Similarly, HSBC used open forums and feedback mechanisms to help the employees understand how digital tools were going to empower them to serve customers better—thereby taking away the concern of job loss at the start ". HSBC, 2023

Sullivan and Albrecht, 2020, mention that the lack of explicit communication during this period of DT can cause disorientation, drift, and therefore, resistance to change. Hence, organizations should approach communication from a holistic perspective in which the message is coherent, clear, and continuous. When leaders communicate the what, how, and—most importantly—why of transformation, employees are much more likely to align with organizational goals, sense a shared purpose, and execute change more smoothly.

In the final analysis, regular communication and dedicated feedback platforms are the keys to fostering a culture of inclusivity and trust during transformation efforts. By addressing concerns directly, acknowledging employees' contributions, and reinforcing the benefits of new technologies, organizations can go a long way in reducing resistance and fostering a more positive mindset toward change.

Employee Engagement through Feedback Mechanisms

The second aspect of employee engagement during digital transformation is structured feedback channels, which provide leaders with a direct line to the concerns, suggestions, and feelings of employees. Indeed, as Brown and Bessant (2013) put it, feedback mechanisms like anonymous surveys, focus groups, and suggestion boxes give employees a chance to put forward their views without fear of retaliation, therefore creating a safe space for open dialogue. These mechanisms can be critical in the identification of underlying issues that may not be immediately visible to leadership. According to Harter et al. (2020), feedback channels allow organizations to identify specific barriers to successful change implementation, such as inadequate training programs or technology gaps, so that targeted solutions improve readiness among employees. Similarly, Sullivan and Albrecht (2020) posit that the sense of being heard and acted upon empowers employees' engagement and thereby motivates them to commit to change.

For instance, a bank in Europe, through feedback, realized that its front-line staff were suffering from the breakneck speed of digital transformation and were experiencing stress and burnout as a result. In realizing the problem, leadership revised the transformation timeline by phasing in new technology gradually so that staff could become proficient in one before taking on the next. This was in line with the phased approach suggested by McKinsey & Company (2020) in dealing with employees' well-being during times of transformation—something that would allow the company to bring in more long-term engagement and retention. In this scenario, slowing down the digital adoption rate translated into a much better reception toward the changes among employees, since they felt their management was responsive to their needs. The bank allowed staff time to adjust to new systems; it was an action that gave the bank its commitment to sustainable transformation with regard for the welfare and development of employees.

More recent literature by Rigby et al., (2016) suggests that such feedback-driven adjustments can also improve organizational resilience, as companies that seek and act on feedback tend to adapt more easily to challenges that may be unexpected. Such adjustments become all the more necessary in today's rapidly changing technology environment, in which people will be slow to adopt new tools unless there is a sense of gaining mastery or control over them. Moreover, as Lichtenstein and Jordan (2021) state, the practice of refining strategies based on employees' feedback would increase the potential for gaining their trust in leadership—one of the imperative components of a collaborative culture in the midst of change. Moreover, as Garvin (2022) points out, feedback channels represent an early warning system through which an organization can take precautions against obstacles likely to grow big enough to block the process of transformation.

In the final analysis, structured feedback channels not only bring valuable insight into the employee experience but also demonstrate an organization's commitment to creating an inclusive and adaptable transformation process. When feedback is actively used to make adjustments, it further solidifies a culture of mutual respect and collaboration, ensuring the transformation journey is both effective and sustainable.

Positive Impact on Transformation Success

This means that the study now concludes that the banks concerned with open communication and employees' feedback attained a higher rate of success in their digital transformation processes. Organizations that engaged the employees as active participants faced few disruptions and fast adoption of new technologies. The employees working in such banks felt valued and respected; this was mirrored in the high levels of engagement and willingness to contribute toward the organizational digital goals.

These findings underline the fact that, beyond technological change, digital transformation has to be viewed as a people-centric process. A two-way communication policy, coupled with the encouragement of managed feedback, kept the workforce of banks aligned with the objectives of workforce transformation, making the actual transition both smooth and effective. This would help to underline the importance of making staff part and parcel of the transformation journey in understanding that their acceptance and adaptation are fundamental for the long term in a competitive, digital-first financial sector (Brown & Bessant, 2013).

Examples of Companies That Failed to Manage Resistance in Digital Transformation

While digital transformation offers significant advantages, many organizations have faced failures due to poor change management and stakeholder resistance. Below are ten notable examples, along with lessons learned:

1. **General Electric (GE) – Predix Failure (2011-2018)**
    - GE launched Predix, an ambitious industrial IoT platform, but struggled with adoption due to internal resistance and external skepticism. Employees were not adequately trained, and customers found the platform complex.
    - **Lesson:** Digital transformation requires phased implementation, stakeholder training, and realistic expectations.
2. **Nike – Supply Chain System Failure (2000)**
    - Nike implemented a new supply chain management system but faced resistance from suppliers and internal teams. The rushed deployment led to inventory mismatches, costing the company over $100 million.
    - **Lesson:** A technology-driven transformation must align with operational realities and undergo proper testing before full-scale deployment.
3. **Hewlett-Packard (HP) – ERP Implementation Failure (2004)**
    - HP's $160 million ERP upgrade faced severe resistance from employees and partners, leading to communication breakdowns and delays in order fulfillment, resulting in a $400 million revenue shortfall.
    - **Lesson:** Digital transformation requires strong internal buy-in and thorough risk assessment before large-scale rollouts.
4. **BBC – Digital Media Initiative (2008-2013)**

- BBC invested £100 million in a digital content management system but failed due to leadership misalignment, lack of employee engagement, and failure to integrate the system with existing workflows.
- **Lesson:** Leadership commitment and cross-functional collaboration are essential for transformation success.

5. **J.C. Penney – Digital Pricing Strategy Failure (2011-2013)**
    - J.C. Penney removed discounts in favor of an everyday low-price strategy, but customers resisted the abrupt change, leading to massive revenue losses.
    - **Lesson:** External stakeholder (customer) resistance must be managed through gradual transition and clear communication.
6. **Target Canada – ERP and Supply Chain Breakdown (2013-2015)**
    - Target's rapid expansion into Canada failed because of a poorly executed digital supply chain system, resulting in stock shortages and pricing errors. Employees were unprepared, leading to a $2 billion loss.
    - **Lesson:** Large-scale digital rollouts require extensive employee training and contingency planning.
7. **Kodak – Digital Transformation Delays (1990s-2012)**
    - Kodak developed digital camera technology but resisted fully embracing digital photography due to fear of cannibalizing its film business, ultimately leading to its bankruptcy.
    - **Lesson:** Fear of disruption should not prevent necessary digital transformation efforts.
8. **Ford – In-house Digital Business (2014-2018)**
    - Ford invested in a Silicon Valley-style digital business, Ford Smart Mobility, but faced resistance from internal teams, leading to inefficiencies and restructuring.
    - **Lesson:** Legacy employees must be integrated into digital initiatives rather than sidelined.
9. **Metro Bank – IT System Failures (2019-2020)**
    - Metro Bank's new IT system faced repeated technical failures, with poor change management leading to customer dissatisfaction and regulatory scrutiny.
    - **Lesson:** Digital transformation must be carefully planned with risk mitigation strategies in place.
10. **McDonald's – AI-Driven Ordering System (2021-2022)**
    - McDonald's introduced AI-driven voice ordering in drive-thrus but faced customer resistance due to errors and dissatisfaction with automated interactions.

- **Lesson:** Digital adoption should prioritize user experience and address external stakeholder concerns proactively.

These failures highlight that resistance to change—whether from employees, customers, or leadership—can derail digital transformation. Organizations must proactively manage stakeholder engagement, provide adequate training, and ensure clear communication to drive successful transformation efforts.

Cultural Differences in Global Digital Transformations: Key Examples

Cultural diversity significantly impacts the success of digital transformation in multinational organizations. Failure to acknowledge regional differences in work ethics, communication styles, and technology adoption can lead to resistance, inefficiencies, and even failure. Here are five notable examples where cultural differences influenced digital transformation efforts:

1. **Walmart's Failure in Germany (1997-2006)**
   - Walmart entered the German market with a U.S.-style retail approach, enforcing American customer service norms like mandatory greetings and smiling at customers. German employees and consumers found this approach unnatural and intrusive, leading to high employee turnover and resistance to corporate policies.
   - **Lesson:** A standardized corporate culture may not work across borders. Digital and organizational transformations must be localized to fit cultural expectations.
2. **McDonald's Digital Expansion in China (2017-Present)**
   - McDonald's introduced self-service kiosks and mobile app ordering in China, but adoption lagged initially due to cultural preferences for face-to-face service and concerns over digital payment security. It had to collaborate with platforms like WeChat Pay and Alipay to gain local trust.
   - **Lesson:** Digital transformation must align with regional consumer behavior and integrate with local technology ecosystems.
3. **IBM's Workforce Restructuring in India (2010-2015)**
   - IBM faced challenges implementing a remote and flexible work culture in India, where employees traditionally valued structured hierarchies and in-person collaboration. The lack of clear adaptation strategies led to disengagement and productivity declines.

- **Lesson:** Workplace transformation should consider cultural attitudes toward hierarchy, flexibility, and remote work.

4. **Airbnb's Struggles in Japan (2018-Present)**
   - Airbnb's digital expansion faced regulatory and cultural roadblocks in Japan. The country's emphasis on privacy and community harmony led to resistance from locals, with authorities enforcing stricter home-sharing laws. Airbnb had to revise its strategy to comply with cultural norms.
   - **Lesson:** Regulatory and cultural sensitivities must be factored into global digital transformation efforts.
5. **eBay's Exit from China (2002-2006)**
   - eBay's failure in China was partly due to cultural misunderstandings in online consumer behavior. Chinese consumers preferred relationship-based transactions, which Alibaba's Taobao facilitated through chat features, while eBay maintained a rigid, Western-style auction model.
   - **Lesson:** Digital platforms must adapt to cultural expectations and user interaction preferences.

CULTURAL DIFFERENCES IN GLOBAL TRANSFORMATIONS

Cultural differences play a crucial role in the success or failure of global transformations. While leadership, communication, and technological advancements are often prioritized, organizations must also recognize the influence of cultural norms, values, and behaviors on transformation initiatives. Hofstede's Cultural Dimensions Theory (1980) provides a framework for understanding how national and organizational cultures impact business practices. Key factors such as power distance, individualism vs. collectivism, and uncertainty avoidance significantly shape how employees and external stakeholders respond to change. Organizations expanding into global markets or implementing digital transformation initiatives across diverse regions often face resistance due to cultural misalignment. For instance, hierarchical cultures (e.g., Japan, India) may resist decentralized decision-making processes, while highly individualistic cultures (e.g., the U.S., Australia) may expect greater autonomy in adopting change. Misinterpretation of communication styles can also hinder transformation efforts. Direct communication, common in Western cultures, may be perceived as too aggressive in high-context cultures like China or the Middle East, where indirect and relationship-driven communication is preferred.

Several companies have faced challenges due to cultural misunderstandings in their transformation efforts. Walmart's failure in Germany (1997-2006) stemmed from an inability to adapt American corporate culture to German business practices, leading to employee dissatisfaction and resistance. Similarly, eBay's exit from China (2006) was largely due to its failure to accommodate local consumer behaviors and trust-based transaction preferences. These cases highlight the importance of cultural sensitivity when implementing transformation strategies. To navigate cultural differences effectively, organizations must adopt localized change management strategies. This includes involving regional leadership, customizing communication styles, and ensuring that transformation efforts align with local work ethics and consumer behaviors. Cross-cultural training and stakeholder engagement can further enhance the success of global transformation initiatives. By acknowledging and integrating cultural diversity into transformation strategies, organizations can improve acceptance, reduce resistance, and achieve sustainable change across different markets.

THE ROLE OF TECHNOLOGY IN MANAGING RESISTANCE AND ENHANCING COLLABORATION

One of the primary reasons for resistance is uncertainty and lack of information. Digital platforms like Microsoft Teams, Slack, and Trello help in real-time communication and project tracking, ensuring that employees stay informed and engaged throughout the transformation process. By fostering open dialogue, these tools reduce anxiety and resistance stemming from a lack of clarity. Collaboration tools also address concerns related to workflow disruption. Cloud-based solutions such as Google Workspace and SharePoint enable employees to access information from anywhere, making transitions smoother for remote and hybrid teams. Additionally, AI-powered chatbots and virtual assistants can provide instant support, helping employees adapt to new technologies more effectively.

Another major factor in resistance is a perceived lack of involvement in decision-making. Crowdsourcing platforms and employee feedback tools like Mentimeter, Yammer, and Workplace by Meta allow organizations to gather input from employees, making them feel heard and valued. When employees actively participate in the transformation process, their resistance decreases, and ownership of change increases. Gamification is another technology-driven approach that enhances engagement. Platforms like Kahoot!, Miro, and Coursera for Business integrate learning and development modules, helping employees upskill in a fun and interactive manner. This reduces resistance by addressing fears of obsolescence and skill gaps.

Furthermore, data analytics and AI-driven sentiment analysis tools such as Qualtrics and Tableau help organizations identify resistance patterns and proactively address concerns. These technologies provide insights into employee sentiment, allowing leadership to tailor their change management strategies effectively. By leveraging digital tools, organizations can make transformation more transparent, engaging, and inclusive, ultimately reducing internal resistance and fostering a collaborative culture. The strategic integration of technology into change management ensures smoother transitions and higher adoption rates for new processes and systems.

Addressing the Weakness: Incorporating Emerging Transformation Trends

As organizations undergo rapid digital transformation, integrating Generative AI, Blockchain, Digital Twins, No-Code Platforms, and Metaverse technologies can significantly enhance change management and reduce resistance. These emerging tools provide automation, transparency, and personalization, enabling smoother transitions for both employees and external stakeholders. Generative AI facilitates personalized training, predictive resistance analysis, and real-time AI chatbots, ensuring employees adapt to new systems with minimal disruption. Blockchain enhances trust and security through immutable record-keeping, smart contracts for automation, and transparent supply chain management, reducing skepticism and misinformation during transformation.

Additionally, Digital Twins allow organizations to simulate change and optimize performance, mitigating risks before real-world implementation. No-Code and Low-Code platforms empower employees by allowing them to build applications and workflows independently, reducing reliance on IT teams and easing resistance to digitalization. The Metaverse and AR technologies enable immersive collaboration and training, helping remote teams engage in interactive learning experiences. Together, these innovations foster greater transparency, inclusivity, and efficiency, making transformation efforts more seamless and widely accepted.

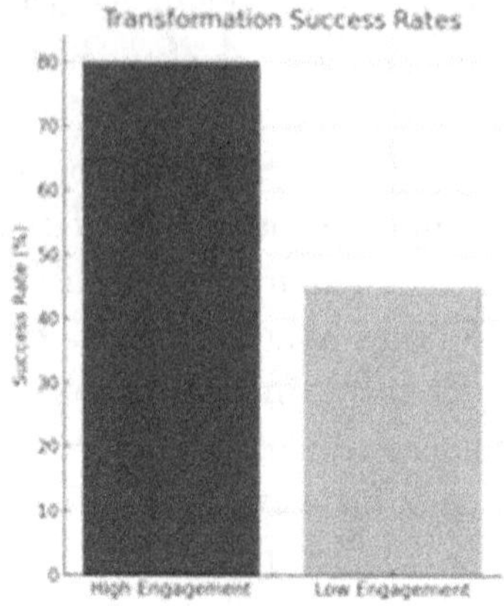
Transformation Success Rates
Success Rate (%)
High Engagement
Low Engagement

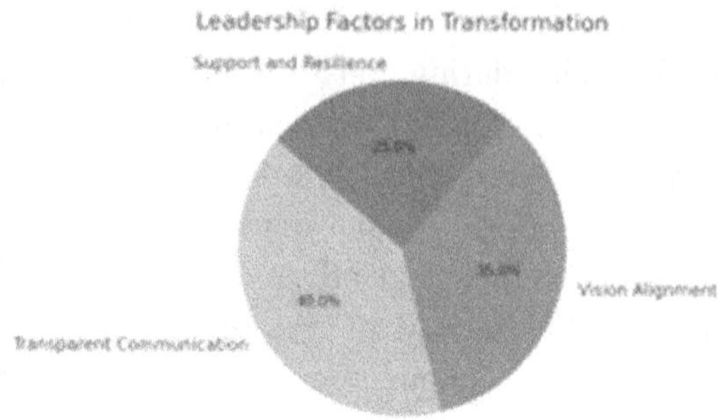
Leadership Factors in Transformation
Support and Resilience
25.0%
35.0%
Vision Alignment
40.0%
Transparent Communication

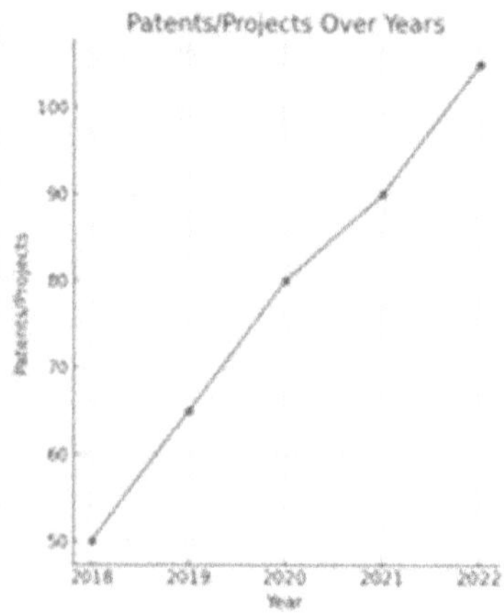
Patents/Projects Over Years
Patents/Projects
Year

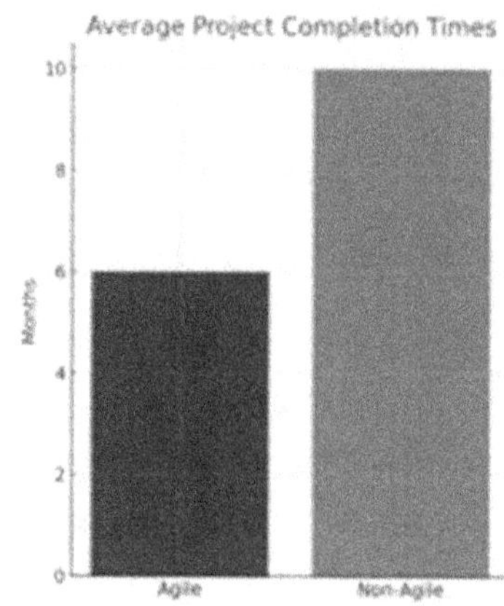
Average Project Completion Times
Months
Agile
Non-Agile

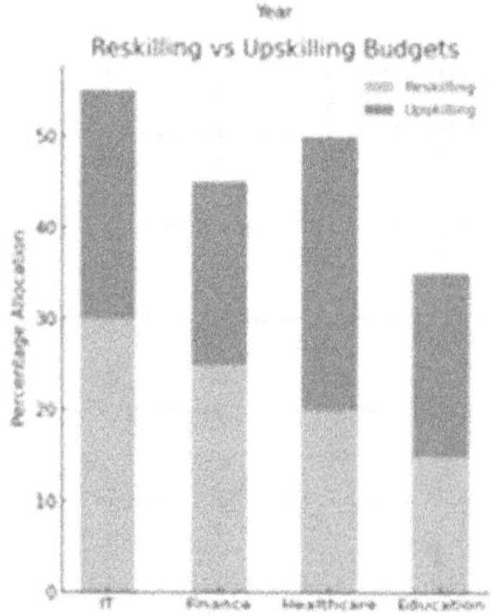
Reskilling vs Upskilling Budgets
Reskilling
Upskilling
Percentage Allocation
IT
Finance
Healthcare
Education

REFERENCES

Anderson, D., & Ackerman Anderson, L. (2010). *Beyond Change Management: How to Achieve Breakthrough Results Through Conscious Change Leadership*. Jossey-Bass.

Bain & Company. (2021) *Overcoming resistance to change: The role of leadership*. https://www.bain.com (Accessed: 20 December 2024).

Bharat, K. (2003). *Google News: How and Why*. Google Blog. https://googleblog .blogspot.com

Brown, T., & Bessant, J. (2013). Case study findings on digital transformation in European banks illustrate the role of open communication in reducing resistance.

Brown, T., & Katz, B. (2015). *Change by Design: How Design Thinking Transforms Organizations and Inspires Innovation*. Harper Business.

Buchheit, P. (2009). *The Origin of Gmail*. [Blog post on Gmail's development] https://paulbuchheit.blogspot.com

Deloitte. (2023) *2023 Global Human Capital Trends: Leading in a Boundaryless World*. Available at: https://www2.deloitte.com

Forbes Insights. (2023) *Transformational leadership and organizational success*. https://www.forbes.com

Gallup. (2022) *State of the Global Workplace Report*. https://www.gallup.com

Gartner. (2022) *The critical role of managers in organizational transformation*. https://www.gartner.com

Hamel, G. (2007). *The Future of Management*. Harvard Business School Press.

Harvard Business Review. (2022) *Supportive managers and employee adaptability in transformation*. https://www.hbr.org

Hess, E. D. (2014). *Learn or Die: Using Science to Build a Leading-Edge Learning Organization*. Columbia Business School Publishing.

Kotter, J. P. (2012). *Leading Change*. Harvard Business Review Press.

Levy, S. (2011). *In the Plex: How Google Thinks, Works, and Shapes Our Lives*. Simon & Schuster.

McKinsey. (2020). The impact of a clear digital strategy on transformation success.

McKinsey & Company. (2020) *The Science of Organizational Change*. https://www.mckinsey.com

Microsoft Corporation. (2023) *Annual report 2023: Reinventing productivity and business processes*. https://www.microsoft.com

Newport, C. (2016). *Deep Work: Rules for Focused Success in a Distracted World*. Grand Central Publishing.

Prosci (2021) *Top Reasons for Resistance to Change*. https://www.prosci.com

PwC. (2022) *2022 Global Culture Survey*. https://www.pwc.com

Rigby, D., Sutherland, J., & Takeuchi, H. (2016). Embracing Agile. *Harvard Business Review*.

SHRM. (2022) *The impact of leadership on employee retention during transformations*. https://www.shrm.org

Starbucks Corporation. (2023) *Starbucks Annual Report 2023: Reinventing the customer experience*. https://www.starbucks.com

Tesla, Inc. (2023) *Tesla Annual Report 2023: Accelerating the world's transition to sustainable energy.* [REMOVED IF= FIELD]https://www.tesla.com

Towers Watson. (2019) *The Role of Communication in Change Management*. [REMOVED IF= FIELD]https://www.towerswatson.com

Chapter 4
Leadership for Digital Transformation

Dinesh Nilkant
Jain University, India

S. Hemanth Kumar
https://orcid.org/0000-0003-2490-1690
Jain University, India

Prabha Kiran
Westminster International University in Tashkent, Uzbekistan

Sumithra Sreenath
https://orcid.org/0009-0001-1441-6520
Independent Researcher, India

ABSTRACT

The chapter explores the critical role of leadership in navigating the complexities of digital transformation, emphasizing the integration of technology, culture, and strategy to drive sustainable organizational change. Through comprehensive case studies, it highlights successful transformation journeys at Microsoft, Tesla, and Walmart, alongside challenges faced by organizations like General Electric. The discussion delves into frameworks such as Transformational Leadership Theory, Agile methodologies, and Digital Maturity Models, underscoring their importance in fostering adaptability, innovation, and inclusivity. Key themes include addressing resistance to change, bridging digital skills gaps, and embedding ethical practices in leadership strategies. The chapter concludes by providing actionable recommendations for future leaders to harness emerging technologies, balance innovation with sustainability, and build resilient organizations in an interconnected global economy.

DOI: 10.4018/979-8-3373-1320-7.ch004

1. INTRODUCTION

In the digital era, transformation is no longer a choice but an imperative for organizations seeking to remain competitive and resilient. Digital transformation encompasses the integration of advanced technologies such as artificial intelligence (AI), blockchain, and cloud computing into core business operations, redefining how value is created and delivered. Gartner (2022) emphasizes that this process extends beyond technology adoption to encompass strategic, operational, and cultural changes. The COVID-19 pandemic underscored the urgency of digital transformation, with businesses worldwide accelerating adoption timelines by three to five years (McKinsey & Company, 2020). A vivid example is the rapid rise in telemedicine during the pandemic, which surged by 78 times in the U.S. alone (World Economic Forum, 2021), showcasing the critical role of digital tools in ensuring continuity during global crises.

The pace of technological disruption in recent years has further heightened the need for digital transformation. Emerging technologies like IoT, 5G, and data analytics are driving fundamental changes in industries, enabling businesses to innovate, scale, and adapt. McKinsey & Company (2023) predicts that digital transformation could add $13 trillion to global GDP by 2030, underscoring its economic significance. Moreover, digital tools are addressing pressing global challenges such as climate change and cross-border collaboration. For instance, IoT-enabled smart grids have reduced carbon emissions by optimizing energy usage (International Energy Agency, 2022), while blockchain enhances transparency in global supply chains, ensuring ethical sourcing (Deloitte, 2023). These developments highlight digital transformation as a multifaceted driver of progress, making it indispensable for organizations aiming for sustainable growth.

Leadership as the Linchpin for Successful Transformation

Effective leadership is the cornerstone of successful digital transformation, ensuring that technological goals align with organizational culture and strategic objectives. Leaders are tasked with navigating complexities, fostering innovation, and driving sustainable change. Forbes (2023) aptly describes leadership in the digital context as the "engine that drives change, innovation, and value creation." A prime example is Satya Nadella's leadership at Microsoft, where his vision of a cloud-first strategy transformed the company into a global leader, achieving a market capitalization of over $2.5 trillion by 2023 (Harvard Business Review, 2022).

Leadership during digital transformation requires a multifaceted approach. Visionary thinking is critical to articulating a clear strategy that aligns with both technological advancements and business objectives. Leaders must also serve as change agents, fostering a culture of trust and collaboration to overcome resistance. Kotter International (2021) found that organizations emphasizing change management principles—such as creating urgency and sustaining momentum—are 70% more successful in achieving transformation outcomes.

Fostering digital literacy within leadership teams is another pivotal aspect. Deloitte (2022) highlights that organizations with digitally competent leaders are 15% more likely to implement transformation projects successfully. This competence enables leaders to bridge gaps between technical teams and business units, ensuring seamless integration of digital initiatives. Moreover, emotionally intelligent leaders excel in managing workforce anxieties and maintaining morale during periods of uncertainty. Harvard Business Review (2022) notes that such leaders are 37% more effective in driving employee engagement and performance during transformational changes.

Ethical stewardship is equally critical in the digital age, where concerns about data privacy, algorithmic bias, and environmental sustainability loom large. Leaders must establish governance frameworks that prioritize transparency and inclusivity while fostering innovation responsibly. The World Economic Forum (2023) emphasizes that balancing these considerations is key to achieving long-term societal and organizational benefits. Real-world examples like Adobe's shift to a subscription model under Shantanu Narayen, which increased market capitalization tenfold between 2010 and 2022 (McKinsey & Company, 2021), underscore the transformative power of visionary leadership.

Importance of Balancing People, Processes, and Technology in Digital Initiatives

The success of digital transformation hinges on achieving a harmonious balance among people, processes, and technology—the three pillars of organizational success. McKinsey & Company (2022) found that organizations integrating these components strategically are 2.5 times more likely to achieve their digital transformation goals. Neglecting any one pillar can lead to inefficiencies, resistance, or outright failure.

People represent the cornerstone of digital transformation, as the workforce determines how effectively technologies are utilized. The World Economic Forum (2020) projects that by 2025, 50% of employees will require significant reskilling due to the rapid adoption of automation and AI. Leadership must prioritize upskilling initiatives, foster collaboration, and address resistance to change to cultivate a culture that embraces innovation. Emotional intelligence and transparent communication are vital in minimizing workforce anxieties and building trust.

Processes form the second pillar, ensuring operational efficiency and scalability. Redesigning workflows to align with digital tools is critical for eliminating redundancies and enhancing performance. Deloitte (2022) identifies poorly defined processes as a leading barrier to successful transformation. Adopting frameworks such as Lean Six Sigma and Agile methodologies can optimize workflows and facilitate seamless integration of technologies. For instance, robotic process automation (RPA) reduces manual errors and accelerates turnaround times, enhancing customer satisfaction.

Technology, the enabler of transformation, must be strategically selected and implemented. Advanced tools like AI and blockchain offer unparalleled opportunities for innovation but must align with broader organizational objectives to deliver value. Gartner (2022) warns against a "technology-first" approach, advocating for technologies to be integrated into well-defined strategies. Amazon exemplifies this balance through its use of AI for demand forecasting (technology), optimization of supply chains (processes), and extensive employee training programs (people). This holistic approach has enabled Amazon to maintain its competitive edge and deliver exceptional customer experiences.

The interdependence of these pillars underscores the need for a systemic approach to digital transformation. Neglecting people can lead to cultural resistance, while disregarding processes can result in inefficiencies. Similarly, overlooking technology can render transformation efforts obsolete in a competitive market. A Harvard Business Review study (2021) found that organizations balancing these elements report a 45% improvement in operational efficiency and a 30% increase in customer satisfaction. Leaders who adopt a holistic approach, integrating people, processes, and technology, are better positioned to achieve their digital goals and create sustainable value for stakeholders.

2. DIGITAL LEADERSHIP DEFINED

In the age of rapid technological change, leadership has taken on a new dimension, evolving from traditional paradigms of stability and control to a dynamic model characterized by adaptability, innovation, and collaboration. Digital leadership emphasizes the ability to navigate complex, rapidly changing environments while integrating technology with people and processes. This section delves into the distinctions between traditional and digital leadership, the essential traits required for effective digital leaders, and the evolving role of leaders in the digital economy.

Differentiating Between Traditional and Digital Leadership

Traditional leadership has historically focused on stability, efficiency, and hierarchical decision-making, emphasizing control and consistency in achieving organizational goals. In contrast, digital leadership thrives in environments of uncertainty and rapid change, requiring leaders to prioritize innovation, flexibility, and decentralized decision-making. While traditional leaders might rely on established frameworks and incremental improvements, digital leaders must be comfortable with experimentation, disruption, and iterative progress.

Frameworks such as Kotter's Change Model underscore the shift from hierarchical to agile decision-making. Digital leaders facilitate collaborative ecosystems, breaking down silos and fostering cross-functional teams. For instance, during Adobe's transition to cloud services, CEO Shantanu Narayen demonstrated digital leadership by encouraging decentralized decision-making and aligning teams with a shared vision of innovation (McKinsey & Company, 2022). Similarly, Satya Nadella's leadership at Microsoft redefined the company's culture to embrace a growth mindset, enabling rapid digital transformation and a market capitalization exceeding $2.5 trillion in 2023 (Harvard Business Review, 2022).

The hallmark of digital leadership lies in the ability to adapt organizational structures to support decentralized workflows. Leaders such as Elon Musk at Tesla exemplify this by creating flat hierarchies that empower teams to innovate independently while maintaining strategic alignment. This approach has been instrumental in Tesla's breakthroughs in electric vehicle production and autonomous driving technologies, showcasing how digital leaders leverage flexibility to drive success in volatile markets.

Essential Traits of Digital Leaders

Digital leaders must embody traits that enable them to navigate the complexities of the modern business landscape. Resilience, adaptability, foresight, and emotional intelligence are among the most critical attributes for success in digital transformation.

- **Resilience:** The ability to persevere through challenges and setbacks is a cornerstone of effective digital leadership. For example, Satya Nadella's resilience was evident in Microsoft's pivot to cloud computing, overcoming initial skepticism and fostering a culture of continuous learning and innovation (Harvard Business Review, 2022).
- **Adaptability:** In an era where technologies evolve rapidly, adaptability is non-negotiable. Shantanu Narayen's leadership during Adobe's transition from traditional software licensing to a subscription-based model highlights the importance of adaptability. This bold move not only disrupted Adobe's legacy systems but also positioned the company as a leader in digital creativity tools, increasing its market valuation tenfold between 2010 and 2022 (McKinsey & Company, 2022).

- **Foresight:** Visionary leaders anticipate market trends and position their organizations to capitalize on future opportunities. Elon Musk's foresight in pioneering sustainable energy solutions through Tesla exemplifies how a forward-thinking approach can disrupt industries and set new benchmarks for innovation.
- **Emotional Intelligence (EI):** Digital transformation often triggers workforce anxiety due to the fear of job displacement and changes in workflows. Leaders with high EI can address these concerns effectively by fostering trust, maintaining open communication, and ensuring employee engagement. Harvard Business Review (2022) found that leaders with strong emotional intelligence drive 37% higher employee satisfaction during organizational change.

The Evolving Role of Leaders in the Digital Economy

In the digital economy, leaders must integrate technology with human-centric strategies to create value while navigating ethical considerations. This dual responsibility requires a balanced approach that prioritizes both technological innovation and stakeholder well-being. Sundar Pichai, CEO of Google, exemplifies this leadership model by driving advancements in AI and cloud computing while emphasizing inclusivity and ethical governance. Under Pichai's leadership, Google launched AI tools to enhance accessibility for individuals with disabilities, showcasing how technology can be leveraged to address societal challenges (World Economic Forum, 2022).

Ethical leadership has become increasingly critical as organizations adopt advanced technologies like AI and blockchain. Leaders are tasked with ensuring that these tools are used responsibly, addressing issues such as data privacy, algorithmic bias, and environmental sustainability. The World Economic Forum (2023) underscores the importance of ethical frameworks in maintaining trust and transparency in digital initiatives.

Furthermore, digital leaders must embrace the principles of inclusivity and diversity to drive innovation. Research by Deloitte (2022) indicates that diverse teams are 20% more likely to develop creative solutions, making DEI (Diversity, Equity, and Inclusion) a strategic imperative in digital transformation efforts. Leaders who prioritize inclusivity create environments where diverse perspectives thrive, enabling organizations to adapt to the complex demands of global markets.

Digital leadership redefines traditional paradigms by emphasizing adaptability, innovation, and collaboration. As businesses navigate the complexities of the digital economy, leaders must embody resilience, foresight, and emotional intelligence to drive transformation successfully. Real-world examples, such as Satya Nadella at Microsoft, Shantanu Narayen at Adobe, and Sundar Pichai at Google, highlight the transformative potential of visionary leadership in aligning technological advancements with ethical and human-centric strategies. By fostering inclusive cultures, empowering decentralized decision-making, and balancing innovation with responsibility, digital leaders can ensure sustained success in an era defined by rapid change and technological disruption.

3. KEY LEADERSHIP COMPETENCIES

In the digital era, leadership competencies have evolved to address the multifaceted challenges of technological disruption, organizational change, and global competition. Three critical competencies stand out: digital literacy combined with an innovation-driven mindset, strategic foresight integrated with effective change management, and emotional intelligence (EI) paired with fostering a collaborative organizational culture. Together, these competencies empower leaders to drive digital transformation by aligning technology with human-centric strategies, anticipating future trends, and cultivating resilient and cohesive teams.

Digital Literacy and an Innovation-Driven Mindset

Digital Literacy as a Foundational Competency

Digital literacy has emerged as a prerequisite for leadership in the digital age, enabling leaders to navigate complex technological landscapes effectively. It encompasses the ability to understand, evaluate, and leverage digital tools, technologies, and data to achieve organizational objectives. Gartner (2022) highlights that organizations with digitally literate leaders are 3.5 times more likely to succeed in their transformation initiatives. Leaders proficient in technologies such as artificial intelligence (AI), blockchain, and the Internet of Things (IoT) can bridge the gap between technical teams and business units, ensuring seamless integration of digital initiatives. For instance, Piyush Gupta, CEO of DBS Bank, leveraged digital technologies to transform DBS into a leading digital bank, integrating AI and machine learning to enhance customer experiences and operational efficiency (Harvard Business Review, 2022).

Innovation-Driven Leadership

Beyond technical proficiency, an innovation-driven mindset is crucial for sustaining competitiveness in an increasingly digital economy. Such a mindset involves fostering a culture of experimentation, encouraging risk-taking, and embracing bold ideas to address emerging opportunities. McKinsey & Company (2021) emphasizes that innovation-driven leaders enable teams to challenge traditional paradigms and explore transformative solutions. Jeff Bezos exemplifies this leadership philosophy. By prioritizing customer obsession and long-term thinking, Bezos transformed Amazon from a bookstore into a technology powerhouse, pioneering innovations such as Amazon Web Services (AWS), AI-driven logistics, and one-click purchasing (MIT Sloan Management Review, 2022).

Synergizing Digital Literacy and Innovation

The combination of digital literacy and innovation is transformative. Leaders like Elon Musk at Tesla demonstrate how technical expertise, coupled with visionary thinking, can disrupt industries. Musk's deep understanding of engineering and digital technologies, aligned with his vision for sustainable energy, enabled Tesla to develop groundbreaking solutions like autonomous vehicles and energy storage systems (Deloitte, 2022). This synergy highlights that leaders must balance technical acumen with a forward-thinking approach to maximize the impact of digital transformation.

To nurture these competencies, organizations must invest in leadership development programs that combine technical training with innovation-focused workshops. For example, Adobe's innovation programs empower employees to pitch creative ideas, fostering a culture of continuous learning and experimentation (Harvard Business Review, 2021).

Strategic Foresight: Charting the Path Forward

Strategic foresight involves anticipating future trends, disruptions, and opportunities to craft long-term strategies. Unlike reactive planning, foresight emphasizes proactive decision-making based on emerging technologies, market dynamics, and societal shifts. The World Economic Forum (2023) underscores that leaders with foresight can navigate uncertainty, positioning their organizations for resilience and growth. Leaders like Sundar Pichai at Google exemplify foresight by prioritizing AI and cloud computing investments, ensuring Google remains at the forefront of technological innovation (Harvard Business Review, 2022). Similarly, Mary Barra's leadership at General Motors demonstrates foresight in transitioning toward electric vehicles (EVs), addressing both consumer demand and regulatory trends (McKinsey & Company, 2021).

Effective Change Management

Strategic foresight must be complemented by effective change management to translate vision into actionable outcomes. Change management addresses the structural, cultural, and human dimensions of transformation, ensuring that stakeholders align with organizational goals. Kotter International (2021) reveals that 70% of change initiatives fail due to inadequate management strategies, underscoring the importance of leadership in this domain. Leaders must communicate the vision clearly, build coalitions, and foster employee buy-in to overcome resistance. Agile frameworks further enhance adaptability, enabling organizations to respond to evolving demands efficiently (Deloitte, 2022).

Integrating Foresight and Change Management

The interplay between foresight and change management is evident in Adobe's transition to a subscription-based cloud model. While foresight enabled leaders to recognize cloud computing's potential, robust change management practices ensured seamless implementation through workforce reskilling and transparent communication (Harvard Business Review, 2021). Microsoft's transformation under Satya Nadella is another compelling example. Nadella's foresight in prioritizing cloud computing and AI, coupled with his emphasis on cultural change, propelled Microsoft's market capitalization to over $2.5 trillion by 2023 (Gartner, 2022).

Leaders must adopt a balanced approach, leveraging foresight to shape strategy and change management to execute it effectively. This dual competency ensures that organizations remain agile, competitive, and well-positioned for sustained success.

Emotional Intelligence and Collaborative Culture

The Role of Emotional Intelligence in Leadership

Emotional intelligence (EI) is a critical leadership attribute in managing the human aspects of digital transformation. As defined by Goleman, EI involves self-awareness, empathy, and interpersonal skills, enabling leaders to build trust, resolve conflicts, and inspire confidence. Harvard Business Review (2021) reports that emotionally intelligent leaders improve employee engagement by 37%, enhancing overall performance during change initiatives. For instance, Shantanu Narayen, CEO of Adobe, used empathy and active listening to address employee concerns during Adobe's transition to a subscription model, fostering trust and minimizing resistance (McKinsey & Company, 2021).

Building a Collaborative Organizational Culture

Collaboration is the cornerstone of successful digital initiatives. The interdisciplinary nature of transformation requires cross-functional teams to work cohesively. Deloitte (2022) notes that organizations with collaborative cultures are 1.8 times more likely to achieve their transformation goals. Leaders play a pivotal role in creating environments where teamwork thrives. Strategies include promoting inclusivity, investing in collaborative tools like Slack and Microsoft Teams, and recognizing team achievements. Leaders with high EI excel in fostering collaboration by addressing tensions and aligning teams around shared objectives.

Overcoming Resistance Through EI and Collaboration

Resistance to change is a common challenge in transformation. Leaders can mitigate this by involving employees in decision-making, addressing concerns transparently, and creating a sense of ownership. Satya Nadella's leadership at Microsoft exemplifies this approach. By emphasizing empathy, inclusivity, and continuous feedback, Nadella transformed Microsoft's culture into one that values collaboration and innovation, driving sustainable success (Harvard Business Review, 2022).

Key leadership competencies—digital literacy paired with an innovation-driven mindset, strategic foresight integrated with change management, and emotional intelligence combined with fostering collaboration—form the backbone of effective digital transformation. Leaders who embody these traits can navigate the complexities of technology, align organizational goals with emerging opportunities, and inspire teams to achieve collective success. By prioritizing these competencies, organizations can thrive in an era defined by rapid change, ensuring resilience, adaptability, and long-term growth.

4. LEADERSHIP FRAMEWORKS FOR DIGITAL TRANSFORMATION

In the age of digital transformation, leaders are tasked with guiding organizations through complex, technology-driven changes. Traditional leadership frameworks, often hierarchical and linear, fall short in addressing the agility, innovation, and collaboration required in digital ecosystems. This section explores three key frameworks that empower leaders to navigate and excel in these challenging environments: the application of Transformational Leadership Theory, the use of Digital Maturity Models to assess readiness and capability, and the integration of Agile methodologies with iterative leadership practices.

Transformational Leadership Theory in Digital Ecosystems

The "Four I's" of Transformational Leadership

Transformational Leadership Theory, conceptualized by James Burns and expanded by Bernard Bass, provides a robust framework for leading in rapidly evolving digital ecosystems. Its principles—Idealized Influence, Inspirational Motivation, Intellectual Stimulation, and Individualized Consideration—offer actionable strategies for fostering innovation, resilience, and engagement in teams.

- **Idealized Influence** emphasizes leaders acting as role models, inspiring trust and commitment through integrity and a compelling vision.
- **Inspirational Motivation** entails articulating a transformative vision that aligns organizational goals with digital imperatives, motivating employees to strive for excellence.
- **Intellectual Stimulation** encourages creative problem-solving, empowering employees to challenge norms and explore new ideas.
- **Individualized Consideration** ensures tailored support and mentorship, addressing the diverse needs of team members to foster growth and adaptability.

Transformational Leadership in Practice

Digital ecosystems thrive on interconnected networks of technologies, stakeholders, and processes, requiring a departure from rigid hierarchical structures. Transformational leaders champion decentralized decision-making, continuous learning, and human-centric approaches to technology adoption. A Gartner (2022) study found that organizations led by transformational leaders are 2.8 times more likely to achieve their digital transformation objectives, underscoring the effectiveness of this approach.

For instance, Satya Nadella's leadership at Microsoft illustrates the power of transformational principles. Nadella's focus on culture, collaboration, and a growth mindset revitalized Microsoft, positioning it as a global leader in cloud computing and AI (Harvard Business Review, 2021). Similarly, Shantanu Narayen of Adobe demonstrated intellectual stimulation and inspirational motivation by steering the company's transition to a subscription-based cloud model, fostering innovation and employee engagement (McKinsey & Company, 2021).

Benefits and Challenges

Transformational leadership enhances employee engagement, fosters innovation, and builds resilience to change. However, its application requires significant emotional intelligence, a balance between visionary thinking and practical execution, and the ability to address both technological and cultural shifts simultaneously. Leaders must actively navigate these complexities to fully harness the transformative potential of their organizations.

Digital Maturity Models: Assessing Readiness and Capability

Understanding Digital Maturity Models

Digital maturity models provide structured frameworks for organizations to evaluate their current capabilities, identify gaps, and create actionable roadmaps for digital transformation. These models assess dimensions such as strategy, technology, culture, processes, and workforce capabilities, offering insights into the organization's readiness for change. According to McKinsey & Company (2021), digital maturity significantly predicts an organization's success in achieving transformation goals.

Key Models and Frameworks

1. **Deloitte's Digital Maturity Model**: Focuses on five dimensions—strategy, culture, technology, operations, and insights—offering comprehensive recommendations for improvement (Deloitte, 2022).
2. **Capgemini and MIT Sloan Digital Maturity Model**: Categorizes maturity into four stages (Beginners, Fashionistas, Conservatives, and Digital Masters) and emphasizes the integration of digital technologies (MIT Sloan Management Review, 2022).
3. **BCG's Digital Acceleration Index**: Provides a quantitative assessment by benchmarking against industry peers, helping organizations prioritize initiatives and allocate resources effectively (Boston Consulting Group, 2021).

Real-World Applications

Digital maturity models have been pivotal in guiding transformation initiatives for several organizations:

- **Unilever** used Deloitte's model to prioritize AI-driven supply chain solutions, achieving a 20% increase in operational efficiency (Deloitte, 2022).

- **Siemens** applied BCG's Digital Acceleration Index to implement Industry 4.0 strategies, optimizing manufacturing processes and reducing costs (Boston Consulting Group, 2021).
- **ING Bank** leveraged Capgemini and MIT Sloan's framework to transition to a customer-centric digital banking platform, improving customer satisfaction by 15% (MIT Sloan Management Review, 2022).

Leadership's Role in Leveraging Maturity Models

Leaders play a critical role in interpreting the findings from digital maturity assessments and integrating them into organizational strategies. By using these models, leaders can prioritize resources, align initiatives with long-term goals, and measure progress over time. Moreover, thcsc frameworks provide a common language for collaboration across departments, enhancing alignment and fostering a shared vision for digital transformation.

Agile Methodologies and Iterative Leadership

Agile Methodologies in Leadership

Agile methodologies, initially developed for software development, have become a cornerstone for managing digital transformation across industries. Agile emphasizes adaptability, cross-functional collaboration, and iterative progress through short cycles known as sprints. Leaders employing Agile frameworks prioritize experimentation, feedback integration, and customer-centric solutions, ensuring that transformation initiatives remain responsive to changing demands (Deloitte, 2022).

Principles of Iterative Leadership

Iterative leadership complements Agile practices by fostering a mindset of continuous improvement and adaptability. Key principles include:

1. **Feedback Integration**: Leaders actively incorporate stakeholder input to refine strategies.
2. **Experimentation and Risk-Taking**: Encouraging calculated risks to drive innovation.
3. **Empowerment and Collaboration**: Building trust and fostering cross-functional teamwork.
4. **Continuous Learning**: Adapting strategies based on data insights and lessons learned from each iteration.

Applications and Success Stories

1. **Spotify's Agile Culture**: Spotify adopted Agile principles to maintain a customer-centric focus, implementing autonomous teams to rapidly develop and deploy features, ensuring continuous innovation (Harvard Business Review, 2021).
2. **GE's Industrial Internet**: GE leveraged Agile and iterative leadership to accelerate IoT-enabled product development, reducing time-to-market and enhancing customer satisfaction (MIT Sloan Management Review, 2022).
3. **ING Bank**: ING's Agile transformation empowered cross-functional teams, enabling faster, more personalized digital banking solutions (McKinsey & Company, 2021).

Advantages and Challenge

Agile and iterative leadership enable faster time-to-market, enhanced innovation, and improved stakeholder alignment. However, their successful implementation requires cultural shifts, adequate training, and consistent leadership commitment to overcome resistance and maximize the benefits of these approaches.

Leadership frameworks for digital transformation—Transformational Leadership Theory, digital maturity models, and Agile methodologies with iterative leadership—provide comprehensive strategies for navigating the complexities of the digital age. By fostering innovation, aligning capabilities, and promoting adaptability, these frameworks empower leaders to drive meaningful change and ensure long-term success in an increasingly dynamic and competitive global environment.

5. CASE STUDIES AND LESSONS LEARNED IN DIGITAL TRANSFORMATION

Digital transformation, a cornerstone of contemporary organizational evolution, involves not just the adoption of technology but a deeper integration of culture, leadership, and strategy. The success or failure of such initiatives often rests on the ability of leaders to inspire, adapt, and innovate. By examining notable successes, such as Microsoft's cultural overhaul under Satya Nadella, and cautionary tales like General Electric's missteps in its Industrial Internet of Things (IIoT) initiative, we gain a nuanced understanding of the factors shaping digital transformation. Additionally, case studies from Tesla, Walmart, ING Bank, and Mayo Clinic provide insights into how different sectors address challenges, fostering innovation and resilience in a rapidly changing digital landscape.

Microsoft's Cultural Transformation Under Satya Nadella

When Satya Nadella took the helm at Microsoft in 2014, the company faced declining relevance in the tech industry, a rigid internal culture, and stagnation in its traditional Windows-based business. Nadella's vision extended beyond financial performance, aiming to reinvent Microsoft's identity as a leader in cloud computing, artificial intelligence (AI), and workplace collaboration tools. His leadership exemplifies the interplay of empathy, innovation, and strategic foresight.

A key element of Nadella's approach was cultural transformation. Recognizing that Microsoft's "fixed mindset" was a barrier to innovation, he embraced Carol Dweck's concept of a "growth mindset," encouraging employees to experiment, collaborate, and embrace failure as a learning opportunity. This cultural shift was instrumental in dismantling silos within the organization, fostering cross-functional collaboration, and inspiring creativity. For example, Nadella promoted transparency by sharing progress and setbacks, ensuring employees felt connected to the company's broader mission (Harvard Business Review, 2021).

Another hallmark of Nadella's leadership was his commitment to customer-centric innovation. The launch of Microsoft Azure positioned the company as a leader in cloud computing, with flexible, scalable solutions tailored to the diverse needs of businesses. Similarly, Microsoft Teams emerged as a pivotal tool during the global shift to remote work, highlighting the company's ability to anticipate and address market demands (McKinsey & Company, 2022). Furthermore, Nadella's emphasis on empathy and inclusivity reshaped both workplace practices and product design, making Microsoft a champion of accessibility and purpose-driven innovation.

Microsoft's transformation yielded remarkable results. The company's market capitalization rose from $300 billion in 2014 to over $2.5 trillion by 2023, underscoring its resurgence as a tech giant (Gartner, 2023). Employee engagement and morale improved significantly, and Microsoft consistently ranked among the best global employers. These achievements illustrate the profound impact of empathetic leadership and a growth-oriented culture on digital transformation.

General Electric's Missteps in Digital Transformation

While Microsoft's success highlights the importance of culture and adaptability, General Electric (GE) offers a cautionary tale of how strategic misalignment and executional challenges can derail digital initiatives. Under CEO Jeff Immelt, GE launched an ambitious plan in 2011 to dominate the Industrial Internet of Things

(IIoT) with its proprietary platform, Predix. Despite its potential to revolutionize industrial analytics, the initiative faltered due to fundamental missteps.

GE overestimated the readiness of industrial customers to adopt Predix, underestimating the infrastructure limitations and cybersecurity concerns that impeded adoption (MIT Sloan Management Review, 2022). Moreover, the company's attempt to position Predix as a universal platform for diverse sectors diluted its focus, preventing it from establishing a strong foothold in any specific industry. Cultural resistance within GE further exacerbated these challenges, as employees struggled to align with the agile, innovative mindset required for digital transformation.

The technical execution of Predix also faced significant hurdles. High development costs, scalability issues, and delays undermined its success. By 2017, GE Digital was incurring nearly $1 billion annually without delivering proportional returns, forcing the company to scale back its ambitions (Harvard Business Review, 2021). Frequent leadership changes during this period compounded these challenges, creating uncertainty and disrupting strategic continuity.

GE's experience underscores critical lessons for organizations embarking on digital transformation. The importance of aligning initiatives with market readiness, focusing on core competencies, fostering cultural adaptability, and maintaining consistent leadership cannot be overstated. By addressing these factors, organizations can avoid the pitfalls that hindered GE's progress.

Tesla's Leadership in Innovation and Automation

Tesla's rise as a leader in electric vehicles (EVs) and renewable energy exemplifies the transformative power of bold vision and innovative leadership. Under Elon Musk, Tesla disrupted the automotive industry by integrating cutting-edge technology with sustainable practices. Musk's vision extended beyond producing EVs; it encompassed a broader mission to accelerate the world's transition to sustainable energy.

Central to Tesla's success is its commitment to vertical integration, controlling key aspects of production, including battery manufacturing and software development. This approach not only ensures quality control but also allows Tesla to innovate at an unprecedented pace. For instance, the company's advancements in autonomous driving technology have redefined industry standards and positioned Tesla as a leader in AI-driven mobility solutions (Deloitte, 2022).

Tesla's customer-centric approach also contributes to its success. The company's direct-to-consumer model fosters strong relationships with its customer base, while its focus on sustainability resonates with environmentally conscious consumers. These strategies highlight the importance of aligning organizational goals with broader societal trends to achieve relevance and impact in the digital age.

Walmart's Digital Reinvention

Walmart's transformation from a traditional retail giant to a digital leader demonstrates the power of omnichannel strategies and data-driven decision-making. Recognizing the competitive threat posed by e-commerce platforms like Amazon, Walmart invested heavily in integrating digital technologies into its operations. Initiatives such as improving inventory management through predictive analytics and creating seamless online-to-offline shopping experiences positioned Walmart as a formidable player in the retail space (MIT Sloan Management Review, 2022).

The success of Walmart's digital transformation underscores the importance of aligning technology with customer experience. By prioritizing convenience and personalization, Walmart enhanced customer loyalty and satisfaction. This case illustrates how traditional organizations can leverage digital tools to remain competitive in an increasingly digital marketplace.

ING Bank's Agile Transformation

ING Bank's adoption of Agile methodologies provides a blueprint for fostering responsiveness and innovation in financial services. By restructuring its operations around cross-functional teams, ING accelerated the development of digital banking solutions tailored to customer needs. This decentralized approach empowered teams to make decisions and adapt quickly to market changes, demonstrating the value of agility in navigating the complexities of digital transformation (McKinsey & Company, 2021).

ING's focus on customer-centricity further reinforced its success. The bank's digital platforms, such as mobile banking apps, prioritized user-friendly interfaces and personalized services, setting new benchmarks for customer engagement in the financial sector. This case highlights the role of organizational agility and customer focus in driving successful digital initiatives.

Mayo Clinic's Digital Patient Care

In the healthcare sector, Mayo Clinic's integration of telemedicine, AI, and predictive analytics exemplifies the potential of technology to enhance patient outcomes. By leveraging these tools, Mayo Clinic personalized treatments and improved diagnostic accuracy, addressing both patient needs and operational efficiency

(Gartner, 2022). The clinic's emphasis on collaboration across medical, technical, and operational teams ensured the seamless implementation of these initiatives.

Mayo Clinic's success demonstrates the importance of aligning technology with human-centric strategies. By prioritizing patient experiences and fostering interdisciplinary collaboration, healthcare organizations can achieve meaningful transformation while maintaining their core mission of care and compassion.

Lessons from Leading Industries

These case studies highlight common themes that drive successful digital transformation. Effective leadership is essential for articulating a clear vision and inspiring teams. Customer-centric innovation ensures that transformation initiatives remain relevant and impactful. Cultural adaptability and collaboration foster resilience and creativity, while strategic focus and alignment prevent resource dilution.

Moreover, leveraging data as a strategic asset enables organizations to optimize decision-making and operations. As digital ecosystems continue to evolve, these lessons provide a roadmap for organizations seeking to navigate the complexities of transformation and achieve sustainable growth.

The journey of digital transformation is multifaceted, encompassing technology, culture, leadership, and strategy. Success stories like Microsoft's resurgence and Tesla's industry disruption offer invaluable lessons on the importance of vision, adaptability, and customer focus. Conversely, failures like GE's IIoT initiative underscore the risks of strategic misalignment and cultural resistance.

By learning from these case studies, organizations can craft resilient strategies that balance innovation with execution, ensuring long-term success in an increasingly digital world. The integration of leadership insights and cross-industry lessons creates a comprehensive framework for navigating the challenges and opportunities of the digital age.

Contemporary Successes and Challenges

In addition to the cases of Microsoft, GE, Tesla, Walmart, ING Bank, and Mayo Clinic, several other organizations provide valuable insights into the complexities of digital transformation. These examples emphasize the importance of sector-specific strategies, global collaboration, and emerging technologies in achieving sustainable growth.

Amazon's Mastery in Supply Chain Automation

Amazon's dominance in e-commerce is a testament to its relentless focus on supply chain automation and customer-centric innovation. The company's implementation of robotics and AI-driven systems in fulfillment centers has redefined operational efficiency. By leveraging AI-powered algorithms for inventory management and last-mile delivery, Amazon consistently exceeds customer expectations in terms of speed and reliability (Harvard Business Review, 2021).

The introduction of Amazon Prime further highlights the company's strategic alignment with consumer demands. By combining fast delivery, exclusive content, and seamless digital experiences, Amazon transformed customer loyalty into a cornerstone of its business model. This case underscores the importance of integrating technology with consumer-focused strategies to achieve competitive advantage.

Alibaba's Integration of Big Data and AI

In the context of global e-commerce, Alibaba's approach to big data and AI offers a compelling case study. The company's use of AI-powered recommendation engines and predictive analytics enhances user experiences by delivering personalized product suggestions. Additionally, Alibaba's "smart logistics" platform integrates real-time data to optimize delivery routes, reduce costs, and minimize environmental impact (MIT Sloan Management Review, 2022).

Alibaba's annual Singles' Day event exemplifies the scalability of its digital infrastructure. By handling billions of transactions within a single day, the company demonstrates the potential of cloud computing and AI in driving operational efficiency. This case highlights the transformative power of data analytics and scalability in sustaining competitive growth.

Netflix's Evolution Through Data-Driven Content

Netflix's transformation from a DVD rental service to a global streaming giant underscores the value of data-driven decision-making. By analyzing user preferences and viewing habits, Netflix produces original content tailored to audience interests, resulting in high engagement and subscriber retention. Shows like *Stranger Things* and *The Crown* exemplify the company's ability to align content strategy with consumer demand (Gartner, 2023).

Netflix's proactive scenario planning also contributed to its success. Anticipating the decline of physical media, the company pivoted to a streaming model, leveraging its digital platform to stay ahead of competitors. This case emphasizes the importance of aligning business models with emerging technologies and consumer trends to ensure long-term relevance.

Zoom's Rise During the Pandemic

Zoom emerged as a critical tool during the COVID-19 pandemic, addressing the global need for virtual communication. The company's user-friendly interface, scalability, and ability to adapt to high demand positioned it as a leader in video conferencing. By focusing on accessibility and seamless integration with other platforms, Zoom enabled organizations, educational institutions, and individuals to maintain connectivity during unprecedented disruptions (McKinsey & Company, 2021).

Zoom's experience highlights the importance of agility and responsiveness in digital transformation. The company's rapid scaling to accommodate millions of new users demonstrates the value of anticipating market needs and investing in infrastructure to support growth.

Key Cross-Industry Takeaways

These additional case studies reinforce several critical themes in digital transformation:

1. **Scalability and Infrastructure**: Companies like Amazon and Alibaba illustrate the importance of investing in robust digital infrastructure to handle high volumes and complexity.
2. **Customer-Centric Innovation**: Organizations that prioritize understanding and addressing consumer needs, such as Netflix and Zoom, achieve higher engagement and loyalty.
3. **Data-Driven Strategies**: The integration of big data and AI enables businesses to personalize experiences, optimize operations, and anticipate market trends.
4. **Agility and Adaptability**: Proactive scenario planning and responsiveness to market changes, as seen in Netflix's pivot and Zoom's scalability, are essential for sustaining relevance.

Emerging Challenges and Future Directions

As digital transformation continues to evolve, organizations must address emerging challenges to maintain momentum and competitiveness. These challenges include navigating regulatory frameworks, addressing ethical concerns related to AI, and balancing technological innovation with sustainability.

Ethical AI and Governance

The rapid adoption of AI raises concerns about transparency, bias, and accountability. Organizations must implement ethical frameworks to guide AI development and deployment, ensuring that these technologies align with societal values. For example, Google's AI principles explicitly prohibit harmful applications, setting a benchmark for responsible innovation (MIT Technology Review, 2022).

Sustainability in Digital Transformation

Sustainability has become a critical consideration in digital strategies. Companies must adopt energy-efficient technologies, reduce carbon footprints, and embrace circular economy principles. Initiatives such as Microsoft's commitment to becoming carbon negative by 2030 demonstrate the integration of environmental goals into digital transformation efforts (Harvard Business Review, 2022).

Leadership for a Global Digital Ecosystem

As businesses operate in increasingly interconnected ecosystems, leaders must develop skills in cross-cultural collaboration, negotiation, and stakeholder management. Building global digital ecosystems requires balancing local relevance with global integration, emphasizing the importance of cultural intelligence and inclusive leadership.

The case studies of Microsoft, GE, Tesla, Amazon, Alibaba, Netflix, and Zoom, along with the lessons drawn from these examples, underscore the multifaceted nature of digital transformation. Successful transformation requires a balance of visionary leadership, cultural adaptability, and technological innovation. By learning from both successes and failures, organizations can craft strategies that foster resilience, drive innovation, and deliver sustainable value.

As the digital landscape continues to evolve, organizations must remain agile, ethical, and consumer-focused, leveraging emerging technologies to navigate complexities and seize opportunities. The integration of lessons from diverse industries provides a roadmap for leaders aiming to thrive in an increasingly digital world.

6. CHALLENGES IN LEADERSHIP FOR DIGITAL TRANSFORMATION

Digital transformation presents a host of challenges for leaders, requiring them to navigate resistance to change, address digital skills gaps, and ensure ethical and sustainable practices in the age of AI and automation. These challenges demand a multifaceted leadership approach that blends emotional intelligence, strategic foresight, and an unwavering commitment to ethical standards.

Addressing Resistance to Change at the Organizational Level

Resistance to change remains one of the most significant barriers to digital transformation. Employees often fear job displacement, distrust leadership intentions, or feel overwhelmed by the complexity of new technologies. Addressing this resistance requires leaders to adopt empathetic, transparent, and inclusive strategies.

1. **Transparent Communication and Visionary Leadership** Leaders must clearly articulate the vision and purpose behind digital transformation. Regular updates, tailored messaging, and storytelling can help employees understand how these changes align with organizational goals and their personal growth. Harvard Business Review (2022) notes that organizations prioritizing transparency in communication are 40% more successful in managing change initiatives. Procter & Gamble (P&G) serves as an example; its leadership engaged employees by co-creating digital solutions, fostering trust and collaboration.
2. **Building a Coalition of Change Agents** Identifying and empowering internal champions who advocate for digital transformation can amplify its acceptance across teams. These change agents influence peers by demonstrating the benefits of transformation and addressing skepticism. According to McKinsey & Company (2021), organizations that involve employees in transformation initiatives reduce resistance by 25%.
3. **Investing in Employee Development** Providing robust upskilling and reskilling opportunities alleviates fears of redundancy and empowers employees to embrace new technologies. The World Economic Forum (2023) estimates that 50% of the workforce will need reskilling by 2025 due to digitalization. P&G's upskilling programs and cultural shift toward innovation ensured its successful transition to digital-first operations.
4. **Cultivating a Culture of Innovation,** a culture that values experimentation and psychological safety encourages employees to adapt and thrive in the digital age. Deloitte (2022) emphasizes that organizations fostering innovation are twice as likely to achieve transformation goals. Leaders can achieve this by

recognizing and rewarding employees who embrace change and contribute to digital initiatives.

Overcoming Digital Skills Gaps Among Leadership and Teams

The digital skills gap is a critical challenge for organizations pursuing digital transformation. Bridging this gap ensures that both leaders and teams possess the technical proficiency and adaptability required to leverage emerging technologies effectively.

1. **Promoting a Learning Culture** Leaders must cultivate an environment where continuous learning is celebrated. Structured training programs, microlearning modules, and peer mentorship opportunities can help employees acquire new skills without feeling overwhelmed. Deloitte (2022) found that organizations with strong learning cultures are 46% more likely to be industry leaders.
2. **Investing in Leadership Development** Leadership must be equipped with the technical literacy needed to make informed decisions and align technology with organizational strategy. Programs like Amazon's "Upskilling 2025," which trains employees in high-demand areas like AI and cloud computing, highlight the importance of preparing leadership and teams for future challenges (Harvard Business Review, 2021).
3. **Leveraging Digital Training Tools** Learning management systems (LMS), gamification, and virtual reality (VR) simulations provide scalable, engaging, and personalized training solutions. These tools not only enhance skill acquisition but also reduce resistance by making learning enjoyable and relevant.
4. **Partnering with External Experts** Collaborating with technology vendors, industry experts, or academic institutions can address internal skill deficiencies. Certifications from platforms like AWS and Google Cloud, combined with tailored workshops, enable employees to stay ahead of technological trends.

Ethical and Sustainable Leadership in the Age of AI and Automation

The integration of AI and automation into digital transformation introduces ethical and sustainability challenges that require careful navigation. Issues like algorithmic bias, data privacy, and environmental impact necessitate a leadership approach rooted in responsibility and transparency.

1. **Addressing Ethical Imperatives**

- **Bias and Fairness:** AI systems must be developed using unbiased data to avoid perpetuating inequalities. Leaders can implement explainable AI (XAI) frameworks to ensure transparency in decision-making processes. Sundar Pichai's focus on inclusive AI at Google demonstrates the importance of ethical leadership (MIT Technology Review, 2022).
- **Privacy and Data Security:** Leaders must enforce robust data governance measures to protect sensitive information. Compliance with regulations like the General Data Protection Regulation (GDPR) ensures accountability and safeguards stakeholder trust.

2. **Embedding Sustainability into Strategy** Automation and AI-driven technologies often have significant environmental impacts, such as high energy consumption in data centers. Leaders should adopt renewable energy solutions and align digital initiatives with the United Nations Sustainable Development Goals (SDGs). Microsoft's commitment to becoming carbon negative by 2030 exemplifies this approach (Harvard Business Review, 2022).
3. **Investing in Workforce Reskilling** Automation disrupts traditional job roles, creating a moral imperative for leaders to reskill employees. By equipping workers with the skills to transition into higher-value roles, organizations can mitigate the social impact of automation. The World Economic Forum (2023) projects that while 85 million jobs may be displaced by automation by 2025, 97 million new roles will emerge in areas like AI and digitalization.
4. **Fostering Trust and Transparency** Transparent communication about the benefits, risks, and implementation of AI and automation is critical for building stakeholder confidence. Leaders should engage employees and customers in open dialogues to ensure alignment with organizational values and societal expectations.

Real-World Examples

1. **Procter & Gamble's Employee-Centric Approach** P&G successfully navigated resistance by involving employees in co-creating digital solutions and offering comprehensive training. This inclusive approach fostered a sense of ownership and eased the transition to digital-first operations.
2. **Amazon's "Upskilling 2025" Initiative** Amazon's large-scale upskilling program demonstrates a commitment to addressing digital skills gaps and preparing employees for automation-driven changes. By investing in its workforce, Amazon ensures sustainable growth and innovation.

3. **Microsoft's Ethical AI Practices** Microsoft established the AETHER committee to oversee the ethical use of AI technologies and committed to reducing its carbon footprint. These initiatives reflect the importance of balancing technological advancement with ethical responsibility and sustainability.

Leadership challenges in digital transformation—resistance to change, digital skills gaps, and ethical concerns—require innovative, empathetic, and forward-thinking strategies. By fostering transparent communication, investing in workforce development, and embedding ethics and sustainability into organizational practices, leaders can navigate these challenges effectively. The success of digital transformation hinges on leaders' ability to align technology with human-centric values, ensuring that the journey benefits all stakeholders while driving long-term growth and innovation.

7. RECOMMENDATIONS AND FUTURE DIRECTIONS

In an era dominated by rapid digital transformation, organizations must equip their leadership with the skills and vision required to navigate unprecedented technological, operational, and cultural changes. Recommendations and future directions for building effective digital leadership include practical guidelines, structured training programs, and strategies for cultivating robust leadership pipelines. These efforts will not only address immediate transformation needs but also ensure long-term resilience and competitiveness.

Practical Guidelines for Developing Digital Leadership Capabilities

1. **Invest in Leadership Development Programs** Leadership development programs tailored to the demands of digital transformation are essential. These programs should integrate technical knowledge, strategic decision-making, and interpersonal skills to prepare leaders for future challenges.

   - **Key Actions:**

- Offer specialized certification courses on AI, data analytics, and automation in partnership with academic institutions or platforms such as Coursera and MIT Sloan.

- Implement experiential learning opportunities, enabling leaders to manage digital transformation projects or participate in cross-functional innovation teams.
- Develop internal digital leadership programs like General Electric's Digital Leadership Development Program, which prepares leaders for digital business (McKinsey & Company, 2022).

2. **Foster a Digital-First Mindset** Digital transformation demands a mindset that prioritizes innovation, experimentation, and adaptability over traditional approaches.

- **Key Actions:**

- Encourage the adoption of Agile methodologies, enabling leaders to respond quickly to dynamic market conditions.
- Establish digital key performance indicators (KPIs) in leadership evaluations, fostering accountability for digital initiatives.
- Promote continuous learning through hackathons, industry events, and digital workshops.
  - **Example:** Adobe's shift to a subscription-based cloud model demonstrates the importance of instilling a digital-first mindset in leadership, enabling the company to adapt to evolving market needs (Harvard Business Review, 2021).

3. **Embed Collaboration Across Functions** Effective digital transformation requires seamless collaboration among technical, operational, and strategic teams.

- **Key Actions:**

- Form cross-functional leadership teams to oversee digital projects.
- Introduce tools like Slack and Microsoft Teams to streamline workflows and foster collaboration.
- Host interdepartmental knowledge-sharing sessions to promote mutual understanding of objectives and challenges.
  - **Example:** Spotify's squad-based structure facilitates cross-functional teamwork, empowering small teams to deliver innovative solutions efficiently (MIT Sloan Management Review, 2022).

4. **Prioritize Ethical and Inclusive Leadership** Leaders must address the ethical implications of digital tools, particularly AI, and promote inclusivity within their organizations.

- **Key Actions:**

  - Develop ethical AI governance frameworks to mitigate biases and enhance accountability.
  - Ensure diversity in leadership roles to integrate varied perspectives into digital strategies.
  - Conduct periodic audits to evaluate the ethical impact of initiatives.
    - **Example:** Microsoft's AI ethics framework and inclusive leadership practices demonstrate a commitment to responsible innovation, fostering trust among stakeholders (Deloitte, 2022).

5. **Build Resilience Through Scenario Planning** Scenario planning equips leaders to anticipate disruptions and develop contingency strategies.

- **Key Actions:**

  - Identify potential risks such as cybersecurity threats and economic volatility.
  - Use data insights to update and refine strategic plans regularly.
  - Engage in resilience training to strengthen leadership adaptability during crises.
    - **Example:** Netflix's proactive scenario planning facilitated its transition from DVD rentals to a leading streaming platform, showcasing resilience in a competitive industry (Gartner, 2023).

Leadership Training Programs and Continuous Development Initiatives

Training programs tailored to the evolving digital landscape are critical for developing resilient and adaptive leaders. These programs should encompass technical, strategic, and ethical dimensions to ensure leaders are prepared to drive meaningful transformation.

1. **Blended Learning Approaches** A combination of traditional and digital learning methods enhances training outcomes.

- **Key Components:**

  - **E-Learning Platforms:** Provide flexible access to courses on digital transformation topics.
  - **Experiential Learning:** Incorporate projects and case studies for practical application.

- **Microlearning Modules:** Offer concise lessons on specific digital tools and strategies.
    - **Example:** Deloitte's Leadership Academy integrates workshops, online learning, and peer mentoring to build digital leadership skills (Deloitte, 2022).

2. **Customized and Continuous Development** Tailored programs ensure training relevance while continuous development maintains leadership competencies.

- **Key Components:**

- Role-based training for diverse leadership levels.
- Ongoing mentorship programs and peer-learning communities.
- AI-driven tools to personalize learning and identify skill gaps.
    - **Example:** Amazon's "Machine Learning University" offers customized AI training programs to enhance technical proficiency among its leaders (Harvard Business Review, 2021).

3. **Emerging Trends in Training** Innovative training methods are reshaping leadership development in the digital age:

- AI-based platforms for real-time feedback and learning analytics.
- Immersive technologies like VR and AR for simulating complex scenarios.
- Focus on sustainability, aligning training with environmental and social goals.
- **Example:** Unilever's leadership programs emphasize digital readiness and sustainability, preparing leaders for future challenges (MIT Sloan Management Review, 2022).

Building Leadership Pipelines for Sustained Transformation

Leadership pipelines ensure a steady supply of capable leaders equipped to sustain digital transformation efforts.

1. **Identifying High-Potential Talent** AI-driven tools and competency frameworks help identify employees with leadership potential.
    - **Example:** IBM's AI-enabled assessments evaluate adaptability, collaboration, and technical skills among employees to build leadership pipelines (McKinsey & Company, 2022).

2. **Fostering Development Opportunities** Structured programs, rotational assignments, and cross-functional projects provide experiential learning for future leaders.
    - **Example:** Procter & Gamble's global leadership programs incorporate real-world problem-solving and rotational roles to prepare emerging leaders (Harvard Business Review, 2021).
3. **Embedding Leadership into Strategy** Leadership development must align with organizational transformation goals.
    - **Example:** Microsoft integrates sustainability and innovation into its leadership pipeline strategies, ensuring alignment with its long-term vision (Gartner, 2023).

Future Directions for Digital Leadership

The role of digital leadership will continue to evolve with emerging technologies and global challenges. Key focus areas include:

1. **AI-Driven Decision-Making:** Leaders must integrate AI tools to enhance strategic planning and operational efficiency.
2. **Sustainability Integration:** Energy-efficient technologies and carbon-neutral initiatives will be integral to future strategies.
3. **Global Ecosystems:** Leaders must navigate increasingly interconnected digital ecosystems, requiring advanced negotiation and cultural intelligence skills.
4. **Commitment to Lifelong Learning:** Continuous education will remain critical as technologies evolve at an unprecedented pace.

Developing digital leadership capabilities is crucial for organizations to thrive in an increasingly digital and interconnected world. By investing in structured leadership programs, fostering inclusive and ethical practices, and building robust leadership pipelines, organizations can empower their leaders to drive transformation. Future leaders will need to adapt to emerging trends such as AI integration, sustainability, and global collaboration, ensuring their organizations remain resilient and competitive in a rapidly changing landscape. Organizations that prioritize leadership development will not only navigate the complexities of digital transformation but also seize opportunities for innovation and growth.

REFERENCES

Adobe. (2021). Transitioning to a subscription-based model: Lessons in digital transformation. Harvard Business Review. https://hbr.org

Amazon. (2021). Upskilling 2025: Preparing for a future-ready workforce. Harvard Business Review. https://hbr.org

Boston Consulting Group. (2021). The digital acceleration index: How to assess digital maturity and prioritize next steps. https://www.bcg.com

Deloitte. (2022). Digital maturity and the agile transformation journey. https://www2.deloitte.com

Deloitte. (2022). Digital maturity assessment framework. https://www2.deloitte.com

Deloitte. (2022). Digital transformation through a learning culture: The importance of upskilling and innovation. https://www2.deloitte.com

Deloitte. (2022). Diversity and innovation in the digital age. https://www2.deloitte.com

Deloitte. (2022). Leadership for the digital era: Fostering innovation and collaboration. Deloitte Insights. https://www2.deloitte.com

Deloitte. (2022). Leadership strategies for digital transformation. https://www2.deloitte.com

Deloitte. (2022). Tesla's innovation ecosystem: Driving sustainability and leadership in the automotive industry. https://www2.deloitte.com

Deloitte. (2023). Blockchain's role in ethical supply chains. https://www2.deloitte.com

Gartner. (2022). Leadership challenges in digital transformation: Addressing skills gaps and resistance. https://www.gartner.com

Gartner. (2022). Predicting digital transformation success: Leadership and market readiness. https://www.gartner.com

Gartner. (2022). The future of healthcare: How Mayo Clinic is leading with digital transformation. https://www.gartner.com

Gartner. (2022). The strategic importance of digital literacy in leadership. https://www.gartner.com

Gartner. (2022). The strategic role of leadership in digital transformation. https://www.gartner.com

Gartner. (2022). Transformational leadership and digital transformation. https://www.gartner.com

Gartner. (2023). The role of data-driven decision-making in digital transformation: Insights from Netflix and beyond. https://www.gartner.com

Gartner. (2023). The role of scenario planning in navigating digital disruptions. Gartner Research. https://www.gartner.com

Gartner. (2023). Top IT and business priorities for 2023–2024. https://www.gartner.com

General Electric. (2022). Digital leadership development program: Building future-ready leaders. McKinsey & Company. https://www.mckinsey.com

Google, A. I. (2021). AI principles: Ethical guidelines for artificial intelligence development. https://ai.google/principles

Harvard Business Review. (2021). Agile and iterative leadership in digital transformation. https://hbr.org

Harvard Business Review. (2021). Amazon's supply chain mastery: Automation and customer focus. https://hbr.org

Harvard Business Review. (2021). Building a growth mindset for digital leadership. https://hbr.org

Harvard Business Review. (2021). Emotional intelligence and collaborative cultures. https://hbr.org

Harvard Business Review. (2021). How Satya Nadella transformed Microsoft into a tech powerhouse. https://hbr.org

Harvard Business Review. (2021). How Satya Nadella transformed Microsoft: Leadership lessons for the digital age. https://hbr.org

Harvard Business Review. (2021). Procter & Gamble's digital transformation journey: Insights from employee engagement strategies. https://hbr.org

Harvard Business Review. (2022). Emotional intelligence and leadership in digital transformation. https://hbr.org

Harvard Business Review. (2022). Emotional intelligence and leadership in transformation. https://hbr.org

Harvard Business Review. (2022). Microsoft's digital transformation under Satya Nadella. https://hbr.org

Harvard Business Review. (2022). Microsoft's leadership under Satya Nadella: Embracing empathy and cultural reinvention. https://hbr.org

Harvard Business Review. (2022). Sustainability in the digital age: A roadmap for responsible innovation. https://hbr.org

International Energy Agency. (2022). IoT in optimizing energy grids. https://www.iea.org

International Energy Agency (IEA). (2022). Data centers and energy consumption: Trends and strategies for sustainability. https://www.iea.org

McKinsey & Company. (2020). Accelerating digital adoption during global crises. https://www.mckinsey.com

McKinsey & Company. (2021). Agile transformation at ING Bank: Building a customer-centric financial future. https://www.mckinsey.com

McKinsey & Company. (2021). Building a coalition of change agents for successful digital transformation. https://www.mckinsey.com

McKinsey & Company. (2021). Digital leadership: Lessons from industry leaders. https://www.mckinsey.com

McKinsey & Company. (2021). Digital transformation in leadership: Strategies and outcomes. McKinsey Insights. https://www.mckinsey.com

McKinsey & Company. (2021). Leadership in digital transformation: A case study of Microsoft and Adobe. https://www.mckinsey.com

McKinsey & Company. (2021). Leadership strategies for digital transformation. Retrieved from https://www.mckinsey.com

McKinsey & Company. (2021). Zoom's scalability during the pandemic: Agility in action. https://www.mckinsey.com

McKinsey & Company. (2022). Achieving scale in digital transformation: Key lessons from industry leaders. https://www.mckinsey.com

McKinsey & Company. (2022). Leadership in digital transformation: Lessons from Adobe and Microsoft. https://www.mckinsey.com

McKinsey & Company. (2022). Microsoft's pivot to the cloud: Lessons in customer-centric innovation. https://www.mckinsey.com

Microsoft. (2022). AI ethics and leadership in the digital age. Gartner Research. https://www.gartner.com

Microsoft AI Ethics Committee. (2022). AETHER principles and the ethical use of AI. https://www.microsoft.com/en-us/ai

MIT Sloan Management Review. (2022). Addressing AI bias and fairness in digital transformation. https://sloanreview.mit.edu

MIT Sloan Management Review. (2022). Alibaba's big data and AI strategy: Lessons from Singles' Day. https://sloanreview.mit.edu

MIT Sloan Management Review. (2022). Case studies in innovation-driven leadership: Lessons from Tesla and GE. https://sloanreview.mit.edu

MIT Sloan Management Review. (2022). Digital maturity models: Frameworks for transformation. https://sloanreview.mit.edu

MIT Sloan Management Review. (2022). Fostering innovation through Agile methodologies. MIT Sloan Management Review. https://sloanreview.mit.edu

MIT Sloan Management Review. (2022). The digital imperative: Aligning transformation strategies with market readiness. https://sloanreview.mit.edu

MIT Sloan Management Review. (2022). Walmart's digital reinvention: Leveraging omnichannel strategies for retail success. https://sloanreview.mit.edu

MIT Technology Review. (2022). Google's AI ethics: Principles for responsible innovation. https://www.technologyreview.com

Procter & Gamble. (2021). Global leadership development and digital transformation. *Harvard Business Review*. https://hbr.org

Spotify. (2022). Driving innovation through squad-based organizational structure. *MIT Sloan Management Review*.https://sloanreview.mit.edu

Unilever. (2022). Leadership training for digital and sustainable transformation. *MIT Sloan Management Review*. https://sloanreview.mit.edu

World Economic Forum. (2021). The acceleration of digital transformation post-COVID-19. https://www.weforum.org

World Economic Forum. (2022). Ethical leadership in the digital economy. https://www.weforum.org

World Economic Forum. (2023). Ethical frameworks for AI and digital innovation. https://www.weforum.org

World Economic Forum. (2023). Reskilling for the future: Workforce implications of automation and AI. *World Economic Forum Reports*. https://www.weforum.org

World Economic Forum. (2023). Strategic foresight in the digital economy. https://www.weforum.org

World Economic Forum. (2023). The future of digital transformation and global leadership. https://www.weforum.org

World Economic Forum. (2023). The future of jobs report 2023: Reskilling and the workforce of the future. https://www.weforum.org

Chapter 5
The Right Talent and Skills for the Digital Economy

Divya Tiwari
https://orcid.org/0009-0004-3688-4511
Amity Global Institute, Singapore

Easwaramoorthy Rangaswamy
https://orcid.org/0000-0003-0222-4764
Amity Global Institute, Singapore

Adrian Raj Angappan
Amity Global Institute, Singapore

ABSTRACT

In the rapidly evolving world of today, the digital economy is changing sectors, occupations, and skill needs. The need for new skills has increased dramatically as technologies like big data and artificial intelligence continue to develop. Businesses must concentrate on gaining the necessary talent and abilities if they want to be competitive in the digital age. Establishing a culture of ongoing learning is essential for both businesses and educational institutions in the quickly changing global economy. Fostering creativity, flexibility, and resilience in the quickly changing world of today requires a learning culture. To help people and organizations achieve continual improvement, this study examines the idea of a learning culture and highlights its significance. The chapter incorporates theoretical frameworks including social learning, andragogy, and transformational leadership. This paper explores the principles, strategies, and outcomes of creating a learning culture that prioritizes continuous development.

DOI: 10.4018/979-8-3373-1320-7.ch005

INTRODUCTION

In the 21st century, the workplace is undergoing a dramatic transformation. As digital technologies rapidly advance, the skills required to thrive in the workforce are evolving. No longer confined to technology specialists, digital competencies have become crucial for workers across all industries and sectors. From automation and artificial intelligence (AI) to data analytics and cybersecurity, digital competencies are reshaping the way businesses operate and how employees contribute to organizational success.

The rapid adoption of digital tools and technologies across industries has given rise to a new trend: digital competencies are now essential for nearly every role within an organization. As businesses increasingly rely on technology to drive efficiency, enhance innovation, and remain competitive, digital literacy and advanced technical skills are indispensable. This chapter delves into what digital competencies are, why they have become a new trend in the workplace, and how they are shaping the future of work.

UNDERSTANDING DIGITAL COMPETENCIES

Digital competencies refer to the knowledge, skills, and abilities required to effectively use digital technologies in a professional context. These competencies go beyond simple computer literacy, encompassing a range of skills needed to navigate and excel in a digitally driven workplace. Digital competencies are categorized into two main areas: basic digital literacy and advanced technical skills.

Basic Digital Literacy

Basic digital literacy includes fundamental skills required to use digital tools and navigate online environments. These skills are foundational and expected of all employees in today's workplace. They include:

1. Proficiency in Common Software: Employees need to be proficient in standard office software, including word processing (e.g., Microsoft Word), spreadsheets (e.g., Microsoft Excel), and presentation tools (e.g., PowerPoint). These skills are necessary for day-to-day tasks such as creating reports, managing data, and delivering presentations.
2. Communication Tools: Effective communication is central to modern work environments, especially as remote and hybrid work models become increasingly common. Employees need to be adept at using communication platforms like

email, instant messaging (e.g., Slack), video conferencing tools (e.g., Zoom, Microsoft Teams), and collaborative platforms (e.g., Google Docs, Trello).
3. Internet Navigation and Research: Employees must be able to efficiently search for information, conduct research, and stay informed about industry trends and developments through the internet. This skill is essential for problem-solving and decision-making.
4. Security Awareness: Basic cybersecurity practices, such as identifying phishing emails, using secure passwords, and understanding data privacy issues, are now critical competencies for all employees. Cybersecurity breaches can have devastating financial and reputational consequences for organizations (Kshetri, 2017).

Advanced Digital Competencies

As businesses become more reliant on digital technologies, more specialized, technical competencies are needed. These competencies often involve a deeper understanding of specific technologies or tools, and they are typically more prominent in roles related to information technology, data science, and innovation management. Some key advanced digital competencies include:

1. Data Analytics: With the explosion of data in virtually every industry, the ability to analyse and interpret data has become a core competency. Data-driven decision-making allows organizations to make informed, strategic decisions. Tools such as Python, R, SQL, and Tableau are used to manipulate and visualize data, helping employees derive actionable insights (Mayer-Schönberger & Cukier, 2013). For example, marketing teams can use data analytics to track customer behaviour and optimize their campaigns for better results.
2. Cloud Computing: Cloud platforms, such as Amazon Web Services (AWS), Microsoft Azure, and Google Cloud, have transformed how businesses store, process, and manage data. Cloud computing has made it easier for businesses to scale operations and collaborate more efficiently. As such, employees in fields such as IT, finance, and engineering must be familiar with cloud-based tools and services (Armbrust et al., 2010).
3. Artificial Intelligence (AI) and Machine Learning (ML): As AI and ML technologies become increasingly integrated into business processes, the demand for employees with the skills to develop and manage these systems is on the rise. These technologies are being used to automate routine tasks, enhance decision-making, and improve customer experiences. Employees working in areas like data science, software engineering, and business analysis are increasingly ex-

pected to have proficiency in AI and ML tools and frameworks (Brynjolfsson & McAfee, 2014).

4. Cybersecurity Expertise: With growing concerns about data breaches and cyberattacks, cybersecurity has become one of the most in-demand digital competencies. Organizations need cybersecurity professionals who are skilled in identifying vulnerabilities, protecting systems, and responding to incidents. For example, cybersecurity analysts and engineers play a critical role in safeguarding an organization's digital infrastructure and ensuring regulatory compliance (Mell & Grance, 2011).
5. Blockchain Technology: Blockchain, the technology underpinning cryptocurrencies like Bitcoin, is being explored for a wide range of applications, from supply chain management to secure digital transactions. Understanding the principles and potential of blockchain technology is becoming a valuable skill, especially in finance, logistics, and legal fields (Catalini & Gans, 2016).

THE DRIVERS BEHIND THE RISE OF DIGITAL COMPETENCIES

Several factors have contributed to the increasing demand for digital competencies in the workplace. The most significant of these drivers include technological advancements, globalization, and the COVID-19 pandemic.

Technological Advancements

The rapid pace of technological innovation is one of the primary drivers of the need for digital competencies. Technologies such as automation, AI, machine learning, and the Internet of Things (IoT) are reshaping industries, creating new opportunities and challenges. To keep up with these advancements, employees must continuously acquire new skills to remain competitive.

For example, in manufacturing, robotics and AI-powered systems are revolutionizing production processes. Factory workers now need to understand how to operate and troubleshoot automated machinery, requiring a higher level of technical competence than in the past. Similarly, AI is transforming the healthcare industry, requiring medical professionals to develop new skills in digital diagnostics and telemedicine (Topol, 2019).

Globalization and Digital Transformation

As businesses expand globally and embrace digital transformation, there is a growing demand for employees who can operate in a digital-first environment. The globalization of business has led to the proliferation of digital tools that enable companies to collaborate, communicate, and operate across borders. For instance, businesses now use cloud-based collaboration tools like Slack and Asana to manage teams and projects spread across different countries and time zones.

Digital transformation is not only about adopting new technologies but also involves rethinking organizational processes, business models, and customer interactions. Employees with strong digital competencies are essential for leading and supporting this transformation within organizations.

The COVID-19 Pandemic

The COVID-19 pandemic has acted as a catalyst for the adoption of digital technologies. As businesses scrambled to shift to remote work, digital tools for communication, collaboration, and productivity became essential. Video conferencing platforms like Zoom and Microsoft Teams saw massive increases in usage, and remote work tools such as Slack and Google Drive became central to daily operations.

The pandemic also accelerated the adoption of e-commerce, telemedicine, and online education, which all require specialized digital skills. As the world continues to navigate the effects of the pandemic, employees' digital competencies will remain critical to the success of organizations in a post-pandemic world.

THE IMPACT OF DIGITAL COMPETENCIES ON THE WORKPLACE

The rise of digital competencies is fundamentally altering the way employees work, interact, and contribute to their organizations. Below are some of the key impacts of digital competencies in the workplace:

1. Increased Efficiency and Productivity

Digital competencies enable employees to leverage technology to streamline workflows, automate routine tasks, and improve productivity. For instance, cloud computing platforms allow teams to collaborate on projects in real time, regardless of location, while data analytics tools enable organizations to optimize their operations and reduce inefficiencies (Brynjolfsson & McAfee, 2014).

2. Enhanced Collaboration and Communication

Digital tools have made it easier for employees to collaborate across departments, locations, and time zones. The widespread adoption of communication platforms like Slack, Microsoft Teams, and Zoom enables seamless interaction between team members, fostering greater collaboration and faster decision-making. This is particularly important in global organizations with remote and hybrid teams.

3. Innovation and Competitive Advantage

Employees with strong digital competencies can drive innovation by identifying new technologies, tools, and processes that can enhance business operations. For example, in the financial sector, employees with expertise in blockchain technology are helping organizations develop secure, transparent systems for transactions and data management (Catalini & Gans, 2016). In the retail sector, employees skilled in data analytics are helping businesses deliver personalized customer experiences that differentiate them from competitors (Verhoef et al., 2021).

4. Employee Development and Career Advancement

As digital competencies become more integral to business success, employees with strong digital skills are better positioned for career advancement. Companies that invest in developing the digital competencies of their employees foster a culture of continuous learning and development. This not only benefits the employees themselves but also enhances the overall competitiveness of the organization.

DEVELOPING DIGITAL COMPETENCIES IN THE WORKPLACE

To remain competitive in the digital economy, organizations must invest in developing the digital competencies of their employees. This can be achieved through a combination of training, mentorship, and continuous learning. Here are some strategies organizations can adopt:

1. Training Programs: Organizations should implement formal training programs that focus on both foundational digital skills (e.g., Microsoft Office, cybersecurity basics) and advanced competencies (e.g., data analytics, machine learning).
2. Mentorship and Peer Learning: Mentorship programs allow experienced employees to share their digital knowledge with junior staff. Peer learning also fosters a collaborative environment where employees can learn from one another.

3. Continuous Learning: Given the rapid pace of technological change, continuous learning is essential. Offering access to online learning platforms such as Coursera, Udemy, or LinkedIn Learning enables employees to stay current with new technologies and trends.
4. Promoting a Digital Culture: Creating a digital-first culture is key to fostering digital competencies across the organization. This includes promoting the use of digital tools, encouraging experimentation, and embracing innovation at all levels of the company.

THE IMPERATIVE OF DIGITAL TRANSFORMATION IN THE MODERN WORKPLACE

In today's fast-paced, technology-driven world, digital transformation is not merely a trend but a strategic imperative. From cloud computing and artificial intelligence (AI) to big data analytics and blockchain, new technologies are reshaping industries at an unprecedented pace. Businesses that fail to adapt risk becoming obsolete, while those that successfully embrace digital transformation position themselves for long-term success. However, digital transformation is not solely about adopting new technologies—it is about equipping employees with the skills and competencies needed to effectively integrate and use these technologies.

Training and developing employees for digital transformation has thus become one of the most critical challenges for modern organizations. Digital skills are no longer limited to a specific IT department; rather, they are essential across all functions and levels of an organization. From front-line workers to executives, every employee needs to possess a degree of digital literacy and proficiency to thrive in this rapidly evolving landscape. This chapter will explore the importance of employee training for digital transformation, identify the key areas of focus, and provide actionable strategies for organizations to develop their workforce.

The Importance of Training for Digital Transformation

Digital transformation encompasses more than just the integration of new technologies; it involves a holistic change in an organization's culture, business processes, and organizational structure. For employees to successfully navigate this transformation, they need continuous training and development in the digital tools, strategies, and methodologies that are reshaping their work environment.

Adapting to New Technologies

The adoption of technologies like AI, machine learning, robotics, and data analytics has brought new capabilities to the workplace. For example, in the healthcare industry, AI algorithms are being used to predict patient outcomes and assist doctors in making diagnostic decisions. Similarly, in the retail sector, data analytics platforms like Tableau and Google Analytics are enabling marketers to make data-driven decisions. Employees need to be proficient in these tools to maintain their relevance and effectiveness in their roles.

Without the right training, employees may struggle to adapt to these technologies, leading to inefficiencies, errors, and resistance to change. Training ensures that employees understand how new technologies can enhance their roles, improve workflows, and contribute to organizational goals.

Fostering Innovation and Agility

Digital transformation is not just about improving existing processes; it is also about fostering a culture of innovation. Employees equipped with the right digital skills are more likely to propose and develop innovative solutions that leverage emerging technologies. For example, a marketing team skilled in data analytics can create personalized customer experiences, while a logistics team with knowledge of AI-driven automation can optimize inventory management.

Furthermore, organizations need to be agile in the face of constant technological advancements. Employees who are adept at using digital tools and platforms can more easily adapt to new technologies, allowing the organization to stay ahead of the competition.

BUILDING A CULTURE OF CONTINUOUS LEARNING

The digital economy is characterized by rapid innovation, and technologies are evolving at an exponential rate. As a result, employee training cannot be a one-time event; it must be an ongoing process. Organizations must build a culture of continuous learning where employees are encouraged and empowered to continuously update their digital skills. This culture helps organizations stay competitive in the digital age while ensuring that employees feel supported and motivated to grow in their careers.

Key Areas of Focus for Employee Training in Digital Transformation

Training for digital transformation must focus on a variety of competencies, ranging from technical skills to softer, leadership-oriented capabilities. The following areas are essential for employee development in the context of digital transformation.

1. Digital Literacy

At the core of digital transformation is digital literacy—the ability to use digital tools and platforms effectively. Digital literacy has become a foundational skill for employees at all levels, as it enables them to interact with and use new technologies in their day-to-day tasks. According to the European Commission (2020), digital literacy includes basic skills such as using computers, accessing and managing digital information, and communicating through digital platforms.

For example, a retail employee should be able to use the company's point-of-sale system, access customer information, and communicate with colleagues through internal messaging platforms. Meanwhile, employees in non-technical roles should be proficient in using digital collaboration tools like Slack, Microsoft Teams, and Google Workspace, which are essential for productivity in the modern workplace.

2. Data Analytics and Data-Driven Decision Making

As businesses generate vast amounts of data, the ability to analyze and interpret that data has become one of the most important skills in the modern workplace. Employees need to be trained in data analytics tools and methodologies so they can use data to make informed, strategic decisions.

For example, marketing professionals must understand how to leverage customer data to drive targeted campaigns and improve customer engagement. Likewise, financial analysts need to analyse market trends and internal performance data to make investment decisions.

Training programs in data analytics should cover a range of tools, such as Excel, SQL, Python, R, Tableau, and Power BI. These programs should also focus on teaching employees how to interpret data visualizations, identify patterns, and draw actionable conclusions from data.

3. Cloud Computing

Cloud computing is the backbone of many digital transformations, enabling businesses to scale their operations, enhance collaboration, and store data securely. Employees across all departments must be familiar with cloud-based platforms and tools to work efficiently and collaborate effectively.

For example, sales teams may use cloud-based CRM systems like Salesforce, while HR professionals use cloud-based payroll systems. Cloud-based collaboration tools like Google Drive, Dropbox, and SharePoint allow teams to collaborate in real time, regardless of location.

Training employees in cloud platforms such as Amazon Web Services (AWS), Microsoft Azure, and Google Cloud can help them utilize cloud services for tasks like file storage, project management, and software development. Additionally, understanding cloud security and data privacy is essential to ensure the safe use of cloud technologies.

4. Artificial Intelligence and Automation

AI and automation technologies are transforming the way businesses operate. Whether it's using AI for customer service through chatbots or automating repetitive administrative tasks, employees must be trained to work alongside AI-driven systems.

For instance, a customer support agent may need to understand how to interact with a chatbot system to address customer queries before escalating issues to human representatives. Similarly, employees in operations may be trained to use Robotic Process Automation (RPA) tools to automate data entry and other repetitive tasks.

Organizations must train employees to embrace these technologies, fostering collaboration between human workers and AI systems to achieve optimal outcomes. Understanding the benefits and limitations of AI can help employees integrate these technologies more effectively into their work processes.

5. Cybersecurity Awareness

As digital technologies become more integrated into everyday business operations, cybersecurity has become a critical concern. Employees must understand the fundamentals of cybersecurity to protect both organizational and customer data.

Training in cybersecurity should cover topics such as password management, phishing prevention, and recognizing suspicious activity. For example, employees should be trained to recognize phishing emails that attempt to steal sensitive information or compromise company systems. Similarly, data protection laws such as

GDPR (General Data Protection Regulation) should be part of the training, ensuring employees understand their responsibilities in safeguarding personal data.

6. Leadership and Change Management

Digital transformation requires strong leadership at all levels of an organization. Leaders must be capable of guiding their teams through the complexities of change, addressing challenges, and fostering a culture of innovation. Training future leaders in digital transformation should therefore include both technical skills and leadership capabilities.

Change management training helps leaders and managers understand how to effectively lead teams through technological transitions. For example, training programs may focus on how to manage employee resistance, communicate the benefits of new technologies, and support employees in adopting new digital tools.

STRATEGIES FOR TRAINING EMPLOYEES FOR DIGITAL TRANSFORMATION

Organizations must adopt a comprehensive and strategic approach to training employees for digital transformation. Here are several strategies to ensure that training is effective and scalable:

1. Develop a Digital Skills Framework

A successful training program begins with a clear understanding of the digital competencies needed for the organization. By developing a digital skills framework, organizations can identify the specific skills required for different roles and tailor training programs accordingly.

For example, a marketing department may require employees to develop competencies in data analytics and digital advertising platforms, while an HR department may need training in cloud-based HR tools and employee engagement platforms. By aligning training with job-specific needs, organizations can ensure that training is relevant and impactful.

2. Offer Blended Learning Solutions

Blended learning combines both online and in-person learning, allowing employees to learn at their own pace while still benefiting from face-to-face interactions with trainers and colleagues. For example, an employee could complete an online module on data analytics and then attend a hands-on workshop to apply what they've learned in real-world scenarios.

Blended learning solutions can be particularly effective for employees with varying levels of digital proficiency. Introductory digital literacy courses can be delivered online, while more advanced technical training can be conducted in workshops or through mentorship programs.

3. Promote Continuous Learning and Upskilling

Digital transformation is an ongoing process, and employee training should reflect this dynamic reality. Organizations must promote a culture of continuous learning where employees are encouraged to update their skills regularly.

Offering access to online learning platforms, such as LinkedIn Learning, Coursera, or Udemy, can help employees acquire new skills at their own pace. Providing incentives for employees to complete certifications or participate in professional development programs can also boost engagement and motivation.

4. Encourage Peer-to-Peer Learning

Peer-to-peer learning can be an effective way to foster a culture of collaboration and knowledge sharing. Employees who are experts in a particular digital skill can mentor others, providing guidance and support. For instance, a senior IT professional might mentor junior employees on cloud computing best practices, or a marketing team member skilled in data analytics could lead a workshop for colleagues.

Peer-to-peer learning encourages employees to take ownership of their professional development and fosters a collaborative work environment.

Conclusion

Training and developing employees for digital transformation is crucial for organizations to remain competitive in today's rapidly evolving business landscape. From basic digital literacy to advanced technical skills, organizations must provide employees with the tools, resources, and training to succeed in the digital economy. By investing in employee development, businesses can foster innovation, improve efficiency, and ensure long-term success in an increasingly digital world. Continuous learning, upskilling, and strategic training initiatives will be the key to unlocking the full potential of digital transformation in the workplace.

ORGANIZING CROSS-ORGANIZATIONAL TEAMS FOR CREATIVITY

The digital economy is an ecosystem in which creativity, innovation, and collaboration drive success. As companies navigate digital transformation, it has become increasingly clear that cross-organizational collaboration is not just desirable—it is essential. The emergence of new technologies, business models, and customer demands has prompted companies to rethink how they organize their workforce and foster creativity.

One of the most effective ways to tap into creativity is by organizing cross-organizational teams. These teams bring together diverse individuals from different departments, functions, and backgrounds to tackle complex challenges and generate innovative solutions. The ability to build and manage such teams is critical in today's fast-paced and competitive digital world.

This chapter will explore how cross-organizational teams are fostering creativity in the digital economy, the benefits and challenges associated with such teams, and practical strategies for organizations to build and manage them effectively.

The Need for Cross-Organizational Teams

In the past, organizations were typically structured in silos, with distinct departments such as marketing, finance, and operations working independently. However, the digital economy has shifted the focus toward more integrated, collaborative approaches. Cross-organizational teams are essential for several reasons:

1. Complexity of Digital Challenges

The challenges facing businesses today are increasingly complex and multifaceted. For instance, when companies seek to implement new technologies, such as artificial intelligence or blockchain, they require input from a variety of departments. IT, legal, marketing, and customer service all need to be involved to ensure the technology is deployed correctly, is compliant with regulations, and addresses customer needs.

By forming cross-functional teams, organizations can pool expertise from multiple perspectives to tackle these issues effectively. For example, in the development of a new software product, engineers may collaborate with product managers, marketers, and customer service representatives to create a solution that meets both technical specifications and user needs.

2. Innovation and Creativity

Innovation is a key driver of success in the digital economy. Research has shown that teams with diverse skills and backgrounds tend to be more creative and innovative (Page, 2007). Cross-organizational teams bring together individuals with different expertise, experiences, and perspectives, which leads to more innovative ideas. When employees from different parts of the organization work together, they challenge each other's thinking, broaden their horizons, and develop more creative solutions to problems.

For instance, a retail company might bring together the IT department, the marketing team, and customer service to design a new e-commerce platform. The IT team provides technical expertise, the marketing team ensures the platform aligns with brand objectives, and customer service helps ensure that the platform is user-friendly and addresses customer concerns. This multi-disciplinary approach fosters creativity by integrating various areas of expertise.

3. Accelerating Digital Transformation

The need for digital transformation is widespread across industries. Cross-organizational teams are critical to accelerating this transformation, as they enable different parts of the business to align and collaborate toward common goals. Digital transformation requires not only technical expertise but also cultural and organizational change. A team made up of individuals from different departments can guide the company through the shift, ensuring that digital initiatives align with the company's overall strategy and objectives.

For example, a financial services company undergoing digital transformation may form a cross-functional team to implement a new customer-facing mobile application. The team might include IT professionals, UX/UI designers, project managers, marketing personnel, and customer support representatives, ensuring that the application is not only technically sound but also aligns with customer expectations and market trends.

The Benefits of Cross-Organizational Teams for Creativity

Cross-organizational teams offer numerous benefits, particularly when it comes to fostering creativity. The diverse expertise and perspectives inherent in these teams often result in innovative solutions that would not have been possible within more siloed structures.

1. Diversity of Thought

One of the most significant advantages of cross-organizational teams is the diversity of thought they bring. Research by Page (2007) highlights that diversity—whether in terms of skills, experiences, or cultural backgrounds—can lead to better decision-making and innovation. By involving people from various departments, organizations ensure that a wide range of perspectives are considered, which can help uncover solutions that may have been overlooked by a more homogeneous group.

For instance, when developing a new product, involving members from research and development (R&D), marketing, sales, and customer support can ensure that the product is not only technically advanced but also meets market demands and solves real customer problems. A team composed solely of engineers may be focused on the product's technical feasibility, but a cross-organizational team can identify ways to make the product more user-friendly, marketable, and profitable.

2. Synergy and Collaboration

When people from different backgrounds collaborate, they bring unique skills and knowledge to the table, which can lead to greater creativity and problem-solving. This synergy can produce results that far exceed the sum of individual contributions. In a cross-organizational team, individuals complement each other's strengths and compensate for weaknesses, resulting in more effective and innovative outcomes.

For example, a cross-functional team working on a new digital marketing campaign might include experts in content creation, SEO, data analytics, and social media strategy. Each member brings their specialized knowledge to the table, but together, they can create a more cohesive and effective campaign than they could individually.

3. Increased Engagement and Empowerment

Employees involved in cross-organizational teams often feel more engaged and empowered because they have a voice in decision-making processes that impact the entire organization. When team members from different functions collaborate, they feel that their expertise is valued, and they are more likely to be motivated to contribute innovative ideas.

For instance, an employee in a customer-facing role may have valuable insights into customer pain points, which they can share with the product development team to improve a new product or service. By giving employees the opportunity to contribute to high-level decisions, cross-organizational teams foster a sense of ownership and involvement, which can lead to increased creativity and productivity.

Challenges of Organizing Cross-Organizational Teams

While cross-organizational teams have many benefits, they are not without their challenges. These teams require careful management and coordination to ensure they are effective in fostering creativity. Some of the key challenges include:

1. Communication Barriers

One of the biggest challenges of cross-functional teams is communication. Employees from different departments may use different terminology, have different priorities, and communicate in different ways. This can lead to misunderstandings, confusion, and inefficiencies.

To mitigate this challenge, it is crucial to establish clear communication channels and expectations from the outset. Regular meetings, progress updates, and collaborative tools (e.g., Slack, Microsoft Teams) can help team members stay aligned and ensure effective communication.

2. Conflicting Priorities and Goals

Each department in an organization has its own set of goals and priorities, and these may not always align with the objectives of the cross-organizational team. For example, the IT department might prioritize technical feasibility, while marketing might focus on user experience and customer acquisition. These competing interests can lead to tensions and conflict within the team.

To address this, it is important to establish a shared vision and common goals for the cross-functional team. Leadership should clearly communicate the purpose of the team and ensure that everyone is aligned on the end goals. Regular check-ins and adjustments to the strategy can also help keep everyone on the same page.

3. Power Dynamics and Hierarchy

In cross-organizational teams, individuals from different departments may have different levels of seniority or authority. This can create power dynamics that affect decision-making and collaboration. For example, an executive from the marketing team may have more influence than a junior employee from IT, which could lead to unequal participation in the team's creative process.

To foster a collaborative and inclusive environment, it is important to create a culture where all team members are encouraged to contribute equally. Leaders should model inclusive behavior and ensure that everyone's voice is heard and valued.

Best Practices for Organizing Cross-Organizational Teams for Creativity

Despite the challenges, cross-organizational teams can be highly effective in fostering creativity if managed properly. The following best practices can help organizations harness the full potential of cross-organizational collaboration:

1. Define Clear Objectives and Roles

At the outset of a cross-functional team, it is important to establish clear objectives and expectations. Every team member should understand the project's goals and how their role contributes to the overall success. By setting clear objectives, teams can stay focused and aligned, ensuring that all members are working toward a common purpose.

Additionally, each team member should be aware of their specific role and responsibilities within the team. This helps prevent confusion and ensures that the right expertise is brought to bear at the right time.

2. Foster Open Communication

Effective communication is the cornerstone of successful cross-functional teams. To foster open communication, organizations should encourage regular check-ins, status updates, and brainstorming sessions. Digital tools such as Slack, Microsoft Teams, and Trello can facilitate real-time communication and collaboration.

Leaders should also promote transparency and openness, encouraging team members to share ideas and concerns freely. A culture of psychological safety—where employees feel comfortable speaking up without fear of judgment—is essential for creativity to flourish.

3. Encourage Diversity and Inclusion

The success of cross-organizational teams hinges on the diversity of their members. To maximize creativity, organizations should actively encourage diversity in terms of skills, backgrounds, and perspectives. A diverse team is more likely to generate novel ideas and approaches.

Moreover, inclusion is equally important. Organizations should create an environment where all voices are heard and valued, regardless of rank or background. This ensures that team members feel comfortable sharing their ideas and contributing to the creative process.

4. Provide Adequate Resources and Support

Cross-organizational teams need the right tools, resources, and support to succeed. This includes access to technology, training, and leadership guidance. Organizations should ensure that team members have the resources they need to collaborate effectively and overcome any obstacles they encounter.

Leaders should also be available to provide guidance and support, helping to resolve conflicts and ensure the team stays on track.

Conclusion: Empowering Creativity in the Digital Economy

Cross-organizational teams are a powerful tool for fostering creativity and innovation in the digital economy. By bringing together diverse talents, perspectives, and expertise, these teams can solve complex problems and drive organizational growth. However, to succeed, cross-functional teams require careful planning, clear objectives, effective communication, and a culture of inclusion and collaboration.

Organizations that prioritize the development of cross-organizational teams will be better positioned to navigate the challenges of the digital economy and unlock new opportunities for innovation and growth. As the business landscape continues to evolve, the ability to organize and manage cross-functional teams will become an increasingly important skill for leaders in every industry.

CREATING A LEARNING CULTURE: CONTINUOUS DEVELOPMENT

The 21st-century landscape is characterized by unprecedented technological advancements and economic shifts, making continuous development essential for survival and success. A learning culture—defined as an organizational environment that prioritizes and values continuous learning—has emerged as a critical determinant of adaptability and innovation. This paper explores the mechanisms, strategies, and challenges associated with cultivating a learning culture.

The concept of a learning culture transcends organizational boundaries, extending to nations, educational institutions, and individuals. It fosters resilience and enhances capacity to navigate uncertainties. Theoretical frameworks such as transformational leadership, andragogy, and social learning theory provide foundational insights into the dynamics of learning cultures. These insights guide the development of actionable strategies that embed learning into organizational processes, promote collaboration, and leverage technology to sustain a competitive edge. Through an analysis of case studies and evidence-based practices, this paper illuminates' pathways to operationalize a learning culture. It also examines barriers, including resistance to change and resource constraints, offering practical solutions for overcoming these challenges. The findings aim to inspire educators, leaders, and policymakers to prioritize learning as a core value, ensuring sustainable growth and development in an ever-evolving world.

Theoretical Frameworks

1. Transformational Leadership and Learning Cultures (60 words) Transformational leaders inspire and motivate individuals to achieve their full potential, fostering an environment conducive to learning and growth. They emphasize vision, intellectual stimulation, and individualized consideration, which are critical for embedding a learning culture within organizations. By encouraging innovation and adaptability, transformational leadership aligns with the principles of continuous development.
2. Andragogy and Lifelong Learning (60 words) Andragogy, the art and science of adult learning, emphasizes self-directed, experiential, and problem-solving approaches. This framework supports lifelong learning by addressing adults' intrinsic motivations and leveraging their prior experiences. Andragogical principles underpin strategies for cultivating learning cultures, as they promote engagement, autonomy, and relevance in professional and personal development.
3. Social Learning Theory (55 words) Social learning theory posits that individuals learn through observation, imitation, and social interactions. This theory underscores the importance of collaborative environments and peer-to-peer learning. In fostering a learning culture, organizations can leverage social learning to encourage knowledge sharing, mentorship, and collective problem-solving, thereby creating a dynamic and interactive learning ecosystem.

Theoretical Frameworks

1. Transformational Leadership and Learning Cultures: Transformational leaders inspire and motivate individuals to achieve their full potential, fostering an environment conducive to learning and growth. They emphasize vision, intellectual stimulation, and individualized consideration, which are critical for embedding a learning culture within organizations. By encouraging innovation and adaptability, transformational leadership aligns with the principles of continuous development.
2. Andragogy and Lifelong Learning: Andragogy, the art and science of adult learning, emphasizes self-directed, experiential, and problem-solving approaches. This framework supports lifelong learning by addressing adults' intrinsic

motivations and leveraging their prior experiences. Andragogical principles underpin strategies for cultivating learning cultures, as they promote engagement, autonomy, and relevance in professional and personal development.

3. Social Learning Theory: Social learning theory posits that individuals learn through observation, imitation, and social interactions. This theory underscores the importance of collaborative environments and peer-to-peer learning. In fostering a learning culture, organizations can leverage social learning to encourage knowledge sharing, mentorship, and collective problem-solving, thereby creating a dynamic and interactive learning ecosystem.

Strategies for Creating a Learning Culture

1. Embedding Learning in Organizational Processes

Embedding learning into organizational processes ensures that professional development becomes an integral part of everyday operations rather than a supplementary activity.

- Integration of Learning Management Systems (LMS): LMS platforms streamline access to training materials, track progress, and provide data-driven insights for continuous improvement. By centralizing learning resources, organizations can standardize training while allowing customization for individual needs. Advanced LMS platforms also enable interactive learning, incorporating quizzes, simulations, and gamified experiences to enhance engagement and retention. Additionally, data analytics tools within LMS platforms provide valuable metrics on learner performance, identifying areas for targeted development and overall organizational learning trends.
- On-the-Job Training: Practical, hands-on experiences enable employees to acquire skills directly relevant to their roles, reinforcing the application of knowledge. Organizations that prioritize experiential learning create opportunities for employees to shadow experts, participate in real-world projects, and engage in role-specific simulations. Such initiatives bridge the gap between theoretical knowledge and practical execution, fostering confidence and competency in the workforce. Regular feedback loops during on-the-job training further reinforce learning outcomes, ensuring alignment with organizational objectives.

• 2. Cultivating Psychological Safety

- Psychological safety—a belief that individuals can voice ideas and take risks without fear of reprisal—is fundamental to a learning culture. When employees feel secure, they are more likely to experiment, admit mistakes, and contribute innovative ideas.
- Leaders play a pivotal role in establishing trust and encouraging open communication. Demonstrating empathy, providing consistent feedback, and leading by example are essential leadership behaviours. Open-door policies, transparent decision-making, and inclusive team discussions foster a supportive environment where employees feel valued and respected.
- Organizations can also implement team-building exercises to strengthen interpersonal bonds and create an atmosphere of mutual respect. Celebrating failures as opportunities for growth and emphasizing collective learning from setbacks shifts the focus from blame to progress. Psychological safety not only enhances individual performance but also promotes team cohesion, innovation, and resilience in the face of challenges.

3. Encouraging Lifelong Learning Mindsets

Cultivating a mindset that values continuous personal and professional development is vital for fostering a learning culture. Organizations can adopt multiple strategies to instil this mindset among employees.

- Recognition and Rewards: Incentivizing learning through awards, promotions, or financial benefits motivates individuals to prioritize personal development. Publicly acknowledging employees who demonstrate exceptional commitment to learning reinforces the importance of continuous improvement and inspires others to follow suit. Reward systems that align with organizational goals, such as certifications in emerging technologies or leadership training, further enhance skill relevance and career growth.

Mentorship Programs: Structured mentorship initiatives provide guidance, support, and opportunities for skill enhancement, fostering a culture of shared growth. Mentors offer insights drawn from their own experiences, helping mentees navigate challenges and identify pathways for development. Reciprocal learning between mentors and mentees strengthens relationships and contributes to a collaborative organizational environment. Formal mentorship frameworks, complemented by peer-to-peer learning opportunities, ensure widespread access to developmental resources.

4. Leveraging Technology for Continuous Development

Technological advancements have revolutionized the way organizations approach learning and development. Leveraging innovative tools can enhance accessibility, engagement, and effectiveness.

- Artificial Intelligence (AI) in Learning: AI-powered tools offer personalized learning experiences, adaptive content delivery, and real-time feedback. By analyzing individual performance data, AI algorithms recommend tailored learning paths that address specific skill gaps. AI-driven chatbots and virtual assistants provide on-demand support, enabling learners to access resources anytime, anywhere. Moreover, AI facilitates predictive analytics, allowing organizations to anticipate future learning needs and proactively address them.
- Microlearning: Bite-sized, focused learning modules cater to modern learners' preferences for accessible and time-efficient education. Microlearning delivers content in short, digestible segments, making it ideal for employees juggling multiple responsibilities. Formats such as videos, infographics, and interactive quizzes engage diverse learning styles. This approach not only enhances retention but also promotes immediate application of knowledge, driving measurable performance improvements.

5. Promoting Collaborative Learning

Collaboration amplifies the collective knowledge within an organization and fosters a sense of community among employees. Structured approaches to collaborative learning maximize its impact.

- Communities of Practice: Groups of individuals with shared interests collaborate to solve problems, exchange knowledge, and innovate. These informal networks encourage continuous learning by facilitating discussions on best practices, industry trends, and emerging challenges. Communities of practice thrive on peer-to-peer interaction, enabling members to learn from one another's experiences and perspectives. Regularly scheduled meetings, supported by online forums and resource-sharing platforms, sustain engagement and productivity.
- Cross-Functional Teams: Diverse teams foster cross-pollination of ideas and promote holistic learning, enriching organizational capabilities. By bringing together individuals from different departments or disciplines, cross-functional teams stimulate creativity and innovation. Exposure to var-

ied expertise broadens participants' skill sets and enhances their adaptability. Collaborative projects, coupled with clear objectives and performance metrics, ensure alignment with organizational goals and deliver tangible outcomes.

6. Creating a Feedback-Oriented Environment

- A feedback-oriented culture is essential for continuous improvement and sustained learning. Feedback loops provide actionable insights, reinforce learning objectives, and promote accountability.
- Constructive feedback fosters individual growth by highlighting strengths and addressing areas for improvement. Regular performance reviews, supported by 360-degree feedback mechanisms, ensure comprehensive evaluations that consider input from peers, supervisors, and subordinates. Real-time feedback, delivered through digital platforms or one-on-one interactions, enables employees to make immediate adjustments and optimize their performance.
- Recognition of achievements through positive feedback builds morale and motivates employees to maintain high standards. Establishing open channels for feedback encourages employees to share ideas and express concerns, enhancing transparency and trust. When feedback is framed as an opportunity for growth rather than criticism, it empowers individuals to learn from successes and failures alike, driving a culture of continuous development.

CHALLENGES IN ESTABLISHING A LEARNING CULTURE

Resistance to Change

Resistance to change is a significant hurdle in fostering a learning culture within organizations. Employees often fear that new learning initiatives might expose their inadequacies, leading to a reluctance to participate. Long-standing employees may feel threatened by changes that disrupt established routines or demand additional effort to acquire new skills. This fear can manifest as skepticism about the benefits of the proposed changes or outright opposition. Moreover, organizational inertia, where existing systems and processes are deeply entrenched, exacerbates resistance. Overcoming this challenge requires clear communication of the benefits of learning initiatives, ensuring alignment with employees' personal and professional growth goals. Leadership plays a pivotal role in addressing concerns, providing reassurance, and creating an environment that normalizes experimentation and learning from failures. Building trust and demonstrating the value of these initiatives through early success stories can gradually reduce resistance and foster acceptance.

Resource Constraints

Implementing a robust learning culture often requires substantial investment in financial, technological, and human resources. Organizations with limited budgets may struggle to allocate funds for advanced learning management systems, training programs, or hiring skilled facilitators. Smaller companies, in particular, may find it challenging to compete with larger corporations that can afford extensive learning and development programs. Technological constraints, such as outdated infrastructure or lack of access to modern tools, further hinder the adoption of innovative learning strategies. Human resource limitations, including insufficient trainers or overwhelmed staff, can also impede progress. Addressing these constraints necessitates strategic planning and prioritization. Organizations can explore cost-effective solutions like microlearning platforms, online resources, or partnerships with educational institutions. Securing external funding, such as government grants or industry sponsorships, can provide additional support. Leaders must also foster a culture of resourcefulness, encouraging teams to leverage available resources creatively and efficiently to achieve learning objectives.

Measuring Impact

Assessing the effectiveness of learning initiatives presents a complex challenge for organizations aiming to establish a learning culture. Unlike tangible business metrics, the outcomes of learning programs are often intangible and long-term, making them difficult to quantify. Traditional performance indicators may not adequately capture improvements in employee engagement, innovation, or adaptability. Furthermore, the diversity of learning activities—ranging from formal training to informal knowledge sharing—complicates the evaluation process. Organizations may also lack the expertise or tools to track and analyze data effectively. To address this challenge, organizations should adopt a multi-dimensional approach to measurement. This includes using a combination of quantitative metrics, such as completion rates and knowledge assessments, and qualitative methods, like employee feedback and case studies. Advanced analytics and AI-driven tools can provide deeper insights into learning patterns and outcomes. Regularly revisiting and refining evaluation methods ensures alignment with organizational goals and helps demonstrate the value of learning initiatives.

Case Studies

1. Google's Culture of Innovation Google's commitment to a learning culture is evident in initiatives like "20% time," which allows employees to pursue projects outside their core responsibilities. This approach has led to innovations such as Gmail and Google Maps (Bock, 2015). Google's emphasis on experimentation and knowledge sharing exemplifies the benefits of a learning-oriented workplace.
2. Singapore's SkillsFuture Initiative SkillsFuture encourages Singaporeans to embrace lifelong learning through financial incentives, industry partnerships, and accessible training programs. The initiative addresses workforce adaptability in a rapidly evolving economy (Chia, 2018). SkillsFuture also demonstrates the importance of government support in fostering a national culture of learning.
3. Amity Global Institute's Unified Video Content Management System (UVCMS) By implementing the UVCMS, Amity has enhanced multimedia teaching and learning, fostering a tech-enabled learning culture that benefits both students and faculty. The system supports a range of pedagogical approaches, from flipped classrooms to interactive simulations.
4. IBM's Skills Academy IBM's Skills Academy offers employees and students access to training in emerging technologies such as AI, blockchain, and cloud computing. This initiative illustrates how corporate investment in learning infrastructure can address skill gaps and drive innovation.

Implications for Practice

1. For Educational Institutions: Prioritize technology integration and collaboration to create engaging learning environments. Institutions should leverage advanced tools, such as learning analytics, to personalize education and strengthen inter disciplinary partnerships for addressing complex, real-world challenges.
2. For Organizations: Embed learning in strategic goals to foster innovation and competitiveness. Organizations must establish clear pathways for employee development, invest in upskilling initiatives, and create an organizational culture that celebrates continuous improvement and knowledge sharing.
3. For Policymakers: Design policies that incentivize lifelong learning and address systemic barriers to education. Policymakers should ensure equitable access to digital resources, promote partnerships between education providers and industry leaders, and implement funding mechanisms to support underserved communities.

A learning culture is a cornerstone of continuous development in the modern era. By embracing theoretical insights and implementing evidence-based strategies, organizations and societies can unlock their full potential. Collaboration among educators, leaders, and policymakers is essential to sustain learning cultures and drive long-term growth and innovation.

Talent Acquisition and Management of Web-Generation Employees The Web Generation, commonly referred to as Generation Z, comprises individuals born between 1997 and 2012. As the first generation to grow up with ubiquitous internet access and advanced technology, their entry into the workforce brings unique challenges and opportunities. Their digital fluency, entrepreneurial mindset, and strong desire for purpose-driven work differentiate them from previous generations. This paper aims to address the evolving dynamics of talent acquisition and management strategies tailored for the Web Generation.

Generation Z has been shaped by a world characterized by rapid technological advancements, social media proliferation, and significant global events. Unlike Millennials, who witnessed the transition to the digital age, Gen Z has never known a world without smartphones, high-speed internet, and social networking platforms. This exposure has cultivated a cohort of individuals who are quick to adapt to new technologies, highly connected, and skilled at multitasking. However, it has also introduced challenges, such as shorter attention spans and high expectations for seamless digital experiences.

Their entry into the workforce coincides with the ongoing transformation of the global economy towards digitalization. Traditional industries are being disrupted, and the demand for digital skills is at an all-time high. Organizations must adapt their talent acquisition and management strategies to attract, retain, and nurture this tech-savvy generation. Employers who fail to understand the unique attributes and expectations of Gen Z risk losing out on a critical talent pool essential for driving innovation and growth in the digital economy.

Moreover, this generation prioritizes values such as diversity, inclusion, and environmental sustainability. They are drawn to companies that align with their personal values and offer opportunities for meaningful work. To attract and engage Gen Z talent, organizations need to craft employer brands that emphasize these attributes while leveraging digital platforms to communicate their message effectively.

This paper explores the specific strategies organizations can employ to optimize talent acquisition and management practices for the Web Generation, ensuring alignment with their unique characteristics and expectations.

Characteristics of Web-Generation Employees

1. Digital Natives

Web-Generation employees are inherently comfortable with digital tools and platforms. Born into a world saturated with technology, they have grown up using smartphones, social media, and cloud-based tools, making them highly proficient in leveraging digital resources for both personal and professional purposes. This fluency enables them to adapt quickly to technological advancements, often outperforming previous generations in adopting and mastering new technologies. However, this digital fluency also comes with elevated expectations. Web-Generation employees anticipate seamless technology integration in their workplaces, including up-to-date software, user-friendly systems, and digital collaboration tools. Organizations that fail to meet these expectations may struggle to attract and retain top talent from this cohort. Furthermore, their digital savviness often leads them to expect continuous upskilling opportunities, such as training in artificial intelligence, data analytics, or emerging technologies. To successfully engage these employees, businesses must prioritize digital transformation, ensuring that their operations and tools align with the capabilities and expectations of this tech-savvy generation.

2. Value-Driven and Purpose-Oriented

Web-Generation employees prioritize value-driven work environments and are deeply purpose-oriented. Numerous studies indicate that this cohort places significant emphasis on working for organizations that reflect their personal values, particularly in areas such as sustainability, diversity, equity, and social justice. Unlike previous generations, who may have been more focused on financial compensation or job security, Web-Generation employees are motivated by a sense of meaningful contribution. This generation seeks transparency and authenticity, expecting companies to demonstrate genuine commitment to causes they care about. Greenwashing or performative diversity initiatives are unlikely to impress them; instead, they value measurable and impactful efforts. Additionally, Web-Generation employees often evaluate potential employers based on corporate social responsibility (CSR) initiatives and their alignment with global challenges such as climate change and ethical governance. For organizations aiming to attract and retain this generation, embedding purpose into their mission and clearly communicating it through branding, policies, and practices is essential. By fostering an authentic connection to these values, employers can build trust, enhance engagement, and create a loyal workforce.

3. Entrepreneurial Spirit

Web-Generation employees often exhibit an entrepreneurial spirit, characterized by a preference for autonomy, flexibility, and innovation. Many in this cohort view traditional employment structures as restrictive and instead gravitate toward roles that offer opportunities to exercise creativity and leadership. This entrepreneurial mindset is reflected in the growing popularity of gig work, freelancing, and side hustles among Gen Z workers. Even within traditional employment, they seek roles where they can experiment with new ideas, take ownership of projects, and make meaningful contributions to the organization. This presents both opportunities and challenges for employers. On the one hand, their entrepreneurial tendencies can drive innovation, as these employees are likely to identify gaps, propose solutions, and contribute to growth. On the other hand, retaining such talent may be difficult if rigid hierarchies or lack of flexibility hinder their ability to thrive. Employers should consider creating intrapreneurial programs that allow employees to explore their entrepreneurial interests within the organization. By fostering a culture of innovation and autonomy, companies can leverage the entrepreneurial spirit of Web-Generation employees while keeping them engaged and committed.

4. Focus on Well-being

For Web-Generation employees, mental health and work-life balance are not just preferences but essential elements of their professional lives. Unlike earlier generations, who may have accepted demanding work schedules and limited benefits as the norm, this generation prioritizes their well-being and expects their employers to do the same. Factors such as rising awareness of mental health issues, the global impact of the COVID-19 pandemic, and the increasing visibility of burnout have shaped their attitudes toward work. They actively seek out employers who offer comprehensive well-being programs, such as mental health counseling, wellness stipends, and gym memberships. Additionally, they value flexible work arrangements, including remote work options and flexible hours, as these contribute to a healthier work-life balance. Organizations that neglect these priorities risk facing higher turnover rates and lower employee engagement. To attract and retain Web-Generation employees, companies must embed well-being into their organizational culture, providing not only tangible benefits but also creating a supportive and empathetic work environment. By addressing these needs, employers can foster a resilient and motivated workforce that thrives in both personal and professional domains.

Talent Acquisition Strategies for the Web Generation

Leveraging Technology

Technology-driven recruitment tools, such as AI-powered applicant tracking systems and chatbots, streamline the hiring process. Gen Z expects a seamless and transparent application experience, complete with real-time updates and interactive elements.

Social Media Recruitment

Social media platforms, particularly LinkedIn, Instagram, and TikTok, have emerged as critical channels for attracting Web-Generation talent. Creative and authentic content, such as behind-the-scenes videos and employee testimonials, can showcase company culture and opportunities.

Employer Branding

To resonate with Gen Z, employer branding must emphasize authenticity, inclusivity, and purpose. Highlighting real employee stories and demonstrating social responsibility can enhance appeal. Companies with a strong online presence and positive reviews on platforms like Glassdoor gain a competitive edge.

Gamification

Gamified elements in the hiring process, such as challenges, quizzes, and virtual reality (VR) simulations, engage candidates and provide insights into their skills and potential. This approach not only evaluates capabilities but also creates a memorable experience for applicants.

Managing Web-Generation Employees

Onboarding and Training

An engaging onboarding process is crucial for retention. Interactive e-learning modules, virtual mentorship programs, and team-building activities can help Gen Z employees acclimate to the workplace. Continuous training opportunities, such as microlearning and gamified courses, cater to their preference for bite-sized, on-demand content.

Feedback and Recognition

Frequent and constructive feedback fosters growth and satisfaction among Web-Generation employees. Digital platforms like Slack, Microsoft Teams, and dedicated employee recognition tools can facilitate real-time feedback and acknowledgment. Peer-to-peer recognition programs further enhance workplace morale.

Flexible Work Models

Remote and hybrid work arrangements align with Gen Z's preference for flexibility. Providing the autonomy to design their work schedules can enhance productivity and loyalty. Flexible workspaces, such as co-working environments, also appeal to their collaborative nature.

Career Development

Gen Z values continuous learning and clear career progression. Employers should offer upskilling opportunities through workshops, certifications, and personalized development plans. Platforms like LinkedIn Learning and Coursera can be integrated into development programs to provide access to a wide range of courses.

Diversity, Equity, and Inclusion (DEI)

Inclusive policies and practices are non-negotiable for Gen Z. Organizations must ensure representation, equity, and belonging at all levels. Regular DEI training, diverse hiring practices, and employee resource groups can create an inclusive culture that resonates with Web-Generation employees.

Retention Strategies

Building a Strong Culture

A strong organizational culture that prioritizes collaboration, transparency, and innovation is fundamental to retaining Web-Generation employees. This generation thrives in environments where their contributions are acknowledged, and their voices are heard. Organizations can foster such a culture by implementing regular team-building activities that encourage camaraderie and mutual respect among employees. Open communication channels, such as town halls, suggestion boxes, or anonymous feedback platforms, provide employees with opportunities to share their thoughts and concerns without fear of reprisal. Transparency in leadership demonstrated through clear communication about company goals, challenges, and achievements, builds trust and a sense of belonging. Additionally, integrating modern workplace tools that enable seamless collaboration and flexibility, such as project management software and hybrid work policies, ensures employees feel supported and valued.

Competitive Compensation

While Gen Z and Millennials often emphasize purpose and values, competitive compensation remains a cornerstone of talent retention. Financial incentives must align with market standards to attract and keep top talent. Beyond salary, organizations can offer benefits tailored to the needs of Web-Generation employees. Programs like student loan repayment assistance address a significant pain point for many young professionals. Mental health resources, such as counselling services and stress management workshops, demonstrate a commitment to employee well-being. Wellness programs, including gym memberships, mindfulness sessions, and nutrition plans, also resonate strongly. Furthermore, non-monetary perks like additional vacation days, sabbaticals, or flexible work arrangements provide employees with the work-life balance they desire, fostering loyalty and long-term engagement.

Employee Advocacy

Transforming employees into brand ambassadors is a powerful strategy for retention and employer branding. When employees take pride in their workplace, their advocacy on social media and professional platforms can significantly enhance the company's reputation. Encouraging employee advocacy involves creating structured programs that reward and recognize those who promote the organization's culture, achievements, or job opportunities. Referral bonuses, exclusive networking events, or public recognition can serve as incentives. Additionally, providing employees with branded content, such as shareable graphics or success stories, simplifies the process and encourages participation. Advocacy programs also strengthen internal morale by showcasing the organization's confidence in its workforce. When employees feel valued and empowered to represent the brand, their loyalty to the organization deepens, further reinforcing retention efforts.

Case Studies and Best Practices

Company A: Tech-Driven Onboarding

Company A, a leading tech firm, revamped its onboarding process to include VR tours of the workplace and AI-driven mentorship matching. This initiative resulted in a 25% improvement in employee engagement during the first six months.

Company B: Purpose-Driven Branding

Company B, a sustainability-focused organization, highlighted its commitment to environmental causes through social media campaigns and employee testimonials. This approach led to a 30% increase in Gen Z applications over two years.

Company C: Flexible Work Policies

Company C implemented a hybrid work model with personalized schedules. By accommodating employees' preferences, they achieved a 20% reduction in turnover among Gen Z staff.

Challenges and Future Directions

Balancing Digital and Human Interaction

Over-reliance on technology can lead to isolation. Organizations must strike a balance between digital tools and meaningful interpersonal connections. Encouraging in-person meetings and fostering a sense of community can mitigate the risks of digital fatigue.

Navigating Ethical Dilemmas

Gen Z is highly attuned to ethical practices. Misalignment between organizational actions and stated values can erode trust and engagement. Transparent communication and accountability are essential for maintaining credibility.

Continuous Evolution

The rapid pace of technological advancements necessitates ongoing adaptation of talent management strategies. Employers must remain agile and forward-thinking to address the changing needs of the workforce.

Addressing Mental Health Stigma

While mental health is a priority for Gen Z, stigma persists in many workplaces. Providing access to resources, normalizing conversations about mental health, and training managers to support employees can create a supportive environment.

CONCLUSION

As the digital economy continues to evolve, digital competencies have emerged as a key trend in the workplace. From basic digital literacy to advanced technical skills, the ability to leverage digital tools and technologies is essential for success in the modern workforce. Organizations that invest in developing the digital competencies of their employees will be better positioned to innovate, enhance productivity, and maintain a competitive edge in an increasingly digital world. Also, Effectively managing Web-Generation employees requires an in-depth understanding of their values, preferences, and expectations. These employees, characterized by their technological fluency, desire for inclusivity, and emphasis on work-life balance, demand a workplace that resonates with their digital-first nature. By leveraging cutting-edge technology, creating diverse and inclusive environments, and prioritizing mental and physical well-being, organizations can not only attract but also retain and empower this digitally native generation. Furthermore, providing opportunities for continuous learning, fostering transparent communication, and offering flexible work arrangements are key to engaging these employees effectively. As the workplace continues to evolve amidst rapid technological and societal shifts, businesses that embrace these forward-thinking strategies will position themselves as employers of choice for the Web Generation. In doing so, they will not only secure a competitive edge but also build a resilient and innovative workforce ready to meet the demands of the digital economy.

REFERENCES

Anderson, H. J. (2021). Managing Gen Z: Work Expectations and the New Workforce. *Journal of Business Research, 124*, 464–475.

Armbrust, M., Fox, A., Griffith, R., Joseph, A. D., Katz, R., Konwinski, A., Lee, G., Patterson, D., Rabkin, A., Stoica, I., & Zaharia, M. (2010). A View of Cloud Computing. *Communications of the ACM, 53*(4), 50–58. DOI: 10.1145/1721654.1721672

Bandura, A. (1977). *Social Learning Theory*. Prentice Hall.

Bass, B. M. (1999). *Transformational Leadership: Industry, Military, and Educational Impact*. Erlbaum.

Bersin, J. (2022). "The Future of Work: Understanding Gen Z in the Workplace." *Harvard Business Review*. https://www.hbr.org

Bock, L. (2015). *Work Rules!* Insights from Inside Google That Will Transform How You Live and Lead. Twelve.

Brynjolfsson, E., & McAfee, A. (2014). *The Second Machine Age: Work, Progress, and Prosperity in a Time of Brilliant Technologies*. W.W. Norton & Company.

Cappelli, P. (2021). Reshaping Talent Acquisition Strategies for a Digital Generation. *HRMagazine*, *66*(4), 32–39.

Chia, A. (2018). SkillsFuture: Enhancing Singapore's Workforce Adaptability. *Journal of Asian Public Policy*, *11*(3), 292–305.

Deloitte. (2023). "Global Gen Z and Millennial Survey 2023." https://www.deloitte.com

Edmondson, A. C. (1999). Psychological Safety and Learning Behavior in Work Teams. *Administrative Science Quarterly*, *44*(2), 350–383. DOI: 10.2307/2666999

Ellis, R. K. (2009). *A Field Guide to Learning Management Systems*. ASTD Learning Circuits.

European Commission. (2020). Digital Skills and Jobs Coalition: What is Digital Literacy? https://ec.europa.eu/digital-strategy/our-policies/digital-skills

E&Y. (2023). "Gen Z at Work: Expectations and Engagement in a Digital World." https://www.ey.com

Knowles, M. (1980). The Modern Practice of Adult Education: From Pedagogy to Andragogy. Cambridge.

Kotter, J. P. (1996). *Leading Change*. Harvard Business Review Press.

Kshetri, N. (2017). 1 Blockchain's roles in meeting key supply chain management objectives. *International Journal of Information Management*, *37*(6), 100–108.

Luckin, R., Holmes, W., Griffiths, M., & Forcier, L. B. (2016). *Intelligence Unleashed: An Argument for AI in Education*. Pearson Education.

Page, S. E. (2007). *The Difference: How the Power of Diversity Creates Better Groups, Firms, Schools, and Societies*. Princeton University Press.

Pew Research Center. (2020). "Generation Z and the Future of Work." https://www.pewresearch.org

Topol, E. (2019). *Deep Medicine: How Artificial Intelligence Can Make Healthcare Human Again*. Basic Books.

Twenge, J. M. (2017). iGen: Why Today's Super-Connected Kids Are Growing Up Less Rebellious, More Tolerant, Less Happy. New York: Atria Books.

Wenger, E. (1998). *Communities of Practice: Learning, Meaning, and Identity*. Cambridge University Press. DOI: 10.1017/CBO9780511803932

Westerman, G., Bonnet, D., & Ferraris, P. (2014). *The 2014 Digital Transformation Review*. MIT Center for Digital Business.

World Economic Forum. (2020). The Future of Jobs Report. https://www.weforum.org/reports/the-future-of-jobs-report-2020

Chapter 6
Managing Resistance to Change

Keai Lim
https://orcid.org/0000-0002-1131-1635
Amity Global Institute, Singapore

Chen Bao
Amity Global Institute, Singapore

Ya Li
https://orcid.org/0009-0000-4189-4686
Amity Global Institute, Singapore

Ruizhe Wang
Amity Global Institute, Singapore

Xiang Zhou
Amity Global Institute, Singapore

ABSTRACT

This chapter explores the key factors contributing to resistance to digital transformation and provides strategies to overcome them effectively. It underscores the importance of stakeholder involvement in fostering trust, collaboration, and alignment with organizational goals, ensuring smoother transitions and greater commitment to change. Effective communication is highlighted as a vital tool in addressing uncertainties, enhancing transparency, and building confidence among employees. The chapter also investigates the critical role of organizational culture in enabling transformation by promoting adaptability, continuous learning, and cross-functional collaboration. Through case studies of organizations such as General Electric (GE) and IBM, practical insights are provided on how leadership-driven

DOI: 10.4018/979-8-3373-1320-7.ch006

communication, targeted upskilling programs, and participatory approaches can successfully mitigate resistance. These insights offer actionable strategies to enhance change readiness and drive sustainable transformation.

1. LEARNING CAUSES OF RESISTANCE

Addressing resistance is crucial for successful organisational transformation. Resistance often stems from exclusion, fear of redundancy, or difficulties adapting to new technologies. By understanding these root causes and applying frameworks like BANI, organisations can contextualise resistance in today's complex environments. This section examines resistance dynamics, its causes, and strategies to mitigate it, highlighting the importance of stakeholder involvement in fostering adaptive and inclusive change.

1.1 Understanding Resistance to Digital Transformation

Definition of Resistance

Resistance to change is a common challenge in organisational transformation, often driven by cognitive biases such as loss aversion, where employees perceive potential losses—like job security and familiarity—as outweighing perceived benefits such as career growth (Wang & Su, 2023). Addressing these psychological barriers requires framing transformation in ways that align with employees' values. Resistance can manifest in passive hesitation, non-compliance, or active opposition, such as open criticism (Anderson & Sun, 2021). Network theory suggests that resistance spreads through informal relationships, with social influencers playing a key role in shaping sentiment. Engaging these influencers can help foster acceptance and mitigate opposition. Additionally, resistance may indicate deeper systemic issues, such as leadership's failure to address perceived risks, while unclear messaging can amplify uncertainty through informal communication channels (Jones & Smith, 2022). Cultural inertia—rooted in established norms and behaviours—further complicates change efforts by reinforcing past practices. Recognising resistance as a signal of broader organisational challenges enables leaders to proactively address concerns and cultivate a culture of adaptability and continuous improvement.

Expanded Theoretical Roots

Resistance to digital transformation stems from various factors, and understanding its root causes through theoretical frameworks is essential for effective management. Lewin's Change Model (1947), which outlines change in three stages—unfreeze, change, and refreeze—provides a foundational approach but is criticised for its linear nature, which fails to capture the iterative complexity of modern digital transformations (Burnes, 2004). Contemporary frameworks such as Kotter's Eight-Step Model (2021) and Adaptive Space Theory (Uhl-Bien & Arena, 2018) address these gaps by emphasising agility, continuous adaptation, and iterative feedback, making them more suitable for today's fast-evolving business landscape. Resistance also varies across cultural contexts, highlighting the need for region-specific strategies (Hofstede, 2021; Silva & Santos, 2020).

Behavioural economics insights (Wang & Su, 2023) reveal that cognitive biases, such as loss aversion, influence resistance, as employees tend to focus more on perceived risks than benefits. Leaders can mitigate this by employing framing techniques that highlight long-term stability and growth opportunities. Additionally, Lewin's assumption that disrupting routines fosters change may no longer hold in an era of constant digital disruption, where repeated changes can intensify resistance (Burnes, 2004). Kotter's model treats resistance as an ongoing challenge, while Anderson & Sun (2021) emphasise the role of social networks in spreading resistance, suggesting that engaging key influencers can mitigate challenges.

Adaptive Space Theory identifies structural barriers to innovation and recommends creating dedicated spaces for experimentation while balancing short-term operational demands with long-term adaptability. The Psychological Flexibility Model (Hayes et al., 2020) offers insights into individual resistance where subconscious fears about job security and unfamiliar technologies fuel opposition. Interventions such as behavioural nudges, targeted training, and clear communication can enhance resilience and alleviate apprehension. However, organisations often underestimate the psychological impact of transformation, leading to disengagement and resistance.

Systems Thinking (Meadows, 2020) takes a holistic view, recognising that digital transformation causes ripple effects, such as workflow disruptions and increased workloads, which can amplify resistance if not managed comprehensively. Tackling systemic barriers, such as organisational silos, through cross-functional teams and digital collaboration platforms fosters cohesion and adaptability. Integrating behavioural economics and network theory helps organisations implement targeted interventions that reduce resistance and support sustainable transformation.

Levels of Resistance

Resistance to change occurs at individual, group, and organisational levels, each presenting distinct challenges that require targeted strategies for effective mitigation. These levels are interconnected, with resistance at one level often reinforcing challenges at another. Addressing these dynamics is crucial to fostering an adaptive environment that supports successful transformation efforts.

At the individual level, resistance often stems from emotional responses, insecurities, or skill deficiencies. Employees may fear redundancy or struggle to adapt to new demands, particularly in technology-driven transformations such as digitalisation (Wang & Su, 2023). Remote and hybrid work models have amplified resistance by fostering isolation and reducing informal communication, weakening connections to organisational culture (Smith et al., 2022). Structured virtual engagement strategies, including regular check-ins and collaboration tools, can help address these challenges. IBM's cloud computing transition illustrates how poor communication and limited skill development fuelled fears of job obsolescence, highlighting the importance of clear communication, empathy, and reskilling initiatives to enhance engagement (Hayes et al., 2020).

Group-level resistance arises from collective dynamics such as peer pressure and groupthink, where teams perceive change as a threat to their cohesion and identity (Uhl-Bien & Arena, 2018). For instance, software development teams transitioning from waterfall to agile methodologies often resist due to fears of losing structure and predictability. Leaders must engage teams early, foster a sense of ownership, and align new practices with team goals. Transparent communication and sustained engagement, as Kotter (2021) emphasises, are crucial to overcoming group-level resistance and preventing organisational fragmentation.

At the organisational level, resistance often results from systemic issues such as outdated infrastructure, rigid hierarchies, and misaligned objectives. Traditional banks, for example, have struggled to integrate fintech solutions due to legacy systems and conflicting strategic goals (Meadows, 2020). Cultural inertia, where entrenched norms hinder adaptability, also presents significant barriers. Addressing these challenges requires phased implementation, iterative feedback, and adaptive spaces for experimentation (Uhl-Bien & Arena, 2018). Leaders must treat resistance as a symptom of deeper systemic misalignment and take proactive steps to align technical and cultural foundations with strategic objectives.

A systems approach helps organisations identify and address resistance drivers effectively, fostering resilience, collaboration, and sustainable change while overcoming challenges such as organisational silos and remote work complexities.

1.2 Root Causes of Resistance to Change in a BANI Context

The BANI framework (Brittle, Anxious, Non-linear, and Incomprehensible) provides a valuable lens for understanding resistance to change in today's volatile and interconnected organisational environments. Unlike traditional models that assume predictable change, BANI acknowledges the fragility, uncertainty, and complexity of modern transformations (Schlegelmilch, 2021). Brittle systems heighten perceptions of instability, anxiety fuels fear of the unknown, and the non-linear nature of change leads to unexpected disruptions. Viewing resistance through the BANI lens enables organisations to implement adaptive strategies that address root causes rather than superficial symptoms. This approach fosters resilience and equips organisations to navigate rapid technological and systemic change effectively (Klein, 2022).

Brittleness of Systems

Brittleness in the BANI framework refers to the fragility of systems that struggle to adapt to change, often due to legacy IT systems, outdated processes, and rigid infrastructures. These weaknesses heighten organisational vulnerability during transformation efforts, increasing employee resistance as they perceive such systems as unreliable and disruptive (Klein, 2022). Deutsche Bank faced significant challenges in adopting digital banking solutions, as its ageing infrastructure was ill-equipped to support emerging technologies like blockchain, causing concerns about stability and increased workloads (Meadows, 2020). To address brittleness, organisations should adopt phased implementation strategies, gradually introducing new technologies to minimise disruptions and support adaptation. Deutsche Bank's approach—upgrading back-end operations first to improve reliability before transitioning to customer-facing services—proved effective in fostering trust and enhancing adoption rates (Deutsche Bank, 2021). Transparent communication, role-specific training, and clear demonstrations of benefits alleviated resistance, showing how incremental changes and active engagement can transform brittle systems into resilient frameworks.

Anxiety About the Unknown

Uncertainty in BANI environments heightens employee anxiety about job security, career growth, and evolving organisational roles, particularly with rapid advancements in automation and artificial intelligence (AI). Employees often perceive such changes as threats rather than opportunities, leading to emotional and psychological resistance that can hinder transformation efforts (Hayes et al., 2020). Tata Motors faced considerable resistance during its automation initiatives, as employees feared redundancy and lacked clarity about their future roles. Insufficient communication regarding the transformation's objectives further diminished trust and engagement (Kumar et al., 2022). To mitigate these concerns, Tata Motors introduced targeted retraining and upskilling programmes aligned with automation processes, reinforced by transparent communication, regular updates, and leadership commitment. These efforts reassured employees of their continued relevance and fostered greater acceptance. Similarly, IBM's "SkillsBuild" programme provides personalised learning pathways and career development opportunities, helping employees align with technological changes (IBM, 2021). Addressing uncertainty through proactive reskilling and clear communication fosters trust, turning resistance into an opportunity for professional growth and organisational innovation.

Non-linearity of Change

Non-linearity in organisational change occurs when small adjustments lead to disproportionately large and unforeseen consequences, disrupting workflows and generating resistance. Traditional linear models often fail to capture these complexities, necessitating a dynamic, interconnected approach (Dawson, 2021). TelCo, a Swedish telecommunications firm, faced such challenges and adopted a complexity-informed strategy, leveraging complexity theory to anticipate and manage disruptions. This proactive approach reduced resistance and improved transformation outcomes (Styhre, 2022). Managing non-linear change effectively requires organisations to embrace systems thinking to understand interdependencies across structures. Scenario planning and pilot testing can identify potential disruptions early, while involving employees foster ownership and minimise resistance. Key strategies include establishing feedback loops for continuous learning, forming cross-functional teams to address challenges, and providing training to equip employees with skills to navigate complexity. Breaking down silos and adapting to evolving environments, such as remote work, enhances resilience, fosters collaboration, and ensures sustainable transformation (Meyer et al., 2022).

The Incomprehensibility of New Systems

In BANI environments, the complexity of emerging technologies—such as blockchain, AI, and advanced analytics—can overwhelm employees, leading to disengagement and resistance when they struggle to understand their practical applications. This disconnect can make employees perceive change as unnecessary or intimidating (Brown & Duguid, 2021). A global automotive company, for example, faced delays in blockchain adoption due to inadequate communication and training, resulting in confusion and scepticism about its relevance to supply chain operations

(Kumar et al., 2022). To address such resistance, organisations must simplify technical concepts and align them with employees' roles through interactive workshops and targeted training programs. The automotive company successfully mitigated resistance by offering hands-on experience and real-world applications. Clear communication and leadership engagement are essential, with leaders articulating benefits and addressing concerns through regular feedback loops. IBM's Watson AI deployment underscores the importance of tailored training and case studies in driving technology adoption (IBM, 2021). Simplifying complex technologies and providing contextual learning fosters adaptability, enhances employee confidence, and supports successful transformation.

1.3 Strategies to Diagnose Resistance Early

Identifying resistance early is crucial for the success of transformation initiatives. Effective diagnostic strategies help organisations address concerns proactively, minimise disruptions and create a supportive environment for change. Emerging technologies such as machine learning and predictive analytics enhance traditional diagnostic methods by detecting resistance patterns before they escalate. These tools analyse data trends to provide actionable insights, allowing leaders to intervene effectively. Furthermore, digital twin technology improves readiness by simulating organisational responses to change, enabling leaders to refine strategies before full-scale implementation. This section explores three key diagnostic tools and their practical applications in early resistance detection, supported by real-world examples demonstrating their effectiveness.

AI-Driven Sentiment Analysis:

AI-driven sentiment analysis enables organisations to detect early signs of resistance by analysing employee communications, such as emails and feedback forms. Machine learning algorithms identify patterns like declining engagement or increased absenteeism, while predictive analytics uses historical and real-time data to anticipate resistance hotspots. Digital twin technology further enhances readiness by simulating organisational scenarios, allowing leaders to test mitigation strategies before full implementation. These tools provide a proactive, data-driven approach to diagnosing resistance and improving responsiveness (Krugmann & Hartmann, 2024). A multinational bank successfully employed AI-driven sentiment analysis during its digital transformation, identifying employee concerns about job security and operational changes. Targeted interventions, such as role-specific training and reassurance initiatives, improved adoption rates and reduced apprehension (Sharma et al., 2024). However, organisations must address ethical concerns, such as employee privacy, and complement AI insights with qualitative feedback for a holistic understanding. Integrating AI sentiment analysis and digital simulations can proactively enhance engagement and facilitate smoother transformations.

Predictive Technologies for Resistance Management

Emerging technologies such as machine learning, predictive analytics, and digital twin technology are valuable tools for identifying and addressing resistance to change. Predictive analytics leverages historical and real-time data to detect behavioural patterns and forecast resistance, enabling targeted, data-driven interventions (Jabbour et al., 2023). Machine learning algorithms analyse diverse data sources—such as employee performance metrics, communication trends, and sentiment data—to identify resistance factors and support timely interventions. Digital twin technology enhances change readiness by simulating operational environments, allowing organisations to test scenarios, identify vulnerabilities, and refine mitigation strategies before full-scale implementation (Fuller et al., 2020). For instance, a global retail firm used digital twins to model employee responses to new supply chain processes, enabling proactive adjustments and a smoother rollout (Gartner, 2023). This controlled environment reduces disruptions and fosters organisational adaptability. Integrating these predictive technologies allows organisations to detect resistance early, engage stakeholders effectively, and refine strategies in real-time. A data-driven approach facilitates informed decision-making, ensuring smoother transitions and improved transformation outcomes (Wilson & Daugherty, 2021).

Stakeholder Mapping:

Stakeholder mapping is a critical tool for identifying key influencers and resistance points within an organisation. By analysing relationships, interests, and power dynamics, organisations can prioritise communication strategies and tailor interventions to address specific concerns. AI-driven stakeholder mapping enhances this process by providing faster, more accurate insights into stakeholder sentiments and potential areas of resistance. Advanced analytics help predict challenges and enable proactive resource allocation to mitigate resistance before it escalates (Dart Project Management, 2024). For example, Reliance Jio used stakeholder mapping during its digital transformation to identify sceptical teams and customise communication stratcgics, resulting in smoother adoption and better alignment with organisational goals (Twimbit, 2021). Stakeholder mapping facilitates targeted engagement by pinpointing individuals and groups likely to resist change, fostering collaboration and trust. However, organisations must balance data-driven insights with qualitative engagement to ensure a comprehensive understanding of stakeholder needs. By integrating AI-driven stakeholder mapping with participatory approaches, organisations can enhance alignment, minimise resistance, and support successful transformation efforts.

Feedback Mechanisms:

Surveys, focus groups, and anonymous pulse checks are essential tools for identifying and addressing resistance to organisational change. They empower employees to voice concerns, providing valuable qualitative and quantitative insights that help organisations identify resistance drivers and implement targeted interventions. Feedback mechanisms foster transparency and trust, promoting employee engagement and inclusivity in the change process. The research underscores their effectiveness, with Taherdoost and Madanchian (2023) highlighting how sentiment analysis, traditionally used in market research, can assess employee perceptions and resistance within organisations. Similarly, The Intelligent Project Manager (2023) emphasises the role of data-driven insights in refining engagement strategies. Singapore's GovTech effectively used anonymous surveys and focus groups to identify concerns, such as inadequate training and introduced tailored interventions to boost confidence and technology adoption. To maximise impact, organisations should integrate traditional and digital feedback approaches, using AI tools to analyse sentiment trends alongside diverse focus groups for broad representation. Crucially, visible action on feedback builds trust and encourages ongoing participation.

Strategies to Diagnose Resistance Early

Managing organisational change effectively requires tools to proactively identify and address potential barriers. Resistance heatmaps provide a visual representation of disengagement or reluctance by analysing employee feedback, performance metrics, and engagement levels, enabling managers to allocate resources strategically and implement targeted interventions (WalkMe, n.d.). Stakeholder engagement checklists further support change efforts by categorising stakeholders based on their influence and potential resistance. Assessing attitudes, communication preferences, and past responses helps organisations tailor engagement strategies to foster buy-in and align stakeholders with organisational goals (OCM Solution, n.d.). Additionally, readiness assessment frameworks offer a structured evaluation of an organisation's preparedness for change by examining leadership commitment, employee capabilities, cultural alignment, and resource availability. These frameworks help identify gaps and support measures, increasing the likelihood of a smooth transition (Prosci, n.d.). Integrating these diagnostic tools allows organisations to adopt a proactive, data-driven approach to resistance management, ensuring a more efficient and sustainable change process.

Practical Examples

DBS Bank: During its digital transformation, DBS Bank utilised real-time dashboards to monitor employee engagement by tracking key metrics such as training participation and feedback submissions. These dashboards helped identify resistance trends early, enabling the bank to implement targeted interventions, such as additional training and clearer communication of transformation objectives. This proactive approach improved employee engagement and alignment with strategic goals. Lim and Goh (2023) highlight that leveraging digital tools enhances the accuracy and timeliness of responses to resistance. By customising support, DBS fostered higher engagement, ensuring employees felt informed and supported throughout the change process.

Reliance Jio: Reliance Jio employed stakeholder mapping during its transition to a digital-first ecosystem to identify resistance hotspots within internal teams. By analysing relationships and sentiments, the organisation tailored communication strategies to address the concerns of sceptical stakeholders. This proactive approach minimised resistance while strengthening trust and collaboration across teams. Rao et al. (2024) emphasise the importance of targeted stakeholder engagement in large-scale transformations. Through stakeholder mapping, Reliance Jio gained deeper insights into resistance drivers and effectively built trust, streamlined communication, and fostered a supportive environment for digital transformation.

2. INVOLVING STAKEHOLDERS IN THE TRANSFORMATION PROCESS

The success of organisational transformation hinges on the effective involvement of stakeholders, who play a pivotal role in aligning objectives, building trust, and addressing resistance. Stakeholder engagement ensures that diverse perspectives are incorporated into planning and implementation, reducing the risks of misalignment and enhancing the sustainability of change initiatives. This section critically evaluates the importance of stakeholder engagement, categorises stakeholder types, and explores effective methods for their involvement in transformation processes.

2.1 Importance of Stakeholder Engagement

Stakeholder engagement is essential for fostering trust, aligning organisational goals, and mitigating resistance to change. Early and meaningful involvement creates an environment of transparency and collaboration, paving the way for smoother transitions.

Fostering Trust

Trust is a cornerstone of successful transformation, as stakeholders are more likely to support changes when their concerns are acknowledged and addressed. Early and meaningful engagement fosters transparency and inclusivity, reducing uncertainty and building confidence in the organisation's vision. This is particularly crucial when implementing disruptive technologies that may create apprehension and resistance. Open dialogue from the outset establishes mutual respect and encourages collaboration, promoting long-term commitment. Infosys effectively built trust during its AI services rollout by conducting co-creation workshops that engaged employees, managers, and clients, allowing them to voice concerns and contribute to decision-making (Infosys, 2023). This participatory approach reduced anxiety and strengthened confidence in the transformation process (Govindarajan & Gupta, 2023). Trust also acts as a bridge between leadership and stakeholders, enabling open communication and minimising resistance. Proactively seeking feedback and integrating it into planning fosters collaboration and adaptability. This iterative engagement process creates a culture of inclusion, where stakeholders feel valued and committed to shared transformation goals.

Aligning Strategic Goals with Operational Realities

Misalignment between strategic goals and operational realities often hinders transformation initiatives, making stakeholder involvement crucial to ensuring that change efforts are both aspirational and practical. Collaborating with stakeholders allows organisations to identify potential obstacles and refine strategies to align with the needs of employees, customers, and external partners. GovTech Singapore exemplifies successful goal alignment through collaboration by engaging representatives from various government departments to co-design digital services that meet public service users' needs (GovTech, 2023). This approach aligned technological advancements with operational requirements, resulting in user-centric solutions that improved public satisfaction and streamlined internal processes (Loo et al., 2022). Aligning strategic objectives with stakeholder priorities reduces friction, enhances feasibility, and fosters a shared sense of purpose by ensuring stakeholder input is reflected in the organisation's vision. Conversely, failing to achieve alignment can lead to overly ambitious strategies that lack practical relevance, ultimately undermining transformation efforts.

Mitigating Resistance

Resistance to change often arises from a misalignment between strategic goals and operational realities, leaving stakeholders feeling disconnected from transformation efforts. When leadership priorities do not resonate with employees, clients, or partners, resistance can hinder implementation and reduce overall effectiveness. Ensuring stakeholder involvement is essential to bridging this gap, making change initiatives both aspirational and practical. Collaborative efforts enable organisations to identify obstacles early and refine strategies to better align with stakeholder needs. Singapore's GovTech exemplifies successful goal alignment by co-designing digital services with input from government agencies and public service users. This participatory approach ensured technological advancements addressed operational realities while enhancing public satisfaction and internal efficiency (GovTech, 2023). Achieving alignment fosters a shared sense of purpose and increases stakeholder commitment. Conversely, failing to align strategic goals with operational realities can result in overly ambitious plans that lack feasibility, leading to disengagement and resistance. Regular engagement through workshops, transparent communication, and continuous feedback allows stakeholders to see their input reflected in the organisation's vision, building trust and promoting commitment. Prioritising alignment reduces resistance, enhances relevance, and ensures transformation efforts are practical, inclusive, and sustainable.

2.2 Types of Stakeholders

Effectively managing resistance to change requires a nuanced understanding of the diverse roles and influences of internal and external stakeholders. Each group holds unique perspectives that can significantly impact the success or failure of transformation initiatives. Engaging stakeholders through tailored strategies is essential to overcoming resistance and ensuring alignment with organisational goals.

Internal Stakeholders

Internal stakeholders—employees, managers, and leadership—are crucial to organisational transformation, as their engagement influences whether resistance is mitigated or exacerbated. Employees, directly impacted by change, may resist due to concerns over job security, evolving roles, or skill gaps. Transparent communication and early involvement alleviate apprehension and foster ownership. For instance, Zappos' Holacracy model empowered employees with greater autonomy, reducing resistance and enhancing engagement (Kelley, 2020). Neglecting employee concerns, however, can lead to disengagement and hinder transformation efforts. Managers play a key role in translating strategic goals into actionable steps while addressing employee concerns. Their involvement aligns leadership's vision with operational realities. Providing managers with training and authority enhances their ability to manage resistance and build trust (IMD, 2021). Leadership teams set the vision and allocate resources, influencing stakeholder perceptions. Transparent and inclusive leadership fosters collective effort and reduces resistance, while inconsistent communication can create uncertainty and impede change (Anderson & Sun, 2021).

External Stakeholders

External stakeholders—clients, regulators, suppliers, and community partners—play a crucial role in shaping transformation outcomes by ensuring relevance, compliance, and acceptance. Clients help validate changes, with early engagement through feedback mechanisms aligning initiatives with market demands and reducing resistance (Kelley, 2020). Regulators define compliance requirements, making proactive collaboration essential to avoid risks and delays. Incorporating regulatory feedback early helps align transformation efforts with legal requirements, reducing external resistance (IMD, 2021). Suppliers and partners must adapt supply chains to accommodate change, yet resistance can arise from workflow disruptions and unclear expectations. Open communication and joint planning sessions help align objectives and ensure smooth collaboration. In the fashion industry, engaging manufacturers in sustainability initiatives has improved programme effectiveness and minimised resistance (Reuters, 2024). Community partners are vital in public sector transformations, where participatory design processes, such as Singapore's GovTech initiative, build trust and foster acceptance by addressing community needs (Loo et al., 2022).

2.3 Case-Linked Insights

Internal and external stakeholders are crucial in overcoming resistance and ensuring successful organisational transformation. Although their roles differ, their interdependence means that resistance within one group can influence the other. For instance, disengagement among employees or managers can lead to external resistance, impacting clients or suppliers through operational inefficiencies. Tailored engagement strategies that address the unique concerns of each group are vital to turning resistance into opportunities for innovation. By fostering trust, encouraging alignment, and prioritising inclusivity, organisations can cultivate an environment where stakeholders collaborate effectively. This holistic approach not only mitigates resistance but also promotes sustainable transformation by ensuring all parties are invested in the change process.

The experiences of Infosys and Singapore's GovTech demonstrate how participatory stakeholder engagement can mitigate resistance and drive successful transformation. During its AI services rollout, Infosys engaged employees, managers, and clients in co-creation workshops to develop user-centric solutions. Employees provided operational insights, managers ensured strategic alignment, and clients contributed market perspectives, fostering trust and ownership across all levels (Govindarajan & Gupta, 2023). Similarly, GovTech Singapore collaborated with government agencies and citizens to refine digital services. Internal representatives ensured operational feasibility, while citizen involvement fostered trust through user-centric design, reducing friction and aligning transformation efforts with public needs (Loo et al., 2022). These cases highlight the importance of incorporating diverse perspectives in engagement strategies. Proactively addressing stakeholder concerns and leveraging their expertise can transform resistance into opportunities for collaboration and innovation, ensuring transformation efforts are practical, widely accepted, and aligned with organisational and societal goals.

While large corporations offer insights into managing large-scale change, SMEs and non-profits face distinct challenges requiring tailored strategies. A mid-sized UK healthcare organisation overcame resistance to a new patient management system by adopting participatory decision-making and phased rollouts, improving staff engagement and operational efficiency despite limited resources (West Africa Civil Society Institute, 2021). Similarly, an African non-profit addressing digital transformation challenges used grassroots engagement to secure stakeholder buy-in. The Climate BRIDGE Programme demonstrated the importance of digital advocacy and collaboration for successful technology adoption in resource-constrained environments. These examples underscore the need for context-specific approaches that consider financial and operational constraints, enabling smaller organisations to achieve transformation goals effectively. Incorporating case studies from SMEs and non-profits provides a comprehensive perspective on resistance management, ensuring strategies are applicable across diverse organisational scales and cultural contexts.

2.4 Methods to Involve Stakeholders

Effective stakeholder involvement is essential for overcoming resistance and ensuring the success of organisational transformation. This requires structured, participatory methods tailored to the unique dynamics of change processes. By fostering collaboration, building trust, and ensuring inclusivity, these methods address stakeholder concerns and encourage active participation.

Participatory Approaches

Participatory approaches, such as design-thinking workshops, effectively engage stakeholders in transformation efforts by providing a structured yet adaptable platform for identifying challenges, brainstorming solutions, and collaboratively prototyping ideas. Direct stakeholder involvement harnesses diverse perspectives and fosters ownership of the transformation process. Singapore's GovTech successfully employed this approach by engaging departmental representatives and end-users to align digital service goals with user expectations. This participatory process fostered innovation, addressed real-world challenges, and increased adoption rates and satisfaction (Loo et al., 2022). The success of design-thinking workshops lies in their ability to simplify complex problems, facilitate open dialogue, and generate practical solutions that align with stakeholder needs. However, their effectiveness hinges on careful facilitation and inclusivity, ensuring the participation of all relevant stakeholders, especially those resistant to change. Moreover, workshops must strike a balance between creativity and feasibility to ensure the realistic implementation of proposed solutions within organisational constraints.

Continuous Feedback Loops

Continuous feedback loops are vital for monitoring stakeholder sentiment and addressing concerns during transformation efforts. Tools such as anonymous surveys, pulse checks, and focus groups provide actionable insights, enabling timely interventions and building trust. DBS Bank effectively used pulse checks to identify employee concerns about training and role adaptation, leading to targeted interventions that improved engagement and reduced resistance (Chan et al., 2021). Feedback loops also reinforce transparency and accountability by demonstrating that stakeholder input drives meaningful action, fostering a collaborative environment. To maximize effectiveness, organisations should combine qualitative methods like focus groups with quantitative surveys for a comprehensive understanding of concerns. Advances in AI-driven sentiment analysis further enhance feedback mechanisms by identifying patterns in stakeholder responses and enabling proactive resistance management (Sharma et al., 2024). By integrating technology with traditional methods, organisations can improve responsiveness, strengthen engagement, and ensure successful transformation.

Gamification and Incentives

Gamification and incentives effectively engage stakeholders, particularly employees, in adopting new processes by incorporating elements of competition, rewards, and recognition. Gamified training transforms learning into an engaging experience, as demonstrated by IBM's cloud technology training, which uses leaderboards, badges, and performance-based rewards to enhance participation, retention, and morale. This approach simplified complex learning and increased motivation. Similarly, Tata Motors incentivised automation adoption through upskilling certifications, which boosted morale and participation (Rao & Verma, 2023). Both financial and non-financial incentives are crucial for sustained engagement. Financial rewards, such as bonuses for achieving transformation milestones, provide immediate motivation, while non-financial incentives, including career development opportunities and recognition, foster long-term commitment. AI-driven sentiment analysis further enhances these efforts by identifying resistance patterns early, enabling proactive interventions (Sharma et al., 2024). However, to ensure lasting impact, organisations must align gamification and incentives with strategic transformation goals, promoting meaningful engagement rather than short-term compliance.

Practical Implementation Strategies

Effective stakeholder involvement requires strategic implementation through key tools such as stakeholder mapping, digital collaboration platforms, and customised engagement plans. Stakeholder mapping identifies key influencers and resistance hotspots, enabling organisations to prioritise engagement efforts and allocate resources effectively. Reliance Jio employed stakeholder mapping during its digital transformation to identify sceptical teams and tailor communication strategies, minimising resistance and ensuring smoother implementation (Twimbit, 2021). Digital collaboration platforms like Slack, Teams, and Trello facilitate real-time communication and collaboration, aligning stakeholders with transformation goals and centralising updates and concerns. Grab successfully used such platforms to maintain transparency and enhance operational efficiency during its transformation (Tan et al., 2023). Customised engagement plans define stakeholder roles, responsibilities, and timelines, ensuring alignment with organisational objectives. These plans should be dynamic, adapting to feedback and evolving needs. Integrating stakeholder mapping, collaboration tools, and flexible engagement plans enhances stakeholder involvement, fostering trust and ensuring successful transformation outcomes.

3. COMMUNICATION FOR CHANGE MANAGEMENT

Effective communication is key to managing resistance and achieving successful organisational transformation by fostering trust, providing clarity, and aligning expectations. Transparent communication minimises uncertainty, counters misinformation, and encourages participation. While technology-driven tools play a role, human-centric approaches such as mentorship programs, empathy-based leadership, and peer-led advocacy are essential for sustained engagement. Mentorship offers direct support during transitions, empathy-based leadership ensures employees feel valued, and peer-led advocacy empowers influential employees to drive change. Combining these strategies with AI-driven sentiment analysis, feedback mechanisms, and collaboration platforms helps organisations create an inclusive environment that supports adaptability and long-term success.

3.1 Critical Role of Communication

Transparency builds trust by providing clarity and aligning stakeholders with organisational goals, while storytelling fosters engagement by making change relatable and inspiring. This section explores how these strategies enhance stakeholder commitment and drive transformation.

Building Trust Through Transparency

Transparency is crucial for reducing uncertainty, fostering trust, and aligning organisational efforts with strategic goals. It goes beyond sharing information; leaders must engage middle managers to translate strategic vision into actionable steps and ensure employees feel included in the transformation process. Effective transparency minimises fear and misinformation, which often heightens resistance to change. Employees are more likely to trust leadership when they understand the purpose, progress, and expected outcomes of transformation initiatives (Wang & Su, 2023). Haier's Rendanheyi (   ) model exemplifies this by linking employee contributions to organisational goals through real-time data, fostering ownership and alignment (Li & Zhang, 2022). Similarly, Huawei's use of digital tools during restructuring provided regular updates, alleviating anxiety, and strengthening trust. Transparency strategies should include clear objectives, regular updates, and open discussions to address concerns. Middle managers play a crucial role in contextualising messages and engaging frontline employees by bridging leadership vision and operational execution, identifying resistance, and supporting employees with tailored interventions. However, transparency must be managed carefully to avoid overwhelming employees or breaching confidentiality, focusing on relevant, actionable information that builds trust without creating confusion (Anderson & Sun, 2021).

Motivating Through Storytelling

Storytelling is a powerful tool for engaging stakeholders emotionally, making change relatable, and inspiring action by transforming abstract goals into compelling narratives that highlight successes and foster motivation. Effective storytelling provides employees with a sense of purpose and belonging, reinforcing their role in the transformation journey. It appeals to both rational and emotional aspects by humanising change through familiar experiences and tangible benefits (Brown & Duguid, 2021). Alibaba's digital transformation exemplifies this, using employee-led innovation stories to showcase the positive impact of new technologies, inspiring broader engagement (Loo et al., 2022). To be effective, storytelling should incorporate relatable examples in engaging formats such as videos, infographics, and case studies. Leaders should craft narratives that emphasise resilience, collaboration, and innovation to foster a positive perception of change (Rao & Verma, 2023). Consistency and authenticity are key to building credibility, and regularly refreshing narratives with new milestones sustains engagement, fosters adaptability, and aligns stakeholders with transformation objectives.

3.2 Modern Communication Tools and Techniques

Advanced communication technologies have transformed how organisations engage stakeholders during change initiatives. By leveraging AI-driven sentiment analysis, feedback mechanisms, and collaboration platforms, organisations can proactively address resistance, build trust, and enhance collaboration. However, while technology streamlines communication, personal interactions remain crucial. Leaders should balance digital tools with human-centric approaches, such as regular face-to-face meetings, mentorship initiatives, and active listening sessions. These efforts provide employees with opportunities to voice concerns, fostering a sense of value and understanding alongside digital engagement. This section explores key communication tools and strategies that drive effective stakeholder engagement during transformation efforts.

AI-Driven Sentiment Analysis

AI-driven sentiment analysis offers real-time insights into employee sentiment by analysing communication channels such as emails, feedback forms, and internal forums. It helps leaders detect resistance trends early and implement targeted communication strategies to address concerns proactively, fostering employee support during transformation efforts. Unlike traditional feedback methods, AI efficiently processes large data volumes to uncover subtle emotional cues indicating dissatisfaction or uncertainty (Sharma et al., 2024). Huawei's use of AI sentiment analysis during restructuring enabled tailored communication strategies, improving acceptance and easing transitions. However, ethical concerns surrounding transparency and trust are critical. Employees must feel assured that data collection supports their well-being rather than surveillance, necessitating clear policies and human oversight to ensure empathetic responses (Wang & Su, 2023). While AI enhances trend analysis, it should complement—not replace—traditional communication methods. A balanced approach, combining AI-driven insights with direct engagement, ensures meaningful dialogue and facilitates a smoother transformation process.

Two-Way Feedback Mechanisms

Two-way feedback mechanisms are essential for fostering open communication and inclusivity during organisational change. Unlike traditional top-down approaches, these mechanisms empower employees to voice concerns, provide suggestions, and receive timely responses from leadership. When employees feel heard and valued, they are more likely to support change initiatives, reducing resistance and fostering a sense of ownership. Encouraging dialogue allows organisations to identify challenges early, enabling proactive interventions and building a culture of trust and collaboration. Meaningful engagement transforms employees from passive recipients to active participants, enhancing morale and facilitating smoother transitions (Brown & Duguid, 2021). DBS Bank leveraged open forums during its digital transformation, allowing employees to address concerns directly with leadership, reducing uncertainty and improving digital adoption (Chan et al., 2021). Similarly, Tata Motors implemented regular team check-ins to reassure employees about automation, boosting morale and acceptance (Rao & Verma, 2023). To maximise effectiveness, organisations should ensure psychological safety by offering anonymous feedback options and visibly acting on employee input to reinforce trust and credibility.

Collaboration Platforms

Collaboration platforms such as Microsoft Teams, Slack, and Workplace by Meta are critical in driving organisational change, especially in hybrid and dispersed work environments. These platforms facilitate real-time communication, support cross-functional collaboration, and centralise information-sharing, ensuring alignment across stakeholders throughout the transformation process. By breaking down silos and offering structured yet flexible interactions, they promote transparency, inclusivity, and responsiveness, helping mitigate resistance and enhance engagement. Centralised communication channels provide employees with access to updates, facilitate knowledge sharing, and allow direct interaction with leadership, addressing concerns early and preventing misunderstandings (Tan et al., 2023). Grab's adoption of Slack during its digital transformation illustrates its benefits. By creating dedicated project channels, the company improved coordination, minimised miscommunication, and facilitated informal dialogue, ensuring alignment with strategic objectives (Loo et al., 2022). To maximise effectiveness, organisations should invest in comprehensive onboarding and training to enhance platform proficiency and engagement. Integrating workflow management tools like Trello can optimise operations, while feedback channels empower employees to voice concerns and receive timely responses, reinforcing trust. Successful adoption of collaboration platforms requires thoughtful implementation, ongoing training, and integration with broader change management strategies.

4. THE ROLE OF ORGANISATIONAL CULTURE IN ACHIEVING DIGITAL BUSINESS

4.1 Culture as an Enabler of Change

Organisational culture plays a fundamental role in enabling digital transformation by shaping how employees respond to change, adopt new technologies, and align with business goals. A culture that fosters flexibility, collaboration, and continuous learning can significantly accelerate digital adoption, while rigid and siloed cultures can hinder progress and create resistance. This section explores how cultural readiness, overcoming barriers, and building a learning ecosystem contribute to successful digital business strategies.

Cultural Readiness

Cultural readiness is essential for successful digital transformation, as adaptable, collaborative, and learning-oriented cultures embrace change more effectively, while rigid cultures face resistance and disruptions (Cameron & Quinn, 2022). Cultural differences shape resistance across regions. In Western cultures, scepticism toward top-down mandates necessitates transparency and participation, whereas Eastern cultures, valuing collectivism and hierarchy, may exhibit passive disengagement, requiring a balance of authority and consultation (Hofstede, 2021). Economic conditions also influence resistance strategies, with Western models emphasizing individual accountability, while Latin America and Africa favour collective decision-making

(Silva & Santos, 2020). In Europe, German SMEs rely on participatory design to foster consensus and alleviate change-related anxiety (Schneider & Schmid, 2021).

Culturally ready organisations exhibit adaptability, collaboration, and a commitment to continuous learning. However, collaboration varies regionally Western firms favour autonomy, while Eastern firms rely on hierarchical coordination. Continuous learning supports proactive upskilling to align with evolving digital landscapes (Kotter, 2021). Tailored strategies, such as mentorship programs in hierarchical cultures and innovation labs in collaborative environments, further enhance cultural readiness. General Electric's (GE) IoT transition exemplifies this approach, overcoming initial resistance by restructuring workflows, empowering leaders, and promoting digital upskilling, fostering agility and collaboration (Smith et al., 2022).

Beyond approaches in the US, China, and India, valuable insights come from Europe, Africa, and Latin America. A German SME addressed automation resistance through participatory design, while a Latin American retailer leveraged community involvement, and a South African non-profit used stakeholder consultations and culturally relevant training to ensure smooth digital adoption (Schneider & Schmid, 2021; Rugman & Verbeke, 2022; West Africa Civil Society Institute, 2021). These cases underscore the importance of context-specific strategies to manage cultural resistance effectively.

Overcoming Cultural Barriers

Cultural barriers, such as rigid organisational structures, legacy practices, and misaligned values, often impede transformation efforts by restricting information flow, collaboration, and adaptability (Brown et al., 2023). Siloed departments hinder cross-functional engagement, delaying technology adoption, while outdated workflows and resistance to change create friction as employees fear disruption and job insecurity (Kotter, 2023). When leadership's vision fails to align with employee expectations, disengagement increases. Addressing these challenges requires fostering a culture that embraces change through strategic interventions. Transparent communication—via town halls, feedback sessions, and open channels—helps surface resistance early and build trust. Leaders should empower teams by acting as facilitators rather than enforcers, offering continuous support (Anderson & Sun, 2021). Aligning organisational values with innovation, customer-centricity, and agility fosters shared purpose across all levels. Infosys' AI integration exemplifies this, overcoming initial resistance through upskilling initiatives, AI champions, and value realignment, ensuring a smooth transition and industry leadership (Rao & Verma, 2023).

Metrics to Measure Cultural Readiness

Measuring cultural readiness is essential for evaluating transformation effectiveness and identifying improvement areas. Key metrics—such as employee engagement scores, cross-departmental collaboration rates, and technology adoption levels—offer insights into how well an organisation's culture aligns with its digital strategy. High employee engagement scores indicate motivation and alignment with change initiatives, whereas low scores may signal resistance or uncertainty, highlighting the need for targeted interventions (Smith et al., 2022). Collaboration rates assess how effectively teams work across functions, with increased cooperation signifying the breakdown of silos and a more cohesive change approach (Brown et al., 2023). Technology adoption rates reflect employees' willingness to embrace new tools and their confidence in the organisation's vision, pinpointing potential training or engagement gaps (Kotter, 2021). To drive continuous improvement, organisations should utilise digital dashboards and analytics for real-time insights, while complementing quantitative data with qualitative inputs from surveys, focus groups, and regular reviews for a comprehensive understanding of cultural readiness.

4.2 Building a Learning Ecosystem

A well-structured learning ecosystem is crucial for organisations navigating digital transformation, as it fosters continuous learning, collaboration, and adaptability. By cultivating a culture that values knowledge sharing and proactive upskilling, organisations can equip employees with the necessary skills to embrace technological advancements and drive innovation. An effective learning ecosystem not only enhances organisational agility but also helps break down silos, ensuring that all stakeholders are aligned with strategic goals.

What is a Learning Ecosystem?

A learning ecosystem is an organisational framework that fosters continuous learning, knowledge sharing, and collaboration across internal and external boundaries to enhance agility and drive innovation in digital transformation (Garvin et al., 2022). It enables organisations to respond effectively to technological advancements and evolving market demands by integrating knowledge resources and promoting adaptability. Key components include knowledge hubs that centralise best practices and learning materials, cross-industry collaborations that introduce new perspectives through partnerships with academic institutions and industry experts, and proactive upskilling initiatives to equip employees with evolving job requirements (Smith & Brown, 2023). Huawei's partnership with global research institutions exemplifies how integrating academic insights into workforce development enhances agility and competitiveness (Loo et al., 2022). However, building an effective learning ecosystem requires a strong commitment to continuous improvement and investment in digital tools that facilitate seamless learning. Without such commitment, embedding learning into daily workflows becomes challenging, leading to stagnation and resistance to change.

Fostering Collaboration and Motivation

Fostering collaboration and motivation is crucial for building an effective learning ecosystem by breaking down silos and cultivating a culture of shared purpose and innovation. Collaboration encourages employees to work cross-functionally, promoting a holistic approach to problem-solving and knowledge sharing (Tan et al., 2023). Motivated employees are more likely to embrace digital transformation initiatives and contribute proactively to organisational goals. Strategies to enhance collaboration include forming cross-functional teams that bring together representatives from different departments to address challenges collectively and leveraging digital collaboration tools such as Microsoft Teams and Slack for real-time communication and knowledge sharing (Jones et al., 2023). Joint problem-solving sessions provide structured forums for brainstorming and troubleshooting, ensuring alignment with organisational objectives. Singtel's digitalisation efforts highlight the effectiveness of these strategies; by implementing cross-departmental teams and structured training initiatives, the organisation improved collaboration, accelerated technology adoption, and enhanced project success rates (Chong & Lim, 2023). Leadership plays a pivotal role in fostering inclusivity and ensuring collaboration remains integral to transformation efforts.

Motivating Teams Through Learning and Recognition

Motivating teams through learning and recognition is essential for sustaining engagement within a learning ecosystem. Employees who see clear career growth opportunities and receive recognition for their learning efforts are more likely to embrace new technologies and processes. Linking structured career growth pathways to upskilling initiatives, celebrating small wins, and incorporating gamification techniques can make learning more engaging and rewarding (Kumar & Patel, 2022). Tata Motors effectively applied this approach by introducing certification programmes tied to career advancement opportunities, addressing automation-related job concerns and increasing participation in upskilling efforts (Rao & Verma, 2023). Additionally, fostering intrinsic motivation through recognition programmes that publicly acknowledge employee achievements reinforces a culture where learning is valued. By rewarding contributions to digital initiatives, organisations can instil a sense of purpose and belonging, encouraging employees to view continuous development as integral to their professional growth. This approach drives long-term commitment, enhances adaptability, and positions organisations for sustained success in an evolving business landscape.

Measuring the Success of a Learning Ecosystem

Assessing the effectiveness of a learning ecosystem is essential for continuous improvement and alignment with organisational goals. Key performance indicators (KPIs) provide insights into learning outcomes and areas requiring enhancement (Smith et al., 2023). Critical metrics include upskilling rates, which measure employee participation in training programmes and reflect the organisation's commitment to workforce development. High completion rates suggest a strong learning culture, while low participation may signal engagement challenges. Innovation output tracks the quantity and quality of new ideas generated, revealing the effectiveness of collaboration and knowledge-sharing efforts. Adoption speed measures how quickly employees embrace new technologies, identifying potential bottlenecks—faster adoption indicates successful integration, while delays may point to resistance or insufficient support. Employee satisfaction, evaluated through surveys, provides insights into how the learning ecosystem influences professional development and morale. Leveraging digital dashboards and analytics tools offers real-time insights, while qualitative feedback from focus groups and anonymous suggestions enriches understanding and informs strategic improvements (Garvin et al., 2022).

5. CASE STUDIES: OVERCOMING RESISTANCE

Organisational change is often met with resistance due to uncertainties, fear of the unknown, and cultural inertia. Overcoming resistance requires strategic approaches tailored to the specific needs of the organisation and its stakeholders. This section explores three case studies—General Electric (GE), IBM, and selected non-US companies (Alibaba, Reliance Jio, and DBS Bank)—to highlight best practices in managing resistance to change, with a focus on leadership strategies, employee engagement, and technology adoption.

5.1 General Electric (GE): Tackling Resistance to IoT Adoption

General Electric (GE) faced considerable resistance during its integration of IoT technologies, as employees feared job redundancy and disruptions to established workflows. The company's hierarchical culture, which emphasised stability over innovation, further intensified these challenges (Brown & Jones, 2022). To address resistance, GE implemented a multi-faceted strategy focusing on leadership transparency, targeted upskilling programs, and active employee involvement. Leadership played a key role by hosting town halls and open forums to communicate the benefits of IoT, reassure employees, and provide clear timelines, fostering trust and reducing uncertainty (Kotter, 2021). Recognising that skill gaps heightened apprehension, GE introduced tailored training programs that equipped employees with IoT-related competencies, certifications, and career growth pathways, aligning professional development with business objectives (Smith et al., 2023). Additionally, employee participation in pilot projects allowed firsthand experience with IoT, encouraging feedback and a sense of ownership. This participatory approach eased resistance and facilitated smoother adoption (Garvin et al., 2022). As a result, GE achieved a 20% increase in operational efficiency and innovation, demonstrating the critical role of transparent leadership and workforce development in successful digital transformation.

5.2 IBM: Cultural Transformation for Cloud Migration

IBM encountered considerable resistance during its transition to cloud computing, particularly from long-serving employees accustomed to traditional IT practices. The shift disrupted established workflows and raised concerns about job redundancy, especially among those with legacy skill sets. Organisational inertia further complicated the transition, as employees hesitated to leave their comfort zones (Wang & Su, 2023). To address these challenges, IBM implemented strategic initiatives to foster cultural transformation and encourage adoption. A key initiative was a gamified learning platform that incorporated leaderboards, rewards, and competitive elements to enhance engagement, motivating employees through points and badges upon course completion (Rao & Verma, 2023). Leadership-driven communication also played a critical role, with senior leaders sharing their cloud learning experiences through regular Q&A sessions and transparent discussions, reassuring employees about career growth opportunities (Kotter, 2021). Furthermore, IBM established employee advocacy networks by positioning early adopters as change champions to mentor peers address concerns, and showcase the benefits of cloud adoption (Cameron & Quinn, 2022). These initiatives accelerated IBM's cloud adoption, with most teams transitioning ahead of schedule. The gamification approach led to a 75% completion rate for cloud training, embedding a culture of continuous learning and innovation. This case underscores the significance of strategic employee engagement and leadership involvement in overcoming resistance and driving successful digital transformation.

5.3 Non-US Examples: Alibaba, Reliance Jio, and DBS Bank

Alibaba (China): Adaptive Resilience During COVID-19

Alibaba encountered significant resistance during its transition to digital operations amid the COVID-19 pandemic, as employees faced challenges adapting to remote work while managing increased e-commerce demand. Uncertainty about job roles, heavier workloads, and the lack of in-person support heightened stress and productivity concerns (Tan et al., 2023). To address these issues, Alibaba implemented a comprehensive strategy to enhance efficiency and support employee well-being. Agile project management methodologies were introduced to streamline decision-making and reduce bottlenecks, enabling teams to respond swiftly to market demands. Virtual training on collaboration tools such as DingTalk and Alibaba Cloud equipped employees with essential remote work skills. Recognising the emotional toll of the transition, Alibaba also offered free counselling services and stress management programs. These initiatives improved operational responsiveness, facilitated smoother adoption of digital tools, and fostered workforce resilience. Alibaba's experience highlights the importance of balancing agility with employee support to manage resistance effectively (Loo et al., 2022).

Reliance Jio (India): Building a Low-Cost Digital Ecosystem

Reliance Jio encountered significant resistance during its transition to a digital-first business model, with internal concerns over job security and external scrutiny regarding the affordability and accessibility of its services. Employees feared the shift would eliminate traditional roles, while regulators questioned Jio's ability to provide low-cost digital connectivity at scale (Kumar & Patel, 2022). To address these challenges, Jio adopted a collaborative and transparent approach. Internally, a robust communication strategy reassured employees about job security and career growth, emphasising how digitalisation would enhance operational efficiency and create new opportunities. Externally, Jio engaged stakeholders—regulators, suppliers, and customers—through regular consultations, aligning its digital goals with national inclusion objectives and ensuring compliance with regulatory standards. Open dialogue sessions built trust and credibility, helping to mitigate resistance. As a result, Jio successfully rolled out its digital services to millions, with employees embracing their evolving roles and the company strengthening its reputation as an innovative industry leader (Sharma & Gupta, 2023). This case highlights the importance of clear communication and proactive stakeholder engagement in managing large-scale transformation efforts.

DBS Bank (Singapore): Driving Engagement Through Feedback and Dashboards

DBS Bank faced significant resistance from employees reluctant to embrace digitalisation, preferring traditional banking methods due to concerns about the complexity of digital tools and their impact on job roles. Uncertainty and unfamiliarity with digital platforms further exacerbated apprehensions (Brown et al., 2023). To address these challenges, DBS implemented a data-driven and participatory approach. The introduction of real-time dashboards provided employees with insights into how digital adoption positively influenced key performance metrics such as customer satisfaction and operational efficiency. By visualising the benefits, employees developed a clearer understanding of their contributions, fostering accountability and motivation. Additionally, DBS established weekly feedback forums to facilitate two-way communication, enabling employees to voice concerns, seek clarifications, and offer suggestions. These sessions provided leadership with actionable insights, fostering a culture of transparency and inclusivity. As a result, employee confidence in digital tools improved, leading to higher adoption rates and a smoother transition to digital banking services. The initiative also strengthened trust and alignment between employees and leadership, highlighting the critical role of data-driven insights and open communication in overcoming resistance and ensuring successful digital transformation (Wang & Su, 2023).

Summary

These case studies illustrate that overcoming resistance to change requires a combination of leadership transparency, employee engagement, and tailored communication strategies. Organisations such as GE and IBM demonstrate the effectiveness of structured upskilling and employee involvement, while Alibaba, Reliance Jio, and DBS Bank underscore the importance of collaboration and agility in managing transformation efforts. By adopting best practices such as stakeholder involvement, continuous learning, and data-driven decision-making, organisations can navigate resistance effectively and achieve successful transformation outcomes.

6. KEY TAKEAWAYS FOR PRACTITIONERS: THE CLEAR FRAMEWORK

The CLEAR framework, derived from the insights discussed in this chapter, provides a structured and practical approach to managing organisational change. It integrates key themes such as stakeholder engagement, communication strategies, and adaptive leadership, offering actionable steps to address common resistance challenges. The framework's components—communication, leveraging diagnostic tools, empowering middle managers, activating employee engagement, and regular review—are grounded in real-world case studies, theoretical models, and best practices. It serves as a comprehensive roadmap to guide practitioners through complex transformations, ensuring strategies are both aligned with organisational objectives and adaptable to evolving business environments. By applying the CLEAR framework, organisations can drive meaningful, sustainable change while fostering a culture of collaboration and continuous improvement.

The CLEAR framework focuses on five critical elements essential for overcoming resistance and fostering long-term success:

- Communication (C): Establishing trust and aligning stakeholders through transparent communication and storytelling, reducing uncertainty, and encouraging engagement.
- Leveraging diagnostic tools (L): Using tools such as resistance heatmaps and stakeholder engagement checklists to identify potential barriers early and implement targeted interventions.
- Empowering middle managers (E): Providing managers with the skills and authority needed to bridge leadership and frontline employees, ensuring strategic objectives are translated into actionable steps.
- Activating employee engagement (A): Encouraging participatory decision-making and implementing feedback mechanisms to foster ownership and commitment, thereby reducing resistance.
- Regular review and refinement (R): Continuously evaluating change initiatives through feedback loops and performance metrics to maintain effectiveness and responsiveness.

Implementing the CLEAR framework enhances transparency, collaboration, and adaptability, enabling organisations to navigate change with resilience and achieve long-term goals. This structured approach not only mitigates risks but also strengthens stakeholder buy-in and creates a foundation for sustained innovation and growth. As organisations encounter emerging challenges such as generative AI and evolving workplace structures, adopting a flexible and adaptive change management approach will be essential to maintaining a competitive edge.

REFERENCES

Anderson, J., & Sun, P. (2021). Impacts of transformational leadership on organizational change capability: The role of trust and learning. *European Journal of Management and Business Economics, 30*(2), 191–204. DOI: 10.1108/EJMBE-06-2021-0180

Brown, J. S., & Duguid, P. (2021). *The Social Life of Information* (2nd ed.). Harvard Business Review Press.

Brown, P., Tan, W., & Zhang, K. (2023). Digital Banking Transformation: Employee Adaptation and Resistance. *International Journal of Financial Innovation, 8*(2), 133–150.

Brown, T., & Jones, L. (2022). Navigating Change in Manufacturing: Overcoming Resistance to IoT. *Journal of Organizational Change Management, 35*(4), 245–260.

Burnes, B. (2004). Kurt Lewin and the planned approach to change: A re-appraisal. *Journal of Management Studies, 41*(6), 977–1002. DOI: 10.1111/j.1467-6486.2004.00463.x

Cameron, K. S., & Quinn, R. E. (2022). *Diagnosing and Changing Organizational Culture: Based on the Competing Values Framework* (4th ed.). Wiley.

Chan, K., Wong, L., & Tan, S. (2021). Pulse checks in digital transformation: A case study of DBS Bank. *Journal of Change Management, 18*(3), 210–224. DOI: 10.1080/14697017.2021.2023

Dart Project Management. (2024) *AI-Powered Stakeholder Analysis: Improving Project Outcomes*. https://www.itsdart.com/blog/ai-powered-stakeholder-analysis.

Dawson, P. (2021). Organisational change as a nonlinear dynamic. In *Global Encyclopedia of Public Administration, Public Policy, and Governance*. Springer., https://link.springer.com/referenceworkentry/10.1007/978-3-319-31816-5_73-1,

Deutsche Bank. (2021). *Innovation and technology strategy: Key developments and progress*. https://annualreport.deutsche-bank.com/files/documents/other-presentations-and-events/4_IDD_2020_CTO_final.pdf.

Floyd, S. W., & Wooldridge, B. (2021). *Building Strategy from the Middle: Reconceptualizing Middle Management's Role in Strategy Execution*. Routledge.

Fuller, A., Fan, Z., Day, C., & Barlow, C. (2020). Digital twin: Enabling technologies, challenges, and open research. *IEEE Access : Practical Innovations, Open Solutions, 8*, 108952–108971. DOI: 10.1109/ACCESS.2020.2998358

Gartner. (2023). Overcoming silos in the hybrid workplace: Best practices for seamless collaboration. *Gartner Research Insights*.https://www.gartner.com/en/insights/overcoming-silos.

Garvin, D. A., Edmondson, A. C., & Gino, F. (2022). Building a Learning Organisation. *Harvard Business Review*, *100*(2), 45–60.

GovTech Singapore. (2023). *Tech Kaki: Enhancing citizen engagement in digital services*. https://www.tech.gov.sg/.

Hayes, S. C., Luoma, J. B., Bond, F. W., Masuda, A., & Lillis, J. (2020). Acceptance and Commitment Therapy: Model, Processes and Outcomes. *Behaviour Research and Therapy*, *60*, 1–25. DOI: 10.1016/j.brat.2019.03.002 PMID: 16300724

Hofstede, G. (2021). *Culture's Consequences: Comparing Values, Behaviours, Institutions, and Organizations Across Nations*. Sage Publications.

Huy, Q. N. (2022). Emotional balancing of organizational continuity and radical change: The role of middle managers. *Administrative Science Quarterly*, *67*(1), 132–169. DOI: 10.1177/00018392211024671

IBM. (2021). *SkillsBuild programme: Empowering workforce transformation through learning and development*. https://www.ibm.com/skillsbuild.

Inform, I. T. (2024). *Stakeholder Analysis with AI*. https://www.informit.com/articles/article.aspx?p=3192418&seqNum=3.

Infosys (2023) *Building trust through co-creation: ESG sustainability report*. https://www.infosys.com/sustainability/documents/esg-2022-23/soc-story4.pdf.

Jabbour, C. J. C., Jabbour, A. B. L. S., Sarkis, J., & Godinho Filho, M. (2023). Unlocking the potential of Industry 4.0 technologies to address resistance to change in sustainable supply chains. *Journal of Cleaner Production*, *368*, 133017. DOI: 10.1016/j.jclepro.2022.133017

Jones, A., & Smith, R. (2022). A systematic review of organizational culture literature: Lessons from the past and avenues for future research. *Journal of Organizational Studies*, *45*(1), 56–80. DOI: 10.1080/JoS.2022.467853

Klein, S. (2022). Change Management: From Theory to Practice. *TechTrends*, *66*(3), 319–328. https://link.springer.com/article/10.1007/s11528-022-00775-0

Kotter, J. P. (2021). Leading Change: Why Transformation Efforts Fail. *Harvard Business Review*, *99*(2), 59–67.

Krugmann, M., & Hartmann, P. (2024). The role of large language models in sentiment analysis for organisational change. *Journal of Organisational Transformation*, *12*(1), 45–59. https://link.springer.com/article/10.1007/s40547-024-00143-4

Kumar, R., Gupta, S., & Rathore, S. (2021). Workforce transformation and resistance to automation: Insights from Indian manufacturing. *Asian Journal of Business and Management*, *9*(3), 21–35. DOI: 10.1080/ajbm.2021.09.003

Kumar, R., & Patel, S. (2022). Digital Inclusion and Organisational Change: The Case of Reliance Jio. *Asian Journal of Business and Technology*, *11*(3), 78–92.

Lewin, K. (1947). Frontiers in group dynamics: Concept, method and reality in social science. *Human Relations*, *1*(1), 5–41. DOI: 10.1177/001872674700100103

Li, Q., & Zhang, H. (2022). The Rendanheyi model and its implications for organisational change: A case study of Haier. *Asia Pacific Journal of Management*, *39*(4), 1235–1250. DOI: 10.1007/apj.2022

Loo, G. C., Rao, P., & Verma, S. (2022). Employee Engagement Strategies for Digital Transformation Success. *The Journal of Business Strategy*, *43*(1), 98–115.

Meadows, D. H. (2020). *Thinking in Systems: A Primer*. Chelsea Green Publishing.

Meyer, A. D., Gaba, V., & Colwell, K. A. (2022). Organising far from equilibrium: Nonlinear change in organisational fields. *Organization Science*, *16*(5), 456–473. https://www.insead.edu/faculty-research/publications/journal-articles/organizing-far-equilibrium-non-linear-change. DOI: 10.1287/orsc.1050.0135

Nkosi, Z. (2019). Stakeholder engagement in healthcare transformation: A South African perspective. *African Journal of Public Administration and Management*, *30*(2), 112–130. DOI: 10.1177/0312896219871463

Prosci. (n.d.). *When should you use a change management readiness assessment?*https://www.prosci.com/blog/when-should-you-use-a-change-management-readiness-assessment

Rao, P., & Verma, S. (2023). Motivating transformation: Gamification and incentives in industry 4.0. *International Journal of Organisational Psychology*, *27*(2), 112–126. DOI: 10.1016/ijop.2023

Rao, S., Gupta, R., & Mehta, P. (2024). Stakeholder mapping in digital transformation: A case study of Reliance Jio. *Asia-Pacific Journal of Business Strategy*, *18*(2), 112–125. DOI: 10.1111/apjbs.2024.18.2

Rugman, A. M., & Verbeke, A. (2022). *International business strategy: Rethinking pathways for global and regional integration*. Cambridge University Press.

Schlegelmilch, B. B. (2021). BANI vs. VUCA: How Leadership Works in the World of Tomorrow. *WU Executive Academy*. https://executiveacademy.at/en/news/detail/bani-vs-vuca-how-leadership-works-in-the-world-of-tomorrow/.

Schneider, S. C., & Schmid, S. (2021). Managing cultural resistance to automation in European SMEs. *Journal of International Business Studies, 52*(4), 789–812. DOI: 10.1057/s41267-021-00404-6

Sharma, R., Gupta, A., & Patel, S. (2024) 'Sentiment analysis and its impact on understanding employee resistance', *AI and Organisational Change Management*, 18(3), pp. 120–135. https://link.springer.com/article/10.1007/s41060-024-00594-x.

Silva, R. M., & Santos, L. M. (2020). Implementing agile frameworks in public sector reforms: The case of Brazil. *Public Management Review, 22*(3), 345–362. DOI: 10.1080/14719037.2020.1725689

Smith, J., Williams, D., & Clark, R. (2022). Embracing IoT in Manufacturing: Lessons from General Electric. *Industrial Technology Review, 9*(4), 300–315.

Smith, R., Jones, A., & Patel, S. (2023). The impact of remote work on organizational resistance: Strategies for engagement. *Journal of Organizational Change Management, 35*(3), 215–232. DOI: 10.1108/JOCM-2022-0134

Solution, O. C. M. (n.d.). *Stakeholder analysis in change management.*https://www.ocmsolution.com/stakeholder-analysis

Styhre, A. (2022). Non-linear change in organisations: Organisation change management informed by complexity theory. *Leadership and Organization Development Journal, 23*(6), 343–351. https://www.emerald.com/insight/content/doi/10.1108/01437730210441300/full/html. DOI: 10.1108/01437730210441300

Tan, C. W., Su, Y., & Zhang, M. (2023). Cross-Functional Collaboration and Resistance to Change in Financial Services. *Journal of Organisational Change, 28*(2), 205–220.

Uhl-Bien, M., & Arena, M. (2018). Leadership for Organizational Adaptability: A Theoretical Synthesis and Integrative Framework. *The Leadership Quarterly, 29*(1), 89–104. DOI: 10.1016/j.leaqua.2017.12.009

WalkMe. (n.d.). *Change management heatmap: Identifying resistance to change.* Retrieved from https://www.walkme.com/blog/change-management-heatmap

Wang, J., & Su, L. (2023). Cognitive Biases in Organizational Change: An Application of Behavioral Economics. *Organizational Psychology Review, 12*(4), 321–342. DOI: 10.1177/2041386623110248

West Africa Civil Society Institute. (2021). *Digital transformation and African non-profits.* https://wacsi.org/wp-content/uploads/2022/01/DIGITAL-TRANSFORMATION-AND-AFRICAN-NON-PROFITS-2021.pdf

Wilson, H. J., & Daugherty, P. (2021). The future of AI in organizations: Balancing automation with the human touch. *MIT Sloan Management Review*, *62*(4), 28–35.

Wooldridge, B., Schmid, T., & Floyd, S. W. (2021). The middle management perspective on strategy process: Contributions, synthesis, and future research. *Journal of Management*, *47*(7), 1805–1834. DOI: 10.1177/01492063211002587

Chapter 7
Data as a Strategic Asset:
A Study

Shrinivas R. Patil
https://orcid.org/0000-0001-6163-665X
Jain University, India

Pujari Sudharsana Reddy
Jain University, India

K. P. Premalatha
Jain University, India

ABSTRACT

Data has emerged as a critical strategic asset in today's digital economy, and in driving innovation, it has become the source of competitive advantage. Companies are under flood studies with data coming from diverse sources, and effective use of such data is an indispensable activity for market differentiation. Value capture refers to converting created value into profit by strategies such as data monetization, dynamic pricing, and customer segmentation. Examples include Google's targeted ads and Netflix's personalized recommendations. Companies like Amazon and Coca-Cola show how it is done. AI, machine learning, and cloud computing give captured data additional value. On the other hand, ethical considerations and conformity with regulations like GDPR are extremely critical. In sum, organizations have to invest in a robust data infrastructure and a culture of responsible use of data to succeed in this data-driven world.

DOI: 10.4018/979-8-3373-1320-7.ch007

1. INTRODUCTION

In today's digital economy, fast-paced and competitive, data has ceased to be just an operational byproduct but has emerged as one of the most valuable resources for any organization. This chapter investigates the transformational potential of data into a strategic asset that drives innovation, competitive advantage, and decision-making in today's dynamic business environment. With digital technologies now in place, businesses face an onslaught of data from all directions: social media, e-commerce platforms, IoT devices, and enterprise systems. Correct tapping into this massive data will help organizations discover new avenues of driving growth, optimizing operations, and improving customer experiences (McKinsey & Company, 2020). This would mean that the main theme of this chapter entails information and its role in the new economy. Traditional resources have largely been substituted by intangible resources in the form of data, which now redefine the basis of competitive advantage. Companies such as Facebook (now Meta) and Alibaba have shown a new way by using data in transforming nature, creating new streams of revenue, and remapping markets in the world (Alibaba, 2021; Meta, 2022). Their example has highlighted the rising importance of Business Intelligence (BI) within the decision-making process.

Building a culture of business intelligence is important for organizations wanting to derive benefit from their data assets. This means organizations have to adopt technologies, tools, and methodologies that allow real-time data analysis and actionable insights (Gartner, 2021). By creating a data-driven mindset, businesses can empower their workforce to make informed decisions at every level. Business intelligence identifies not only the trends and patterns but also the future behaviors of the organization, thus enabling the organization to lead in the competitive market.

Another significant emphasis of this chapter is capturing value from data. Monetizing data is not about direct financial value but also in the form of increased efficiency, lowered costs, and high levels of customer satisfaction (Deloitte, 2020). Companies must tune their data strategies to the overall business goals to sustain growth in a competitive world. Hence, deriving actionable insights from the data has now become a source of competitive advantage in terms of differentiation in the marketplace. However, there are challenges in using data. In this age of governance and privacy of data, personal and corporate information is more under siege than ever before (Pew Research Center, 2021). As the dependency on data in informing business decisions increases, it needs to be balanced with ethics and respect for regulations relating to data protection. Finding the right balance in exploiting data for insight while preserving privacy is material in earning and sustaining trust and credibility (General Data Protection Regulation [GDPR], 2016; California Consumer Privacy Act [CCPA], 2018).

The discussion also explores how analytics can give meaningful data insights while ensuring confidentiality where needed. Advanced analytic tools and techniques offer organizations a chance to extract value from data without exposing themselves to security vulnerability (Cisco, 2020). The following practical advice would also be useful in supporting strong data governance structures and meeting high demands for analytical functions. In effect, data has turned into one of the important cornerstones of strategic management in the twenty-first century. The businesses able to manage, analyze, and protect their data effectively will have a better foundation for success in an increasingly digitized world (IBM, 2021). This chapter uses case studies and real-world examples to show the importance of data-driven strategies and gives best practices for any organization looking to derive maximum value from their data assets. Companies like Alibaba and Facebook give resonance to the power of using data to create a remarkable trajectory of growth in their respective market influence (Alibaba, 2021; Meta, 2022).

Alibaba Group: Founded in 1999, Alibaba has grown into a global leader in e-commerce, cloud computing, and digital services. Its success is deeply ingrained in its ability to bring a data-driven approach in every business strategy and customer experience enhancement (Alibaba, 2021).

E-commerce Dominance: The domestic commerce retail business of Alibaba contributed about 40.56% to its total revenue in the second quarter of 2024.

Cloud Computing Growth: In the same period, Alibaba's cloud intelligence division saw a 7% increase in revenue to 29.61 billion yuan.

AI Investments: The company aggressively increased capital expenditures on artificial intelligence infrastructure, despite the challenges in the global environment. It more than doubled the investment to 50 billion yuan in the first half of the year (Alibaba, 2021).

Facebook (Meta Platforms) -a leading social media by using data to change the landscape of online interactions and advertising, creating a gigantic user base and generating revenues based on the vast amount of data (Meta, 2022)

- User Engagement: As of the third quarter of 2023, Facebook reported roughly 2.91 billion monthly active users from around the world, reflecting its immense reach and the amount of data created through interactions of users.
- Advertising Revenue: The advertising model of Facebook, driven by sophisticated data analytics, has been a cornerstone of its financial success in enabling highly targeted marketing campaigns, which appeal to advertisers looking for effective audience engagement.

In a nutshell, both Alibaba and Facebook present great examples of the strategic use of data for driving business success. While Alibaba uses data to optimize its e-commerce operations and expand its cloud services, Facebook uses user data to fine-tune its advertising algorithms and improve user engagement. Their heavy investments in technology and data analytics hint at the critical role that data plays as a strategic asset in the modern economy.

2. THE ROLE OF INFORMATION IN THE NEW ECONOMY

Information has emerged as one of the most potent sources of growth and innovation. In a world of rapid digital technological development, businesses now bank on information more than ever before to drive strategic decisions, create a competitive edge, and raise efficiency (McKinsey & Company, 2020). Companies such as Alibaba, Facebook (Meta), and Google have shown that with the smart use of information, industries can be transformed and new value created (Alibaba, 2021; Meta, 2022).

2.1 Information as a Key Asset

In the earlier economies, wealth and growth in the economy were founded on physical assets: land, factories, and machinery. On the other hand, the digital revolution placed information at the forefront. Now, information and data stand at the core of resources that help in the development process (McKinsey & Company, 2020). Information helps firms understand better their consumers, anticipate trends, and ameliorate products or services.

A good example is Alibaba. The company operates as one of the largest e-commerce platforms in the world, bringing information to tailor-make shopping experiences. Its recommendation system, developed by its AI algorithms based on customer behavior and purchasing patterns, suggests products that would chime with the preferences of an individual (Alibaba, 2021). In the fiscal year 2024, e-commerce retail business contributed about 40.56% of revenues for Alibaba Inc., mainly driven by data-driven insights (Alibaba, 2021).

2.2 Driving Decision-Making

Information enables swifter and better decision-making in the new economy. When properly equipped with the right tools and technologies, companies can convert reams of raw data into insights that can be acted upon (Gartner, 2021). In the case of Facebook, user data is used to fine-tune its algorithms for better delivery of targeted advertisements. As of the third quarter of 2023, Facebook had 2.91 billion monthly active users. This huge user base generates data, which in turn helps advertisers to reach the intended audience for much better engagement and returns (Meta, 2022). Using the collected information from its search engine and other services, Google improves the user experience. With data analytics, it tries to predict what the user wants or is searching for and accordingly produces relevant search results (McKinsey & Company, 2020). Its present position at the top as a search engine globally and being dominant in digital advertising are thus consolidated.

2.3 Building a Competitive Advantage

Information is a competitive advantage in that it enables businesses to innovate and get ahead of their competitors. Businesses that leverage information either to optimize operations or to create new services seem to outpace their competition. For example, Amazon uses information in its supply chain to ensure deliveries are fast and dependable (Amazon, 2021). Its predictive shipping model, based on forecasts of customer demand, has established new efficiency standards for the retail industry. On the other hand, Alibaba has also invested strongly in its cloud computing wing, with 2024 revenue jumping 7% to 29.61 billion yuan (Alibaba, 2021). Its cloud services analyze massive datasets for businesses, helping them make informed decisions and improve their operations. Hence, by being a pacesetter of cloud computing, Alibaba gets an upper hand in the technology sector (Alibaba, 2021).

2.4 Enhancement of Customer Experience

One of the largest advantages of information in the new economy is its function in making the customer's experience better. Nowadays, organizations use data to understand customer needs and preferences better. This allows them to create products, services, and experiences that are much more relevant to the customers (McKinsey & Company, 2020). For instance, Netflix uses data on viewership to recommend films and series as per the taste of a particular individual. Due to this factor, Netflix became the global streaming service leader (McKinsey & Company, 2020). Similarly, Starbucks uses its loyalty program and mobile app to collect information on customer preferences. This information aids the company in designing promotions and menu options that resonate with its audience (Deloitte, 2020).

2.5 Information Monetization

Information is not just a decision-making tool but an asset that can be gainfully monetized. Firms like Facebook, Google, and Alibaba use data as the asset by offering an advertising platform service to different concerns. Thus, the ad platforms enable business concerns to more effectively reach their customers, ensuring effective marketing campaigns (Meta, 2022; Google, 2020). Facebook's advertisement model is one such case in point: it analyses user data and directs advertisements, therefore becoming one of the profitable companies in the world (Meta, 2022). Advertisers go for Facebook since it enables them to reach specific demographics, locations, and interests. In this way, this has strategic importance to Facebook in the digital economy (Meta, 2022).

2.6 Challenges in Using Information

The immense opportunities which information brings with it also bring challenges. Top on the list of the new economy, data privacy and security are fast becoming major concerns that companies have to ensure they use data in a manner that is ethical and compliant with regulations (Pew Research Center, 2021). Scandals like the Cambridge Analytica misused users' data through Facebook clearly show how responsible handling of data should be. For instance, the Alibaba group has seen an increase in investments related to data protection (Alibaba, 2021). For example, in 2024, it doubled its spending on artificial intelligence infrastructure to improve data security. Companies that respect ethical data practices can earn the trust of their consumers—without which no business can be successful in the long run (General Data Protection Regulation [GDPR], 2016; California Consumer Privacy Act [CCPA], 2018).

2.7 The Future of Information in the New Economy

Information is going to play an even greater role in the new economy. Technologies in the realm of artificial intelligence, machine learning, and blockchain are going to make data analysis and usability much easier. They will allow companies to delve deeper into the insights and come up with creative solutions for some of the complex issues. In that direction, Alibaba's focus on AI-driven infrastructure speaks volumes about company preparedness. Similarly, Google and Facebook continue to invest in AI and machine learning to drive improvements in their platforms and offer new services. Companies adopting the aforementioned new technology developments will be in a good position to succeed in the new economy.

Information is the backbone of the new economy. It provides the backbone for making decisions, building competitive advantage, and improving customer experiences. Firms such as Alibaba, Facebook, Google, and Amazon show how well-harnessed data can lead to great growth and innovation. But with increased dependence on information, data protection and security become key challenges to overcome. Leading practices for ethical data practice contribute much toward ensuring the needed level of trust, securing ongoing operations in business at all times. With these rapid developments happening, information becomes all the more imperative as this new economy starts emerging in a way. Business enterprises adopting the power of data strategies and the impetus provided by emerging technologies may go quite confidently and adeptly in addressing different dynamics. The skill of converting information into actionable insights is no longer optional; it's a requirement for success in today's digital world.

3. UNITED NATIONS SUSTAINABLE DEVELOPMENT GOALS (UN SDGS)

The United Nations Sustainable Development Goals form a framework for tackling some of the world's most serious social, economic, and environmental problems by 2030. The goals—17 in all—are closely linked to ensuring inclusive growth, eradicating poverty, increasing sustainability, and ensuring an equal distribution of resources. Technology and the data-driven industry have played a leadership role in achieving these goals, as innovation and digital transformation produce solutions that can be scaled up. The list includes such large players as Alibaba, Amazon, and Meta; they aligned their strategies with the key SDGs and proved that private enterprise could be a big driver of sustainability worldwide. For instance, SDG 7 (Affordable and Clean Energy) and SDG 13 (Climate Action) are supported through heavy investments in renewable energy and initiatives in carbon reduction. With sustainable practices, these companies not only reduce their ecological footprint but also increase the resilience of their business for the long run, therefore meeting SDG 8 on decent work and economic growth.

Table 1 shows Data analysis of Alibaba, Amazon, and Meta between 2020 and 2024 gives a perspective on the interaction between strategic investments, financial performance, and the United Nations Sustainable Development Goals. These companies give very good examples of how sustainability integrated into core business operations can provide improved competitiveness, innovation, and resilience while contributing to global goals.

Table 1. Investments, Returns and UN SDG alignment

Company	Investments (2020–2024)	Returns (2020–2024)	Link to UN SDGs
Alibaba	- Investments in digital businesses and philanthropy impacting 16 out of 17 UN SDGs. - Long-term investments totaling $173.25 billion	ROI of 9.47% as of September 30, 2024. - Net income of $16.39 billion	SDG 1: No Poverty - SDG 8: Decent Work and Economic Growth - SDG 13: Climate Action
Amazon	More than $20 billion invested in wind and solar energy. - Continued AI technology investment, including funding for MultiOn startup.	Growing market share of renewable energy procurement. - Growing revenue from AI initiatives.	SDG 7: Affordable and Clean Energy - SDG 9: Industry, Innovation, and Infrastructure
Meta (Facebook)	Massive procurement of renewable energy (more than 45 GW globally, combined with other tech companies).	- Positive financial returns from sustainability investments and growing revenue from digital advertising.	SDG 7: Affordable and Clean Energy - SDG 13: Climate Action

Source: Compiled from publicly available data, including corporate sustainability reports, financial statements, industry benchmarks (e.g., REBA and GRI), and independent market research (2020–2024).

Alibaba's investment strategy shows the potential to be aligned with business growth and sustainability. Long-term investments of $173.25 billion, undertaken by the company, have yielded an ROI of 9.47%, thus proving that the targeting of key SDGs, including SDG 1 (No Poverty) and SDG 8 (Decent Work and Economic Growth), has the capability to yield substantial financial and social returns. Its focus on digital transformation and philanthropic initiatives underlines its commitment toward fostering inclusive economic progress.

Amazon has pioneered the buying of renewable energy in more than 20 gigawatts of clean energy projects. This aligns with SDG 7 (Affordable and Clean Energy) and SDG 9 (Industry, Innovation and Infrastructure). Its concurrent investments in artificial intelligence only reinforce the dual focus on sustainability and technological advancement, yielding both economic and environmental benefits.

4. BUILDING A CULTURE OF BUSINESS INTELLIGENCE

In this era of digital revolution, building a culture of Business Intelligence is something which any organization must consider if it has to stay ahead in the race. Business Intelligence is a technology, process, and best practice used to gather, analyze and present data in a way that enables business users to make better decisions. A strong BI culture allows organizations to exploit data for strategic growth, operational efficiency, and innovation. Some of the leading companies, such as Amazon, Alibaba, and Meta, have opted to adapt to BI tools in an effort to improve their decision-making while aligning their operations with sustainable global goals to contribute equally to business success and impact society.

For example, Amazon has deeply integrated its business with BI. It uses data analytics in improving its supply chains, predicting customer behavior, and enhancing its services. AWS, Amazon's cloud computing arm, grew 29% in revenues year over year in 2020—a clear testament to the fact that the firm uses data-driven insights to propel growth. On its path toward sustainability, Amazon also uses BI to track its consumption of renewable energy, ensuring it stays on track to meet its goal of powering all operations with 100% renewable energy by 2025.

Similarly, Meta has a strong BI culture in place. The company performs deep analyses of vast user data to drive improvement in advertising products for better revenue generation and engagement with users. It invests in AI and machine learning at the backbone of its infrastructure in creating tailored user experiences while committing to dealing with climate concerns, as Meta has dedicated itself to net-zero emissions by the year 2030. The BI systems of the company help track the progress toward this goal.

Lastly, the creation of a Business Intelligence culture is no longer an option; it is a must for any organization intent on thriving in a data-driven world. In this sense, companies like Amazon, Alibaba, and Meta indicate how developing Business Intelligence at the core of operations helps drive such aspects as innovation, sustainability, and competitive advantage. Alignment of BI strategies with global goals, such as the UN SDGs, allows organizations not only to achieve financial success but also to contribute to a sustainable future.

4.1 Introduction to Business Intelligence (BI)

Business Intelligence is the application of technologies, tools, and practices in analyzing and presenting data for businesses to make business decisions. The setup of the culture of BI is very important today for an organization to remain competitive in this digital age. BI helps organizations understand their operations, market trends, and customer behaviors in efforts toward driving growth and efficiency. Companies such as Amazon, Alibaba, and Meta have built strategies around business intelligence and aligned them with broader sustainability goals.

4.2 Amazon: BI in Growing and Sustaining

Amazon has been able to smoothly integrate BI throughout its business processes. For instance, its cloud computing division, AWS, which contributes greatly to revenue, uses BI to optimize its services. In the year 2020, AWS recorded 29%

year-over-year revenue increase, which reflects how BI tools enable Amazon to forecast demand and manage supply chains in order to improve customer satisfaction.

Amazon, however, goes beyond growth in using BI for sustainability. It is using data analytics to track its consumption of renewable energy and make progress toward its target of using 100% renewable energy by the year 2025. In a manner driven by BI, Amazon could both improve operational efficiency and reduce environmental footprint.

4.3 Alibaba's Data-Driven Transformation

Alibaba has built a strong BI culture in its e-commerce, logistics, and cloud computing businesses. In 2021, enabled by its use of BI tools and AI technologies in the optimization of business processes and improvement of customer experiences, the company's cloud business expanded by 50%. The ongoing investment in data analytics keeps Alibaba agile, responsive to market demands, and cost-efficient.

Moreover, Alibaba uses BI to underpin its sustainability efforts. Through data analysis, Alibaba could trace and measure the environmental impact of its activities to ensure its operations are in line with the UN SDGs, particularly in areas related to affordable energy (SDG 7) and climate action (SDG 13).

4.4 Meta's Integration of BI for User Experience and Sustainability

Meta, commonly known as Facebook, has built a strong BI infrastructure for analyzing its massive user data. The company makes use of BI tools to drive improvements in its advertising products by better targeting users and increasing engagement. Hence, this leads to a gradually increasing revenue for Meta, as the data-driven model of advertising drives revenue.

On the sustainability front, Meta has committed to becoming net-zero by the year 2030. Its BI systems help track and monitor progress in reducing its carbon footprint. This application of BI in sustainability initiatives is in line with SDG 13 on Climate Action and is another example of how BI can be applied—not only for increasing profits but also for creating environmental change.

4.5 Creating a Data-Driven Mindset Across the Organization

In order for firms to tap into the full potential of BI, there needs to be a data-driven culture inculcated at all levels within an organization. This calls for investment in the right BI tools and systems besides training of employees so that they understand and interpret data. A strong BI culture empowers teams to make faster, more informed decisions, which in turn can lead to improved operational efficiency and customer satisfaction.

For instance, through data analytics, Amazon has been able to forecast its inventory needs and optimize delivery routes for lower costs and improved sustainability. Making data available across all the departments means that Amazon ensures that each team within an organization is working toward attaining wider organizational objectives in profitability and sustainability.

4.6 Interdepartmental Collaboration for Effective BI

Equally, effective BI implementation calls for cross-departmental collaboration, where data becomes transparent and is shared with all teams for proper collaboration in relation to continuous improvement. How Meta does carbon footprint tracking for its operations shows a big aspect that is involved by all departments: sustainability.

This would mean that sustainability initiatives are in every operation of the organization, driving business success and positive environmental impact. For instance, through its focus on integrating BI across its cloud, logistics, and e-commerce operations, Alibaba has been able to enhance business efficiency while at the same time monitoring its environmental impact.

Table 2. Business Intelligence, Investments and Sustainability Efforts of Amazon, Alibaba, and Meta, 2020-2024

Company	Investment in Sustainability (2020-2024)	Revenue Growth (2020-2024)	Sustainability Goals/SDG Alignment	Key BI Use Cases	BI Impact on Performance
Amazon	$11 billion—Amazon investment in sustainability by 2025	29% YoY growth in AWS (2020)	SDG 7: Affordable and Clean Energy; SDG 13: Climate Action	Supply chain, inventory prediction, and carbon tracking	optimization Lower carbon footprint, better operational efficiency, and increased revenue from AWS.
Alibaba	Alibaba $15 billion investment in green energy by 2030	50% growth in cloud business in 2021	SDG 7: Affordable Energy SDG 13: Climate Action SDG 9: Industry, Innovation, and Infrastructure	AI-driven logistics and e-commerce insights; energy consumption tracking	Lowered operational costs and enhanced customer experiences while meeting SDG targets
Meta (Facebook)	$1 billion in RE and carbon offset projects	22% YoY growth in ad revenues (2021)	SDG 13 (Climate Action), SDG 12 (Responsible Consumption and Production)	Monitor user engagement, providing customized advertising, and carbon emissions tracking.	Increased revenue, better-targeted advertisements, and reduced carbon footprint

Data source: Compiled from publicly available sustainability reports, financial statements, and industry reports from 2020–2024.

Table 4.1 presents how Amazon, Alibaba, and Meta—using BI—have strategically used their platforms to increase business performance and sustainability efforts in line with some of the important SDGs set by the United Nations. Amazon, for instance, invested $11 billion in sustainability initiatives, including renewable energy projects, while its AWS division grew 29% year-over-year in revenues in 2020, driven by BI tools integral to optimizing supply chain management, inventory prediction, and carbon footprint reduction. This exemplifies how BI supports operational efficiency and contributes meaningfully to Amazon's commitment to SDG 13 (Climate Action). Similarly, Alibaba's $15 billion green energy investment by the year 2030 exhibits the commitment of this organization to SDG 7 (Affordable and Clean Energy) and SDG 13 (Climate Action), with a cloud business growing 50% in 2021, supported by AI-driven logistics and energy consumption tracking through BI. By using BI, Alibaba could ensure its financial growth, hence much closer to fulfilling its commitment to sustainable goals by reducing costs to work toward a more sustainable future. Meta, on the other hand, has invested in renewable energy and carbon offset initiatives of $1 billion; coupled with BI, it saw a 22% ad revenue increase this past quarter because of the crucial role BI played in personalizing user experiences and optimizing ad targeting. The company's commitment to SDG 13 (Climate Action) and SDG 12 (Responsible Consumption and Production) underscores its commitment to the reduction of the company's carbon footprint while accelerating revenue growth. In other words, the cases of these three organizations show that BI is not just an enabler but one of the very important drivers in business performance improvement and sustainability pursuits. Their investment and strategic use of BI allow balancing financial success with responsible environmental practices, thus setting role models for other firms to follow.

Figure 1. Business Intelligence Impact on Performance

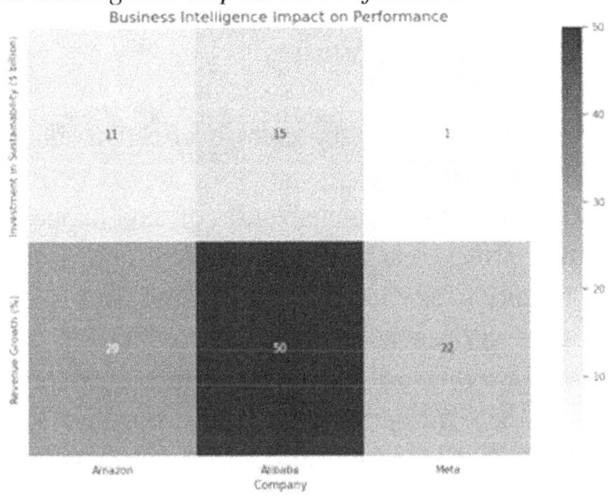

5. CAPTURING VALUE AS THE SOURCE OF COMPETITIVE ADVANTAGE

5.1 Introduction

Value capture from data—probably one of the most relevant themes in the current state of sustaining competitive advantage in business today. As organizations increasingly depend on data-driven insights to drive decision-making, the ability to harness and monetize this data becomes a differentiator in the marketplace. This will go into more detail on value capture from data assets of a company, strategies, examples, and the interplay between value creation and capture.

5.2 The Importance of Value Capture

Value capture is the process of converting created value into profit for a firm. This is generally done through the use of pricing strategies, market positioning, and unique capabilities differentiating a firm from its competitors. The relationship between value creation and value capture is a complex one: superior value creation can augment a firm's ability to capture value by differentiating its offerings and commanding higher prices or greater market share. On the other hand, an excessive focus on capturing value at the expense of creating it can undermine long-term competitiveness and growth prospects.

5.3 Strategies for Capturing Value

Organizations can adopt a variety of strategies for effective value capture from data:

a. Data Monetization: The direct monetization of data happens through its sale or as a source of improved products and services. In this regard, for instance, Google is able to derive tremendous value out of data in the form of highly targeted advertising services based on user behaviors.
b. Pricing Strategies: Dynamic pricing models, which are based on real-time data analysis, will help the firm optimize revenues. For instance, airlines change ticket prices frequently depending on demand forecasts, competitor pricing, and customer booking patterns.
c. Customer Segmentation: The use of data analytics in segmenting customers helps a business target its marketing efforts and product offerings. A good example is Netflix, which uses viewer data to recommend personalized content to its users, increasing their engagement and retention rates.
d. Customer Experience Enhancement: Organizations use data to enhance customer interactions and service delivery. Amazon uses advanced algorithms to analyze buying behavior, which helps them provide personalized recommendations to enhance the shopping experience.
e. Building Ecosystems: The firms build an ecosystem whereby multiple stakeholders could work in concert to drive greater value capture, for example, Apple has built the App Store whereby multiple developers create apps integrated into the Apple devices; these not only contribute to added revenues but further enrich customer experiences.

5.4 Examples of Successful Value Captures

Few businesses have succeeded in deriving benefit from the capture of their data value:

a. **Amazon:** With massive consumer data at its disposal, Amazon created a much more personalized shopping experience that boosts both sales and customer loyalty. Additionally, through predictive analytics, it would be able to foresee customer demand and manage inventory accordingly.
   a. **Netflix:** The streaming giant not only uses viewer data to give content recommendations but also to decide on content creation. Viewed in this light, viewing patterns help Netflix make informed investment decisions in original programming that resonates with audience preferences, maximizing viewership and growing subscribers.
      c. **Coca-Cola:** Coca-Cola captures value through advanced analytics in its supply chain management. They optimize production schedules and distribution routes from analysis of sales across various regions, ensuring product availability at lower costs.

5.5 Balancing Value Creation and Capture

While capturing value is important for profitability, organizations need to balance it with value creation. The sole focus of companies on short-term gains may make them risk alienating customers or stifling innovation. Example:

High Value Creation, Low Value Capture: Most startups deliver high value to their customers but fail to monetize it due to poor pricing models or a lack of market presence. For instance, early-stage tech companies usually offer free services to grow their user bases but struggle to convert them into paid models without losing users.

Creating Low Value but Capturing High Value: Leading firms in attractive markets might be earning returns with little contribution towards the provision of substantial customer benefit or innovation. This sets an organization up for disruption when newcomers enter with disruptive solutions that finally deliver meaningful solutions to customer problems.

5.6 The Role of Technology in Value Capture

Technological progress is an important factor increasing an organization's ability to capture value from its data assets.

a. **Artificial Intelligence (AI) and Machine Learning (ML):** These technologies enable organizations to analyze large datasets quickly and derive actionable insights from them. For example, financial institutions use AI algorithms for fraud detection by analyzing transaction patterns in real time.
b. **Cloud Computing:** Cloud platforms provide scalable infrastructure for storing and processing vast amounts of data. Companies like Microsoft are using cloud services to offer analytics tools that empower businesses to make decisions driven by data in an efficient manner.
c. **Blockchain Technology:** This will help the organizations in sectors like finance and supply chain management make transactions more transparent and secure, hence helping them capture more value from gaining the trust of the customers.

5.6 Ethical Considerations in Data Utilization

With increasing dependence on data for competitive advantage, ethical considerations of data privacy and governance become very important. Balancing between the leverage of data to gain insights and safeguarding customer privacy is the key to maintaining the trust.

Data Privacy Regulations: Organizations operating in multiple jurisdictions must comply with regulations like GDPR. Non-compliance can lead to heavy fines and reputational damage.

Transparency in Data Usage: Companies need to clearly explain to the customers how their data will be used. Transparency can help build trust, which in turn will make customers more loyal and more likely to share information.

5.7 The Future of Data-Driven Business

The future of business will be driven by data exponentially. Since organizational competitiveness in that arena strongly depends on harvesting power from the vast amounts of data, emphasis should continue on investing in the required infrastructure of technology: organizations seriously need to consider investments in strong data infrastructures, rendering them advanced analytics capabilities.

Create a data-driven culture: It will be important to foster a culture where employees are encouraged to use data in their decision-making processes to maximize the potential of available information.

Ensure strong data governance and security: Strong governance frameworks ensure that there is compliance with regulations while protecting sensitive information from breaches.

Upskilling workforce continuously in data analysis and related skills: Providing training opportunities will empower employees to use analytical tools effectively.

In using data power to its fullest effect, these principles can give organizations new opportunities for growth, innovation, and success in an increasingly data-driven world. To sum it all up, capturing value from data is no longer an option but a necessity for any organization to thrive in the current competitive landscape. By monetizing data effectively, creating a data-driven culture, and overcoming the challenges in using data, organizations can realize their full potential and achieve a sustainable competitive advantage. Capturing the value of data isn't about short-term returns; it's about building sustainable competitive advantages through informed decision-making, improved customer experiences, and innovative business models. Organisations have to strike a careful balance in creating value for customers while effectively monetizing their offerings. Such strategic decisions, in an approach that effectively balances technological benefits with ethical considerations, would, therefore, really help an organization to succeed long-term in the growing competitive space. Effective value capture is, therefore, the differentiator in the era of digital transformation and data proliferation. Companies living by these principles won't merely be successful; they will become new standards of excellence in their markets.

Table 3. Key Aspects of Data as a Source of Competitive Advantage

Aspect	Description	Examples
Value of Data	- Strategic asset driving innovation, competitive advantage, and decision-making. - Redefined competitive advantage from physical to intangible resources.	- Facebook, Alibaba, Google: Revolutionized industries through data-driven strategies.
Capturing Value	- Direct Revenue: Selling data directly to other companies. - Operational Efficiency: Predictive maintenance, supply chain optimization, fraud detection. - Improved Customer Experience: Personalized marketing, enhanced customer service, product innovation.	- Predictive Maintenance: Manufacturing companies using sensor data to predict equipment failures. - Personalized Marketing: Companies tailoring marketing campaigns based on individual customer preferences.
Building a Data-Driven Culture	- Data literacy training for employees. - Data democratization: Making data accessible to all relevant stakeholders. - Data-driven decision making integrated into all key processes. - Rewarding data-driven success.	- Training employees on data interpretation, visualization, and basic statistical concepts. - Providing self-service analytics tools and dashboards.
Data Governance	- Data quality management: Ensuring data accuracy, completeness, and consistency. - Data security and privacy: Protecting data from unauthorized access and breaches. - Data compliance: Adhering to data privacy regulations (e.g., GDPR, CCPA).	- Implementing data cleansing, validation, and ongoing monitoring. - Encrypting data, implementing access controls, and conducting regular audits.

continued on following page

Table 3. Continued

Aspect	Description	Examples
Advanced Analytics	- Artificial Intelligence (AI) and Machine Learning (ML): Predictive modeling, fraud detection, customer segmentation. - Natural Language Processing (NLP): Analyzing customer feedback, social media sentiment. - Predictive Analytics: Forecasting future trends and events.	- AI-powered recommendation systems like those used by Netflix. - NLP to analyze customer reviews and identify areas for product improvement.
Challenges	- Data silos: Data fragmented across different departments. - Data integration challenges: Integrating data from diverse sources. - Lack of skilled talent: Shortage of data scientists and analysts. - Ethical considerations: Protecting user privacy and avoiding bias in data-driven decisions.	- Implementing data integration strategies to break down silos. - Investing in training and development programs to address the skills gap.
Future of Data-Driven Business	- Continued investment in data infrastructure and technology. - Further development of data-driven cultures. - Prioritizing data governance and security. - Continuous upskilling of the workforce in data analysis and related skills.	- Embracing emerging technologies like AI, ML, and the Internet of Things (IoT).

Source: Synthesized from provided information and general knowledge on data-driven business strategies.

Table 5.1 does an excellent job of portraying the major aspects related to data as a source of competitive advantage. It brings together how organizations can leverage data to drive operational efficiency, improve customer experience, and foster innovation in an excellent manner. That is, building a data-driven culture is stressed in the table, and the importance of data governance is emphasized within the current research trends. However, there are many areas that need further investigation. For example, the table could be better with more detailed information on the ethical issues pertaining to data usage, such as bias, privacy, and responsible use of AI/ML algorithms. Equally, research is needed that would open the way toward exploring the new role that data plays in view of emerging technologies such as IoT, edge computing, and blockchain. How these technologies work with data is something organizations in the future need to understand and master to gain a competitive edge.

6. PROTECTION OF DATA PRIVACIES AND CONTROL DRIVING THE DISCOVERY OF INSIGHTS

6.1 Introduction

The modern business environment is highly driven by data and offers unparalleled opportunities for insight and innovation. At the same time, the exponential growth of data collection and processing has given way to a strong concern for personal privacy. Striking a balance between harnessing the power of data analytics and safeguarding individual privacy is what organizations need to do in order to thrive in the digital age. Data breaches have been becoming alarmingly frequent, with major incidents affecting millions of people around the world. For example, the 2013 Equifax breach exposed sensitive personal information of 147 million Americans, pointing out the very serious consequences of inadequate data security. These breaches could be corrosive to public trust, damaging to the reputation of brands, and possibly resulting in immense financial penalties for those organizations. Misuse of personal data in targeted advertisement, discriminatory practices, and even social manipulation through personal data pose extremely serious ethical and societal challenges.

6.2 The Importance of Data Privacy

Data privacy deals with individuals' rights concerning how their personal information is collected, used, and shared. In a world undergoing a wild data breach spree and misuse, stringent data protection measures become all the more crucial. According to IBM's report, the average cost of a data breach had already reached $4.24 million in 2021, showing the financial impact of poor data protection. A further note from the Cisco survey says that 84% of consumers take data privacy seriously—a situation business can't leave unheeded if they wish to stay competitive and gain loyal customers. Legislation such as Europe's GDPR, as well as CCPA on the west of the United States, calls upon organizations to establish rigorous controls pertaining to data. GDPR imposes fines of up to €20 million or 4% of annual global revenue for non-compliance, which is a very critical reason why businesses must be compliant with these regulations. CCPA also enforces the rights of the consumers in relation to their personal data; it allows Californians to request details on how their data is collected and used.

6.3 Control Over Data

One of the most critical aspects of modern-day data governance is giving control to individuals over their data. It is very critical that organizations make clear choices for users on how their data is used through mechanisms of obtaining consent for data collection and processing. A Pew Research Center study showed that 79% of Americans are worried about how their data is being used by companies—basically, this explains the need for transparency in data practices. For instance, tech giants like Apple have already made huge strides toward users' privacy—among the steps they have initiated is the introduction of functions giving users easy-to-use capabilities to control tracking and limit app permissions. Apple has its App Tracking Transparency feature—when enabled—demanding that apps seek permission from users before they track activity across other apps and websites reflects a change in control over personal information to users. Now,

6.4 Driving Insights Through Analytics

While data protection is a top priority, equally important is unleashing the power of analytics to drive insights for business strategies. Advanced analytics techniques, such as artificial intelligence and machine learning, can be applied to the analysis of large datasets to arrive at meaningful insights without breaching individual privacy. The use of techniques like anonymization and aggregation may mitigate risks pertaining to personal data while retaining the possibility of valuable analysis. For instance, health care organizations can analyze patient data trends to improve treatment outcomes while safeguarding the identities of the individuals through de-identification methods. According to a report from McKinsey & Company, health care organizations that effectively use analytics can improve patient outcomes by as much as 30%. Similarly, retail companies can use consumer behavior analytics to optimize inventory management and marketing strategies while respecting customer privacy preferences.

6.5 Ethical Considerations

Ethical considerations of using data extend beyond mere regulation compliance but, rather, committing to responsible data stewardship. Clear organizational policies on data collection, usage, retention, and sharing practices must be in place. According to a survey done by Deloitte, 61% of consumers believe organizations should bear accountability for securing personal information entrusted to them. Culturally, awareness of ethical values must prevail with employees. Such data ethics training programs will prepare the staff with the necessary knowledge of how to handle sensitive information responsibly. An organization can thus strengthen its relations with customers in trust and accountability through ethical practices.

6.6 Role of Technology in Privacy Enhancement

Technological enhancement is one factor increasing the ability of any organization to guard privacy while leveraging valuable insights from data. New technologies, such as blockchain, give increased security features to help protect personal information during transactions. Blockchain, for instance, by its very nature, does not have a central point storing sensitive information; hence, the probability of breaches is greatly reduced. In addition, it facilitates tracking compliance in regard to privacy regulations by its automation in procedures related to consent management and requests for data access. Gartner estimated that in the year 2023, 40% of organizations will adopt AI-based solutions in managing their compliance-related activities.

6.7 Future Trends of Data Privacy and Analytics

In many ways, the future of business, as of January 28, 2025, will have to combine strong data privacy measures with advanced analytics capabilities. There are a few areas to which organizations should pay heed:

1. Investment in Data Infrastructure: Companies should invest in strong infrastructure supporting advanced analytics while meeting the requirements of privacy regulations.
2. Building a Data-Driven Culture: A culture within the organization where employees would be encouraged to use data in a responsible manner while making decisions with the aim to maximize the value of available information.
3. Prioritizing Data Governance and Security: Strong governance frameworks ensure that all regulations are followed while safeguarding sensitive information against breaches.
4. Continuously Upskilling the Workforce: This will enable the employees to use the analytic tools effectively and, at the same time, take actions with regard to the ethical issues in data usage.

6.8 Related Companies in Data field

a. Alibaba

Since Jack Ma founded it in 1999, Alibaba has developed into a global powerhouse by using data-driven approaches to guide its growth and innovation. It started off targeting businesses for online trade, though Alibaba has used extensive collection and analysis of data related to consumer behavior, market trends, and business operations as key building blocks in the transformation of its business. Strong data ecosystem: advanced AI and machine learning technologies feed the very powerful systems behind recommendation personalization, optimization in logistics, and improvement in the overall customer experience. Much of the Alibaba strategy lies in investment in cloud computing through its arm, Alibaba Cloud. The division supports not only the company's diverse business ventures but also gives it a competitive edge in the fast-expanding cloud market. With its scalable infrastructure and advanced analytics capability, Alibaba Cloud empowers businesses to innovate and enhance operational efficiency. The company's landmark initial public offering in 2014 raised over $25 billion, in what was the largest IPO ever at that time. This large amount of capital served to boost its further expansion into e-commerce, cloud computing, and a lot of other fields. It put Alibaba in a position where it could enhance its technology infrastructure, expand the breadth of its services, and beef up its competitive position. As of January 28, 2025, Alibaba continues to harness the power of data to maintain its competitive advantage. From supply chain management to customer engagement, Alibaba integrates AI-driven insights into every aspect of its operations, showing what a data-driven strategy looks like in action for continued growth and innovation in the digital economy.

- Data-Driven Ecosystem: Alibaba runs a sprawling e-commerce, cloud computing, fintech, and more, throwing up vast reams of data pertaining to consumer behavior, market trends, and business operations.
- Privacy Concerns: The company has also been under scrutiny for its practices regarding data privacy and security. - 2019 Cyber Attack: A huge cyberattack hit Alibaba Cloud, raising questions about the security of sensitive information stored on it.
- Government Collaboration: Alibaba's close ties with the Chinese government have raised concerns over potential government access to user data.
- Data Localization: China has very strict requirements for data localization, which may create a challenge for how Alibaba will handle user data and international privacy regulations.

b. Facebook

Facebook, founded in 2004 by Mark Zuckerberg and his college roommates, is today one of the most powerful social media platforms in the world, having fundamentally changed the way people communicate and share information. The company's growth trajectory has been largely driven by its innovative use of data analytics and targeted advertising. From the time of its launch, Facebook was focused on developing a user-friendly platform that encourages social interaction. As the user base started growing, so did the amount of data being generated. Facebook sweeps up vast reams of data on user behavior, preferences, and interactions. It analyzes this data to build detailed user profiles, enabling the platform to provide highly tailored content and advertisements. The company is also using advanced algorithms and machine-learning techniques in ad placement optimization so that users can see relevant ads based on their interests and online activities. This advertising model has proved to be enormously lucrative for the company. With targeting options as specific as demographics, interests, and behaviors, Facebook has been a haven for business establishments of all scales intent on reaching their target audience effectively. For instance, in 2020 alone, ad revenue on Facebook crossed the $84 billion mark, which further rings credence to its data-driven approach. Besides the core platform, Facebook expanded its ecosystem through several strategic acquisitions and innovations. Its acquisition of Instagram in 2012 and WhatsApp in 2014 had enabled the company to expand its product portfolio and collect even more user data. It probably brings in better analysis of cross-behavior on platforms and further refines advertising strategies for Facebook. On the other hand, its data-driven nature has also brought about great ethical questions regarding privacy and the security of the data. In 2018, the misuse of personal data by Cambridge Analytica highlighted how personal information could be misused, further calling for greater transparency and regulation within the tech industry. But challenges are not preventing it from still innovating, investing heavily in new technologies like virtual reality with Ocu lus, and exploring new opportunities through the metaverse. And so, keeping such strong focus on data analytics and user engagement, Facebook is going to strive to continue its competitive advantage in this ever-complex digital landscape.

- **Data Collection and Usage**: Facebook sweeps up vast reams of user data, encompassing personal info, browsing history, and social interactions. This is then used to tailor experiences for users, target ads, and fuel its AI/ML algorithms.
- **Cambridge Analytica Scandal:** The high-profile case brought to light the improper access and usage of user data for political purposes, where Facebook's reputation was severely dented and put under extreme scrutiny over its data practices.

- **Privacy Settings:** The criticism of Facebook came about because of its complex privacy settings, which are quite hard for users to understand and control.
- **GDPR Compliance**: Facebook has heavily invested in the process of compliance with GDPR and other data protection regulations. It still faces a big challenge of ensuring user privacy while maintaining its business model.

c. Google:

Founded in 1998 by Larry Page and Sergey Brin, Google grew from a simple search engine to become one of the world's most powerful and influential technology companies. Crystallizing and harnessing the vast reams of data that inform its many services and products have been critical to Google's success. Primarily, at the core of Google lies a data-driven model driven by user experience and targeted advertising. The company sweeps up vast reams of data on user behavior, search patterns, and preferences through its search engine, YouTube, Google Maps, and other platforms. This information is then processed using sophisticated algorithms to deliver personalized content and targeted advertisements, thereby enhancing user engagement while maximizing advertising revenue. Indeed, most of Google's income—more than $147 billion in 2020—comes from targeted advertising services that capitalize on user insights. It also mirrors Google's investments in technologies around AI and ML, which power the business as it continues to rework its search algorithms for better results, enables better ad targeting, and creates new products like Google Assistant. By using AI and ML, Google strengthens not only its current offerings but also forays into new areas such as the Google Cloud Platform, a scalable infrastructure with advanced analytics capabilities for businesses. The company's growth is also due to its strategic acquisitions in large measure. The purchase of YouTube in 2006 allowed Google to tap into the burgeoning video content market, and acquisitions such as Android positioned the company as a leader in mobile operating systems. However, Google's reliance on data has raised ethical concerns regarding privacy and data security. It has faced criticism over practices in collecting personal information and whether it completely complies with regulations such as GDPR. In response, Google has implemented measures that make operations transparent and increase users' control over personal data. As of January 28, 2025, Google continues to move forward in exploring new technologies in quantum computing and augmented reality. With such a strong focus of its innovation efforts on the exploitation of data, Google has retained the competitive advantage it had sought for in this dynamically changing digital economy.

Google sweeps up vast reams of user data through its search engine, Android operating system, and other services. This data is then used for targeted advertising, product development, and AI/ML research. Google has also been questioned about data privacy, especially in relation to user location tracking and sharing data with third parties.

d. Amazon:

Founded in 1994 by Jeff Bezos, Amazon really came to be the whip andiciary in e-commerce and cloud computing, much due to its unique use of data to derive business strategies. Capturing and analyzing the voluminous data from the consumers has always been the key to creating a highly personalized shopping experience, fostering customer loyalty, and driving sales. Advanced data analytics forms the most integral part of Amazon's strategy. The company can decide on its inventory management, product recommendations, and marketing strategies based on consumer behavior data. Using predictive analytics, the company can forecast customers' needs and accordingly optimize logistics so that products reach customers in an efficient and effective manner. This data-driven approach not only enhances the shopping experience but also significantly reduces operational costs. Innovation at Amazon goes beyond e-commerce. Its cloud computing arm, Amazon Web Services, has positioned the company to lead in the cloud market. AWS offers scalable infrastructure and sophisticated analytics tools that help companies make the most of their own data. This strategic investment in cloud technology underpins not just the internal operations of Amazon but also arms it with substantial revenues from external customers. The company has based its focus on customer experience through various initiatives, including Amazon Prime, which gives its members benefits such as free shipping and access to streaming services. The company constantly refines its offering based on user engagement and preferences. Amazon is thus seen to have worked out the demand of the customers. Despite the successes, Amazon faces challenges related to data privacy and ethical considerations. This will need to happen hand in hand with transparency and observance of regulations such as GDPR, because of the massive collection of user data. Up to January 28, 2025, Amazon is continuing to drive growth and innovation on all its platforms through data-driven approaches. With new technologies in a fast-changing digital world, Amazon goes ahead, further building on its competitive advantage by always having customer satisfaction at the top of the agenda.

Amazon sweeps up data on customer purchases, browsing history, and other interactions to make targeted recommendations, optimize its supply chain, and develop new products and services. Others have expressed concern that it will use that data to give itself an unfair competitive advantage at the expense of smaller businesses.

- **Data Privacy is Paramount:** The growing pressure to hold data privacy to the highest regard and keep abreast of new regulations becomes a challenge for such companies, among many.
- **Innovation and Privacy Balance**: The key challenge is in striking a balance between data-driven innovation and protection of individual privacy rights.
- **Transparency and Control:** The best way to obtain trust is by being transparent with regard to the practices of collecting and using data and by giving users meaningful control over their data.
- **Continual Evolution:** This landscape of data privacy is dynamic, hence firms are forced to evolve their practices and invest much in strong security and privacy measures for data.

Table 4. Data Privacy Protection and Insights Discovery through Analytics

Aspect	Details
Importance of Data Privacy	- Consumer Concern: 79% of Americans are concerned about how their data is used (Pew Research Center).
	- Cost of Data Breaches: Average cost of a data breach reached $4.24 million in 2021 (IBM).
Data Privacy Regulations	- GDPR: Fines up to €20 million or 4% of global annual revenue for non-compliance.
	- CCPA: Allows Californians to request information on data collection and usage.
Control Over Data	- User Empowerment: Companies must provide options for users to manage their data preferences.
	- Apple's Approach: App Tracking Transparency requires user consent for tracking across apps and websites.
Driving Insights Through Analytics	- Advanced Techniques: AI and ML enable quick analysis of large datasets for actionable insights.
	- Healthcare Impact: Organizations using analytics can improve patient outcomes by up to 30% (McKinsey & Company).
Examples of Successful Value Capture	- Amazon: Personalized shopping experience through predictive analytics.
	- Netflix: Uses viewer data for content recommendations and creation.
	- Coca-Cola: Optimizes supply chain management with analytics.
Role of Technology in Value Capture	- AI and ML: Used for fraud detection in financial institutions by analyzing transaction patterns.
	- Cloud Computing: Scalable infrastructure provided by companies like Microsoft for data-driven decisions.
Ethical Considerations in Data Utilization	- Accountability: 61% of consumers believe companies should be accountable for protecting personal information (Deloitte).
	- Transparency: Clear communication about data usage enhances customer trust.

continued on following page

Table 4. Continued

Aspect	Details
Future Trends in Data Privacy and Analytics	- Focus on investing in data infrastructure, developing a data-driven culture, prioritizing governance and security, and continuously upskilling the workforce in data analysis and related skills.
Conclusion	Balancing privacy with analytical capabilities is key to unlocking sustainable growth. Companies that prioritize ethical data practices will set new standards for excellence while maintaining consumer trust and fostering innovation.

Source: Compiled from "Capturing Value as the Source of Competitive Advantage

Table 6.1 gives a broad overview of the critical relationship between data privacy and the effective use of analytics in modern organizations. As of January 28, 2025, it is evident that data privacy has been gaining in importance because of growing consumers' concern and increasing regulatory pressures. The table starts by setting the baseline with data privacy and how 79% of Americans are worried about how their data is being used, not to mention the financial implications a data breach had reached an average cost of $4.24 million in the year 2021. This setting helps understand the reasons organizations should have strong measures in place regarding data protection and regulatory frameworks shaping data privacy practices, such as GDPR and CCPA. The GDPR also imposes severe penalties for non-compliance, further enforcing the requirement for organizations to be transparent in data handling. It flags that empowering users to take control of their data is of great importance, and companies like Apple are pioneering in putting mechanisms in place for user consent. Another area of focus would be driving insights through analytics—how organizations can leverage advanced technologies like AI and machine learning to quickly analyze large datasets.

7. CONCLUSION

As we navigate the complexities of the digital economy today, January 28, 2025, the importance of data as a strategic asset has never been more obvious. Organizations in every sector are rapidly realizing that data is not an operational by-product; it's a fundamental resource through which innovation will be driven, customer experiences will be enhanced, and competitive advantages will be built. Companies like Alibaba, Facebook, Google, and Amazon are only some of the things that can be achieved when data is harnessed properly to produce transformational results. Alibaba demonstrates how a data-centric approach can fuel growth and innovation. The company uses data on consumer behavior to enhance services by offering recommendations and optimizing logistics. Its investment in Alibaba Cloud positions it as a leader in cloud computing that supports diverse business ventures. The company's 2014 IPO raised over $25 billion, which provided further expansion into e-commerce and cloud services. Facebook has revolutionized social interactions through its data-driven strategies. Through the collection of extensive user data, Facebook delivers personalized content and targeted advertisements. In 2020, the platform generated over $84 billion in advertising revenue, hence showing the effectiveness of its data utilization. But the privacy issues with Facebook only underline the need for ethical considerations in using data. Google has turned out to be the giant by making use of user data for targeted advertisements. The company captures most of the value from its vast datasets through services like Google Ads, which generated more than $147 billion from advertising in 2020. Its commitment to innovation is reflected in investments in AI and machine learning that enhance its ability to analyze data quickly and derive actionable insights. Amazon transformed e-commerce by using consumer data to create personalized shopping experiences. Through predictive analytics, Amazon tries to envision what the customer needs and optimizes its inventory accordingly. In 2020, Amazon netted sales of about $386 billion, which underlines the success of its data-driven strategies. In spite of such successes, the firms are exposed to data privacy and governance challenges. The compliance is very important, as failure could result in massive fines and reputational damage. For instance, GDPR imposes fines up to €20 million for non-compliance. One of the things companies must institute to establish trust with consumers is the transparency of data practices. In this respect, technology has a very important role in improving privacy while extracting insights from data. Advanced analytics techniques, including AI and machine learning, give organizations the opportunity to analyze large datasets efficiently while ensuring ethical standards are being upheld. Moreover, cloud computing offers scalable infrastructure for processing voluminous information. The organizations also have to work on the balance between value creation and value capture. It calls for dynamic pricing models, customer segmentation, and enhancement of customer experiences through personalized interactions. For instance, Netflix uses data of viewers not only for content recommendations but also for decisions on content creation. In a nutshell, value capture from data is something that organizations must do to thrive in the competitive landscape today. The good handling of the privacy-related issues allows corporations to monetize their data effectively, generating new opportunities to fuel growth and further innovation. Its validity can clearly be shown with the case study of Alibaba, Facebook, Google, and Amazon: balancing value creation with ethics will distinguish leaders from laggards in industry. Adopting these principles not only drives business success but also charts a new way forward for excellence in the digital age. As organizations continue to evolve in this increasingly data-driven world, the ability to unlock sustainable growth and keep consumer trust will indeed be tied to harnessing the power of data in a responsible manner.

REFERENCES

Alibaba. (2021). *Annual report 2021.* Alibaba Group.

Amazon. (2021). *Sustainability at Amazon.* Amazon.com.

California Consumer Privacy Act (CCPA), Cal. Civ. Code §1798.100 et seq.

Cisco. (2020). *Cybersecurity series: Data privacy and security trends in 2020.* https://www.cisco.com

Deloitte. (2020). *The ethics of data: A survey of consumer attitudes towards data privacy and protection.*https://www2.deloitte.com

European Parliament and Council of the European Union. (2016). *General Data Protection Regulation (GDPR), Regulation (EU) 2016/679 on the protection of natural persons with regard to the processing of personal data and on the free movement of such data.*

Gartner. (2021). *AI in compliance: How AI is transforming compliance management.* https://www.gartner.com

IBM. (2021). *Cost of a data breach report 2021.* IBM Security. https://www.ibm.com/security/data-breach

McKinsey & Company. (2020). *The role of data analytics in healthcare: How to improve patient outcomes.*https://www.mckinsey.com

Meta. (2022). *Sustainability report 2022.* Meta Platforms Inc.

Pew Research Center. (2021). *Americans and privacy: Concerned, confused, and feeling a lack of control over their personal information.*https://www.pewresearch.org

Chapter 8
Digital Analytics in Action From Data to Impact

S. Ajitha
https://orcid.org/0000-0002-0459-3208
M.S. Ramaiah University of Applied Sciences, India

K. J. Reshma
M.S. Ramaiah University of Applied Sciences, India

Geevarathna
https://orcid.org/0009-0009-7384-2237
Acharya Bangalore B School, India

S. Huxley
https://orcid.org/0009-0004-7407-1415
Alliance University, India

Chandra Sen Mazumdar
M.S. Ramaiah University of Applied Sciences, India

Divya D. S. Rajan
https://orcid.org/0009-0000-0033-5554
St. Xavier's Catholic College of Engineering, India

ABSTRACT

The chapter "Digital Analytics in Action: From Data to Impact" explores the transformative role of digital analytics in bridging the gap between raw data and actionable business outcomes. It highlights the integration of theoretical frameworks such as decision theory, behavioral analytics, and predictive analytics, which serve as guiding principles for leveraging data effectively. By delving into key areas like customer behavior analysis, conversion rate optimization, and advertising performance tracking, the chapter illustrates how organizations can extract meaningful insights to enhance decision-making, improve user experiences, and drive innovation. Real-world applications are discussed, including the use of analytics in personalized marketing, resource allocation, and operational efficiency, showcasing its interdisciplinary nature across industries. Additionally, the chapter examines challenges such as data scalability, privacy concerns, and the need for continuous refinement, emphasizing the importance of ethical and strategic approaches.

DOI: 10.4018/979-8-3373-1320-7.ch008

1. INTRODUCTION

In today's digital-first world, organizations gather vast amounts of data from multiple sources—websites, mobile apps, and social media platforms. While managing this volume presents challenges, it also offers opportunities to gain actionable insights that drive business success. Digital analytics (Kliestik et al., 2022) serves as the bridge between raw data and impactful decision-making, enabling businesses to enhance operations, enrich customer experiences, and achieve strategic goals.

For instance, global e-commerce companies utilize digital analytics to personalize customer interactions. By examining browsing history, purchase patterns, and feedback, businesses can recommend products that align with user preferences, increasing customer satisfaction and revenue. Logistics companies leverage real-time analytics to optimize delivery routes, reducing costs and improving efficiency.

This chapter explores digital analytics principles, applications, and strategies, illustrating how businesses can harness its power to remain competitive. Covering both foundational and advanced techniques, it provides a comprehensive guide to turning data into actionable insights.

1.1 Understanding the Fundamentals of Digital Analytics

Definition: Digital Analytics refers to the process of collecting, measuring, analyzing, and interpreting data from digital platforms such as mobile apps, websites, email campaigns, social media, and other virtual channels.

Data analytics is the process of examining raw data to uncover patterns, draw conclusions, and make informed decisions. It involves collecting, cleaning, transforming, and analyzing data to extract meaningful insights that support strategic goals and problem-solving. The primary goal is to understand user behavior, improve digital marketing strategies, optimize user experiences, and achieve business objectives through data-driven insights.

Characteristics of Data Analytics

Data analytics is distinguished by several key characteristics, each playing a vital role in deriving actionable insights. It is a data-driven process (El Khatib et.al.,2022), relying on both structured and unstructured data to uncover patterns and trends. For example, an e-commerce platform might analyze user browsing behavior and purchase history to recommend personalized product suggestions. This process is objective-oriented, designed to solve specific problems or achieve goals, such as optimizing a marketing campaign to increase conversion rates.

The process is also iterative, involving repeated cycles of data collection, cleaning, analysis, and refinement to ensure accurate results. For instance, a healthcare organization analyzing patient data may iterate its models to improve diagnosis accuracy. Moreover, data analytics is scalable, capable of handling large volumes of data, as seen in social media platforms analyzing millions of posts daily to identify trending topics.

Another defining feature is its interdisciplinary nature, combining statistical methods, computational techniques, and domain expertise. In the financial sector, data scientists, statisticians, and economists might collaborate to predict stock market trends. Additionally, analytics is often predictive and prescriptive, moving beyond describing past events to forecast future outcomes and suggest optimal actions. For example, ride-sharing apps like Uber use predictive analytics to estimate demand and dynamic pricing.

Lastly, modern data analytics is automation-friendly, leveraging tools and algorithms to automate repetitive tasks. In a manufacturing scenario (Lei et al.,2023), automated analytics might monitor machinery performance in real-time, detecting potential failures before they occur. These characteristics make data analytics a versatile and powerful tool across industries, driving informed decision-making and innovation.

Uses of Digital Analytics

Digital analytics offers valuable insights that enhance various facets of digital marketing, user experience, and business strategies. Some of its uses include:

1. **Customer Behavior Analysis**: Digital analytics helps businesses comprehend how customers engage with their platforms. This includes tracking pages visited, time spent, and where they drop off.
2. **Conversion Rate Optimization (CRO)**: Analyzing user interactions, businesses can optimize the conversion process, such as turning website visitors into paying customers or encouraging newsletter sign-ups.
3. **SEO and Content Strategy**: Identifying high-performing keywords and content, enabling businesses to create targeted SEO strategies and produce content that attracts more traffic and engagement.
4. **Advertising Campaign Performance**: Evaluating the effectiveness of online advertisements, such as pay-per-click (PPC) campaigns or social media ads. This helps in optimizing ad budgets and maximizing return on investment (ROI).

5. **Customer Segmentation**: Digital analytics groups customers based on behaviors, preferences, or demographics. This segmentation allows businesses to provide personalized content and tailored offers, enhancing customer satisfaction.
6. **Product and Service Optimization**: Identifying areas where products or services may be underperforming, allowing for data-driven decisions to improve user experience and overall satisfaction.

Applications of Digital Analytics

1. **Website Analytics**: Digital tools like Google Analytics allow businesses to track website metrics like traffic, bounce rates, session duration, and goals (e.g., form submissions, purchases). Example: A retail website can track how many visitors complete a purchase versus those who leave the site before buying.
2. **Social Media Analytics**: Social media platforms (e.g., Instagram, Facebook,Twitter) offer tools to measure engagement, reach, and sentiment. Example: A brand can analyze which types of posts generate the most interaction and adjust their content strategy accordingly.
3. **E-commerce Analytics**: Monitoring e-commerce activities like cart abandonment rates, product views, and purchase conversions. Example: A fashion retailer might use digital analytics to see which products are getting the most clicks and which categories need promotion.
4. **Email Campaign Analytics**: Tracking open rates, click-through rates (CTR), and conversion rates of email campaigns to gauge their effectiveness. Example: A company might optimize email subject lines, design, and content based on data from previous campaigns.
5. **Mobile App Analytics**: Analyzing app usage patterns, including which features are used the most, session lengths, and user retention rates. Example: An app developer can use analytics to improve user onboarding or identify why users stop using the app.

Examples to Illustrate Digital Analytics

- **E-Commerce Site**: An online clothing retailer might use Google Analytics to track which product pages have the highest bounce rate. By analyzing the data, they may find that users abandon the page due to slow loading times. The company can then optimize the page's performance, leading to better conversions.
- **Social Media Analytics**: A brand uses Facebook Insights to monitor the engagement on its posts. After analyzing the data, they discover that posts featuring user-generated content receive 40% more likes and comments com-

pared to regular promotional posts. The brand shifts its strategy to feature more user content.

- **Email Campaigns**: An online food delivery service might track email open rates and conversion rates using Mailchimp. They find that emails sent on Tuesdays at noon have the highest open rates and orders, helping them schedule future campaigns for optimal performance.

Theories on Digital Analytics

Digital analytics operates on a foundation of several key theories that shape its application and enable impactful outcomes across industries. Decision theory is pivotal in guiding businesses toward rational, data-driven choices (Mariani & Nambisan, 2021) by analyzing available information and determining the optimal course of action. For example, it is commonly employed in dynamic pricing models to maximize revenue or optimize resource allocation. Information theory, on the other hand, delves into the encoding, transmission, and interpretation of data. This theory plays a crucial role in improving data processing efficiency and reducing communication errors, as seen in industries like telecommunications and digital platforms.

Behavioral analytics theory emphasizes understanding user actions, preferences, and interactions. By analyzing this data, businesses can predict customer behavior and create highly personalized experiences, such as tailored content recommendations on streaming platforms. Game theory examines competitive scenarios where businesses strategize to outmaneuver rivals, such as optimizing bidding strategies in digital advertising to balance visibility with cost-effectiveness.

Big data theory addresses the challenges of managing and analyzing vast and complex datasets. It enables organizations to extract meaningful patterns and trends from massive information streams, such as analyzing social media data for sentiment analysis and trend identification. Complementing this, Bayesian theory provides a framework for updating predictions and probabilities dynamically as new data becomes available, making it invaluable for applications like fraud detection and real-time analytics.

Statistical learning theory focuses on creating predictive models based on historical data. It is widely used to forecast market trends, improve inventory management, and support decision-making through machine learning algorithms. Finally, predictive analytics theory leverages past data to anticipate future outcomes, empowering businesses to adopt proactive measures. This is particularly impactful in sectors like healthcare, where patient data can be analyzed to predict potential health risks and provide preventative care.

Together, these theories form the backbone of digital analytics, enabling organizations to transform raw data (Vollrath & Villegas, 2021) into actionable insights, optimize operations, and achieve sustainable growth. They are instrumental in bridging the gap between data collection and meaningful impact, ensuring decisions are both strategic and evidence-based.

Metrics in Digital Analytics

In the context of Digital Analytics, metrics refer to the quantitative data points or key performance indicators (KPIs) that are used to measure the effectiveness of digital marketing efforts, user engagement, and the overall performance of digital channels. These metrics provide insight into how well a website, app, or online campaign is performing and help businesses evaluate progress toward their goals.

Key Types of Metrics in Digital Analytics

- **Traffic Metrics**: Traffic metrics are fundamental indicators that measure the flow and behavior of visitors on digital platforms. They include pageviews, sessions, and unique visitors, each serving distinct purposes.
    - **Pageviews** track the total number of times a webpage is loaded or refreshed. For example, a content website can analyze pageviews to identify trending articles, guiding editorial focus toward high-demand topics.
    - **Sessions** represent a user's activity during a single visit, offering insights into engagement levels. For instance, an online learning platform may evaluate session duration to determine which course materials hold student attention the longest.
    - **Unique visitors** measure the count of distinct users accessing a site, helping businesses assess their overall reach and audience growth. A marketing agency might use this data to evaluate the success of a campaign aimed at attracting new customers. Together, these metrics provide actionable insights for improving website performance, enhancing user experience, and achieving business goals.
- **Engagement Metrics**: Engagement metrics evaluate how users interact with digital platforms and content, providing critical insights into user behavior and satisfaction. These metrics include bounce rate, time on site, and click-through rate (CTR), each offering unique perspectives on engagement.
    - **Bounce Rate** is the measure of the percentage of visitors who leave a site after viewing only one page. A high bounce rate may indicate that the content is not relevant or engaging. For example, a blog can analyze

bounce rate to determine which topics fail to capture audience interest and adjust its content strategy accordingly.

- o **Time on Site**: This metric tracks the average duration users spend on a website. Longer time on site often signals higher user engagement. For instance, an online education platform might analyze this metric to evaluate the effectiveness of its video tutorials or interactive exercises
- o **Click-Through Rate (CTR)**: CTR measures the ratio of users who click on a specific link compared to the total number of users who view it. It is commonly used in digital advertising and campaigns using email. For example, an e-commerce business may monitor CTR for promotional banners to assess the effectiveness of its marketing efforts.

Engagement metrics are vital for understanding the depth of user interactions, identifying areas for improvement, and tailoring strategies to foster meaningful engagement and enhance user experiences.

- **Conversion Metrics**: Measure how effectively a website, app, or digital campaign persuades users to take desired actions. These actions, often referred to as **conversions**, can include purchases, form submissions, newsletter sign-ups, downloads, or any other goal relevant to a business. Conversion metrics help businesses understand user behavior, evaluate performance, and optimize strategies to increase desired outcomes.
    - o **Conversion Rate** is the percentage of visitors who complete a desired action (e.g., making a purchase, signing up for a newsletter, downloading an app). A higher conversion rate indicates better performance in meeting business objectives. For example, an online clothing retailer might have a landing page where visitors can browse products. If a significant portion of visitors proceeds to make a purchase, the retailer can evaluate how effectively the website converts casual visitors into paying customers.
    - o **Goals** are specific user actions that a business wants to measure. These actions can vary depending on the business objectives, such as signing up for a newsletter, downloading a brochure, or submitting a contact form. For instance, an educational website may set a goal for users to download a free eBook. If many users take this action, the business can assess the effectiveness of the eBook as a lead generation tool.
    - o **E-commerce Conversion Rate** is the specific type of conversion rate, measuring the percentage of visitors who make a purchase on an e-commerce site. For example, an electronics store running a promotional

campaign might evaluate the percentage of visitors who end up buying products from the website.

- **Acquisition Metrics**: Measure how effectively a website, app, or marketing campaign attracts users or customers. These metrics focus on tracking the channels, strategies, and costs involved in bringing visitors or leads to a digital platform. They help businesses understand the sources of their audience and assess the effectiveness of different acquisition strategies.
  - **Traffic Sources** is the channels through which visitors arrive at a website or app, such as organic search, paid search, social media, direct traffic, or referrals.
    - **Organic Search**: A small business selling handmade candles notices that 50% of their website traffic comes from people searching for terms like "best scented candles" on Google and clicking on their website in the search results.
    - **Paid Search**: An online clothing retailer runs a Google Ads campaign targeting the keyword "summer dresses." Visitors who click on these ads make up 20% of the website's traffic.
    - **Social Media**: A fitness brand promotes its products on Instagram and Facebook, leading to a 15% increase in website traffic from these platforms.
    - **Direct Traffic**: A popular restaurant chain has customers who directly type their website URL (e.g., www.bestpasta.com) into their browsers, making up 10% of their overall traffic.
    - **Referral Traffic**: A tech blog links to an online store selling gadgets. Visitors clicking this link generate 5% of the store's traffic.
    - **Email Marketing**: A subscription box company sends a newsletter featuring a special discount. The email generates 8% of their monthly website visits.
  - **Cost per Acquisition (CPA)** is the cost incurred to acquire one customer, typically used in paid advertising campaigns. A lower CPA is usually more efficient. For example, an online store spends $5,000 on a Facebook Ads campaign promoting a new product line. The campaign results in 250 purchases. The CPA is $20 per customer.
  - **Click-Through Rate (CTR)** is the percentage of people who click on a link (such as an ad or email link) compared to the total number of people who saw the link. For example, a home decor store runs a Google Ads campaign that is displayed 10,000 times (impressions). Out of these, 300 people click the ad. The CTR for this campaign is 3%.
- **Retention Metrics**: Measure a business's ability to retain users or customers over time, reflecting how effectively it engages and satisfies its audience.

These metrics are critical for assessing customer loyalty, identifying potential issues, and optimizing strategies to encourage repeat interactions or purchases. They are especially important in industries where long-term relationships with customers drive revenue, such as SaaS, e-commerce, and subscription-based models.

- **Customer Retention Rate** is the percentage of customers who return to make additional purchases or interact with a service after their initial visit or interaction. For example, a streaming service might calculate its retention rate over a month. If the platform starts with 10,000 subscribers, gains 2,000 new ones, but ends the month with 9,000 active users, it indicates that a significant portion of customers remained loyal despite some churn. A high retention rate suggests strong customer satisfaction and effective engagement strategies, such as personalized recommendations or exclusive content.
- **Churn Rate** is the percentage of customers who stop using a product or service within a specific time period. A high churn rate suggests issues with user satisfaction or product-market fit. For instance, a SaaS company notices that 10% of its subscribers cancel their subscriptions every month. This could indicate issues like poor onboarding, lack of perceived value, or strong competition. By addressing churn factors such as improving customer support or introducing loyalty discounts the company can reduce its churn rate and boost long-term revenue.
- **Return Visitors** is the number of users who come back to your website or app after their first visit. This is often an indicator of user loyalty and satisfaction. For example, an e-commerce site tracks return visitors and notices that 40% of its monthly traffic comes from people who have visited the site before. This could indicate that users find the platform valuable, whether through compelling content, relevant products, or excellent customer experiences. Encouraging return visits might involve strategies like email reminders, retargeting ads, or loyalty programs. For instance, an online bookstore might send an email recommending books based on a user's browsing history, enticing them to return and make a purchase.

- **Social Media Metrics**: Quantitative measures that help businesses evaluate the performance of their social media strategies. These metrics provide insights into how audiences interact with content, how far a message spreads, and the overall effectiveness of a campaign. By analyzing metrics like engagement rate, impressions, and reach, businesses can refine their strategies to achieve better results and stronger connections with their target audience.

- **Engagement Rate** is the level of interaction (likes, comments, shares) relative to the number of followers or impressions. This metric helps assess how well social media content is resonating with the audience. For example, a beauty brand posts a tutorial video on Instagram and notices a high engagement rate as users comment on the tips, share the post with friends, and click the link to purchase products. This metric indicates that the content is both relevant and compelling.
- **Impressions** is the total number of times your content is displayed on social media, regardless of whether it's clicked or not. For instance, a clothing retailer runs a Facebook ad showcasing its winter collection. If the ad appears 50,000 times across users' feeds, that number represents its impressions. While impressions alone don't guarantee user action, they are useful for measuring the visibility and exposure of content.
- **Reach** is the total number of unique users who have seen your content on social media platforms. For example, a nonprofit organization posts about a fundraising campaign on Twitter, and 15,000 unique users see the post. This reach helps the organization understand how many people its message is potentially influencing.

Digital Analytics Process

The digital analytics process is a systematic approach that businesses and organizations use to collect, measure, analyze, and interpret data from their digital channels (websites, apps, social media, etc.). This process helps organizations make data-driven decisions (Lutfi et al., 2022), optimize marketing strategies, and improve user experience. Here's a detailed breakdown of the digital analytics process:

1. Define Business Objectives and KPIs

The first step in the digital analytics process is to define clear **business objectives** and align them with measurable **Key Performance Indicators (KPIs)**. This phase ensures that data collection and analysis will address the business's specific goals.

- **Business Objectives**: What do you want to achieve? Examples include increasing website traffic, boosting online sales, enhancing customer satisfaction, or growing social media followers.
- **KPIs**: These are specific, quantifiable metrics that directly reflect the progress toward the business objectives. Examples of KPIs are website conversion rate, customer retention rate, click-through rate (CTR), or return on investment (ROI).

Example: If a retail brand's business objective is to increase online sales, the KPIs might include e-commerce conversion rate, average order value, and cost per acquisition (CPA).

2. Identify the Data Sources

Next, you need to identify the relevant **data sources** that will provide the information needed to track the KPIs. Data sources can come from various digital channels, such as:

- **Websites**: Google Analytics, server logs, landing pages
- **Mobile apps**: In-app analytics, event tracking
- **Social media**: Facebook Insights, Twitter Analytics, Instagram Insights
- **Email campaigns**: Email marketing platforms like Mailchimp or HubSpot
- **Paid Advertising**: Google Ads, Facebook Ads, LinkedIn Ads

Understanding where your data is coming from helps in setting up the correct tools for tracking and ensures you capture the relevant data points.

3. Data Collection

Once the sources are identified, the next step is to **collect the data**. This involves using analytics tools and platforms to capture the data from your digital assets. Depending on the platform, you might need to:

- **Set up tracking codes**: For example, Google Analytics requires adding tracking code to your website to capture visitor behavior.
- **Implement tags**: Tools like Google Tag Manager can be used to track specific actions, such as button clicks or form submissions.
- **Integrate APIs**: For integrating data from different platforms like CRM systems or social media platforms to get a unified view of performance.

  Example: If a company wants to track website traffic and user behavior, it would set up Google Analytics with tracking codes and configure goals to measure key actions, like purchases or sign-ups.

4. Data Processing and Organization

Once the data is collected, the next phase involves **processing and organizing** it in a way that makes it easy to analyze. Raw data often requires cleaning and structuring before it can provide meaningful insights. This can involve:

- **Data cleaning**: Removing duplicates, fixing errors (e.g., misreported page views), and eliminating irrelevant data.
- **Data aggregation**: Organizing the data into categories or groups (e.g., grouping user data by demographics or traffic sources).
- **Data transformation**: Converting data into a usable format, such as aggregating daily visits into monthly visits.

5. Data Analysis

Once the data is clean and organized, the next step is **analyzing** the data to extract meaningful insights. This involves:

- **Descriptive Analytics**: Understanding what has happened by examining historical data. This might involve looking at metrics like traffic trends, bounce rates, and conversion rates over time.
- **Diagnostic Analytics**: Identifying the reasons behind certain outcomes. For instance, analyzing why a certain campaign led to a spike in traffic or why a page has a high bounce rate.
- **Predictive Analytics**: Using historical data to forecast future outcomes. For example, predicting sales or customer churn based on past behaviors.
- **Prescriptive Analytics**: Recommending actions based on analysis. This can include optimizing a website's layout based on user behavior or adjusting a paid ad campaign to maximize ROI.

  Example: After analyzing traffic data from different sources, a business might discover that a large percentage of users drop off during the checkout process. This insight can inform efforts to improve the checkout flow.

6. Reporting and Visualization

The results of the analysis need to be presented in a clear and actionable format. **Reporting and visualization** are key to ensuring stakeholders understand the data and its implications.

- **Dashboards**: Tools like Google Data Studio or Tableau are often used to create interactive dashboards that display live data in an easy-to-understand format.
- **Reports**: Regular reports summarizing the findings (e.g., weekly, monthly) are shared with key stakeholders, often including visuals like graphs, charts, and tables.

- **Actionable Insights**: Reports should focus on what the data means and how it can drive future actions. For example, a report might highlight that social media traffic is driving more conversions than paid ads, suggesting a shift in budget allocation.

7. Take Action

The ultimate goal of digital analytics is to **take action** based on the insights gained from the data. This could mean adjusting strategies, launching new campaigns, or optimizing digital assets. Actions could include:

- **A/B Testing**: Running tests on different versions of a webpage or ad to identify the most effective one.
- **Optimizing Campaigns**: Adjusting ad copy, targeting, or budget allocation based on campaign performance.
- **Website Optimization**: Making changes to the website layout, improving content, or enhancing user experience to increase conversions.

8. Monitor and Iterate

The final step is to **continuously monitor** the results and **iterate** on the strategies. Digital analytics is an ongoing process, and it's important to regularly assess how changes are impacting business goals.

- **Ongoing Monitoring**: Continuously track KPIs and adjust strategies as necessary.
- **Iteration**: Based on new data, tweak the approaches, test new hypotheses, and improve performance over time.

2 . FROM RAW DATA TO ACTIONABLE INSIGHTS

In today's digital landscape, businesses and organizations collect vast amounts of data from various sources like websites, mobile apps, social media, and customer interactions. This data, while valuable, is often unstructured and overpowering, making it challenging to draw meaningful conclusions or take informed actions. The key lies in transforming this raw data into actionable insights that can drive strategic decisions (Ochoa & Wise, 2021), improve performance, and optimize customer experiences.

The process of converting raw data into actionable insights typically involves several stages: data collection, data cleaning, analysis, interpretation, and reporting. By following this process, businesses can extract valuable information that informs marketing strategies, enhances product development, and fosters customer loyalty. To gain clarity, we need to break down and explain key concepts involved in this transformation: Data Collection, Data Processing, Data Analysis, Data Visualization, and Actionable Insights.

2.1 Data Collection and Processing Strategies

Data collection and processing are essential steps in converting raw data into valuable insights that drive business decisions (Akter et al.,2022). Efficient data collection strategies and processing techniques ensure that the right data is gathered, cleaned, and transformed into actionable business intelligence.

Sources of Data

Google Analytics is one of the most widely used web analytics tools for tracking and analyzing website traffic. It provides insights into user behavior, page views, bounce rates, traffic sources, conversions, and much more. By tracking user interactions on your website, businesses can measure the effectiveness of marketing campaigns (Sponder et al., 2018), understand user engagement, and identify opportunities for optimization.An e-commerce website uses Google Analytics to track where customers are coming from (e.g., organic search, paid ads, social media) and what actions they take on the site, such as adding items to their cart or completing a purchase. **Customer Relationship Management (CRM)** systems, such as Salesforce or HubSpot, track interactions with customers, including email communications, sales, and customer service activities. These systems provide valuable insights into customer behavior (Choi et al.,2022) preferences, and purchase history, helping businesses tailor their marketing and sales strategies. A SaaS company uses its CRM system to track which features customers use most, their subscription renewal status, and their interactions with customer support. This data helps personalize communication and improve customer retention. The Internet of Things (IoT) refers to the network of physical devices that are embedded with sensors and software to collect and exchange data. IoT devices are used in various industries to monitor everything from manufacturing processes to smart homes. A manufacturing company uses IoT sensors to track machinery performance in real time, collecting data on temperature, pressure, and vibration levels. This data can be analyzed to predict equipment failures and optimize maintenance schedules.

Techniques for Data Processing

Automated data pipelines are processes that streamline the flow of data from multiple sources to a centralized data storage or analysis platform. This automation minimizes the risk of human error and ensures timely, accurate data collection. A marketing team sets up an automated data pipeline that pulls customer interaction data from social media, email campaigns, and website visits into a central analytics platform, where it is processed and analyzed in real time. ETL (Extract, Transform, Load) Processes, the ETL process is a common data integration method used to move and transform data from one or more sources into a destination database or data warehouse. The process consists of: Extract: Collecting data from various sources (e.g., CRM, IoT, web analytics). Transform: Cleaning and transforming the data into a standardized format, including filtering, aggregating, and enriching the data(Mohamed et al., 2024). Load: Loading the transformed data into a data warehouse or analytics platform for further analysis. A retail company extracts customer purchase data from its CRM system, transforms it by filtering out incomplete or outdated records, and loads the clean data into a data warehouse for in-depth analysis to improve inventory management and sales strategies.

2.2 Transforming Data into Valuable Business Intelligence

Once data has been collected and processed, it must be transformed into actionable business intelligence. This involves presenting the insights in a way that is easy to understand and can directly inform business decisions.

Data visualization

Data visualization is the process of presenting complex data in graphical formats, such as charts, graphs, heatmaps, and dashboards. It allows decision-makers to quickly grasp insights and trends without needing to sift through raw data. Effective visualization can highlight key patterns, performance metrics, and outliers that might otherwise go unnoticed.

- **Dashboards**: A dashboard is a real-time, interactive visual interface that aggregates and displays key performance indicators (KPIs) and metrics. Dashboards provide a comprehensive overview of business performance at a glance.

- **Example**: A marketing dashboard might display metrics such as website traffic, conversion rates, social media engagement, and email campaign performance—all in one view, making it easy to track campaign progress.

- **Reports**: Reports are more detailed, often showing historical data and trends, and are typically used to provide insights over a longer period.
  - **Example**: A monthly sales report showing year-over-year growth, best-performing products, and customer acquisition costs allows the sales team to make informed decisions about product focus and resource allocation.

Interpretation

The process of **interpretation** involves analyzing the data in the context of business objectives and goals. Contextualizing the data ensures that insights are not just numbers, but actionable conclusions that can inform strategy, optimization, and decision-making.

- **Contextualization**: To properly interpret data, it's important to understand the business context, such as market conditions, target audience, and specific challenges the organization faces. Without context, data can be misleading or difficult to apply in decision-making.
  - **Example**: If a retailer sees a sudden spike in traffic from a particular region, they need to interpret whether it's a result of a localized promotion, seasonal trends, or something else. If the spike is from a promotion, the retailer can analyze which specific products are being bought and use that information to adjust inventory or plan similar future campaigns.
- **Informing Decision-Making**: Insights drawn from data can be used to guide both tactical and strategic decisions (Niu et al., 2021). Whether it's optimizing a website layout to increase conversion rates, adjusting marketing campaigns to improve customer acquisition, or refining customer service based on feedback, interpreting data correctly can lead to more effective decisions.
  - **Example**: A SaaS company analyzing user data from its platform might find that certain features are being underused. They could interpret this as a need for better onboarding or create targeted content to highlight those features, improving customer retention.

3. ADVANCED ANALYTICS TECHNIQUES FOR BUSINESS OPTIMIZATION

Advanced analytics has become a cornerstone for business optimization, enabling organizations to make informed, data-driven decisions (Song & Yu, 2021) that drive growth, efficiency, and innovation. By leveraging cutting-edge technologies such as machine learning (ML), artificial intelligence (AI), and predictive modeling, businesses can uncover deeper insights, improve operations, and create value across various functions. This comprehensive approach to analytics allows companies to transition from descriptive reporting to proactive and prescriptive decision-making.

Key Advanced Analytics Techniques

1. Machine Learning and Artificial Intelligence (AI)

Machine Learning and AI empower businesses to automate data analysis, identify complex patterns, and make predictions without explicit programming. These technologies learn from historical data to optimize operations, forecast outcomes, and personalize customer experiences.

- **Applications:**
  - **Fraud Detection:** Financial institutions use ML algorithms to detect fraudulent transactions.
  - **Predictive Maintenance:** Manufacturers use AI sensors to predict equipment failures.
  - **Chatbots:** AI-driven virtual assistants improve customer service with real-time responses.
- **Benefits:**
  - Faster decision-making through automation.
  - Cost reduction by automating repetitive tasks.
  - Enhanced customer engagement through personalization.

2. Predictive Analytics

Predictive analytics involves using statistical models and machine learning algorithms to analyze historical data and predict future outcomes. It helps organizations anticipate trends, customer behavior, and potential risks.

- **Applications:**
  - **Sales Forecasting:** Retailers predict demand for products to manage inventory efficiently.

- Customer Churn Prediction: Telecom companies forecast which customers are likely to cancel services.
  - **Risk Management:** Banks assess the likelihood of loan defaults.
- **Benefits:**
  - Proactive risk management.
  - Improved marketing campaign targeting.
  - Better demand and resource planning.

3. Prescriptive Analytics

Prescriptive analytics (Barbosa et al., 2024) goes beyond prediction to recommend actions based on predicted outcomes. It uses algorithms, simulations, and optimization models to suggest the best course of action.

- **Applications:**
  - **Dynamic Pricing:** Airlines adjust ticket prices based on demand and competition.
  - **Supply Chain Optimization:** Logistics companies optimize delivery routes to reduce costs.
  - **Workforce Management:** Retailers allocate staff based on store traffic forecasts.
- **Benefits:**
  - Maximized operational efficiency.
  - Cost reduction and revenue maximization.
  - Enhanced strategic planning.

4. Data Mining

Data mining involves discovering hidden patterns, correlations, and insights within large datasets. It uses algorithms to explore data and extract meaningful information.

- **Applications:**
  - **Market Basket Analysis:** Retailers analyze purchase patterns to cross-sell products.
  - **Customer Segmentation:** Banks classify customers based on spending behavior.
  - **Fraud Detection:** Insurance companies identify unusual claim patterns.
- **Benefits:**
  - Deeper customer understanding.
  - Identification of new business opportunities.
  - Enhanced fraud detection.

5. Natural Language Processing (NLP)

NLP allows machines to understand, interpret, and generate human language. Businesses use NLP to analyze unstructured data from emails, social media, and customer feedback.

- **Applications:**
    - **Sentiment Analysis:** Brands assess public sentiment about their products.
    - **Chatbots:** E-commerce platforms automate customer inquiries.
    - **Voice Assistants:** Devices like Alexa or Siri provide hands-free services.
- **Benefits:**
    - Improved customer engagement.
    - Automation of communication tasks.
    - Real-time feedback analysis.

6. Big Data Analytics

Big data analytics involves processing and analyzing massive volumes of data from diverse sources in real-time. It combines structured and unstructured data to uncover insights that drive business strategies.

- **Applications:**
    - **Social Media Analytics:** Brands track trends and customer engagement.
    - **IoT Analytics:** Smart devices collect and analyze performance data.
    - **Healthcare Data Analysis:** Hospitals analyze patient records to improve care.
- **Benefits:**
    - Faster, data-driven decisions.
    - Real-time performance monitoring.
    - Competitive market intelligence.

7. Optimization Algorithms

Optimization algorithms are mathematical models designed to find the most efficient solutions to complex problems. These models are used in resource allocation, logistics, and process improvements.

- **Applications:**
    - **Route Optimization:** Delivery companies minimize fuel costs with optimized routes.
    - **Inventory Management:** Retailers balance stock levels to reduce storage costs.

- **Production Scheduling:** Manufacturers streamline workflows for maximum output.
- **Benefits:**
  - Cost savings through efficient resource use.
  - Reduced waste and improved sustainability.
  - Higher operational productivity.

Business Optimization through Advanced Analytics

Advanced analytics optimizes business performance by enabling better decision-making across all departments. Here's how it impacts various business areas:

- **Marketing:** Personalizes marketing strategies through customer segmentation and targeted campaigns.
- **Sales:** Improves lead scoring and sales forecasting for better conversion rates.
- **Supply Chain:** Streamlines operations with demand forecasting and logistics optimization.
- **Finance:** Enhances risk assessment and fraud detection for financial stability.
- **Human Resources:** Predicts employee turnover and improves talent management strategies.

Challenges in Implementing Advanced Analytics

While advanced analytics offers significant advantages, it also presents challenges:

1. **Data Quality:** Inaccurate or incomplete data can lead to misleading insights.
2. **Integration Complexity:** Combining data from multiple sources can be technically challenging.
3. **Talent Gap:** Skilled data scientists and analysts are in high demand but short supply.
4. **Cost of Implementation:** Investing in advanced tools and infrastructure can be expensive.
5. **Data Privacy:** Handling sensitive customer data requires strict compliance with privacy laws.

Future of Advanced Analytics in Business Optimization

The future of advanced analytics is shaped by rapid technological advancements:

- **AI and Automation:** Greater use of AI for autonomous decision-making and process automation.
- **Real-Time Analytics:** Instant insights through real-time data processing.
- **Augmented Analytics:** AI-powered tools that automate data preparation and analysis for non-technical users.
- **Edge Computing:** Analyzing data closer to the source (IoT devices) for faster decision-making.
- **Explainable AI:** Models that provide transparency into how AI-driven decisions are made.

Advanced analytics techniques like machine learning, AI (Ajitha & Huxley, 2024), predictive modeling, and data mining have transformed how businesses operate, enabling smarter, faster, and more informed decision-making. By leveraging these tools, companies can optimize operations, personalize customer experiences, mitigate risks, and drive growth. Although challenges exist, businesses that invest in and embrace these analytics techniques will gain a competitive advantage and position themselves for long-term success in the evolving marketplace.

4. REAL-TIME ANALYTICS FOR INSTANT DECISION-MAKING

In a fast-paced digital world, the ability to process and act on data instantly has become a critical component for business success. Real-time analytics allows organizations to analyze and interpret data as it is generated, enabling immediate responses to emerging trends, potential risks, and market opportunities (Zhang et al.,2022). This capability helps businesses stay agile, competitive, and responsive to the dynamic demands of customers and markets.

Case Studies of Businesses Adapting in Real Time

1.Real-Time Monitoring of Social Media Sentiment

Social media platforms generate vast amounts of data every second. Real-time analytics (Petrov M et al.,2021) tools can process this data to gauge public sentiment about brands, products, or services. By analyzing customer opinions, trending topics, and feedback instantly, businesses can identify emerging issues, capitalize on positive trends, and mitigate reputational risks.

Example: Apple Inc., a leading consumer electronics company, launched a new iPhone model. Shortly after its release, Apple's real-time social media sentiment analysis detected a surge in customer complaints about rapid battery drain. Recognizing the potential impact on customer satisfaction and brand perception, Apple's marketing and technical teams collaborated to address the issue swiftly. The company issued a public statement on Twitter and other platforms acknowledging the problem and assured users that a software update was being developed to resolve the battery performance issue. Within days, Apple rolled out the update, effectively mitigating the issue and maintaining customer trust. This prompt and transparent response demonstrated Apple's effective use of real-time analytics to protect its brand reputation and improve customer experience.

2.Adjusting Digital Campaigns Based on Live Performance Metrics

Digital marketing campaigns can be optimized in real-time by continuously monitoring performance metrics such as click-through rates (CTR), conversion rates, and audience engagement (Zhou, C et,al,2021). This allows marketers to adjust ad placements, budgets, and targeting strategies dynamically to maximize campaign effectiveness.

Example: Amazon, a global e-commerce giant, launched a multi-channel marketing campaign to promote its Prime Day sales across social media, search engines, and email marketing. Using real-time analytics, Amazon's marketing team discovered that Instagram and TikTok ads were generating significantly higher engagement compared to search engine ads. Recognizing this trend, Amazon swiftly reallocated a larger portion of the advertising budget to these high-performing social media platforms and fine-tuned its targeting strategy to focus on younger demographics. This immediate adjustment led to a substantial increase in web traffic and sales during the Prime Day event, showcasing Amazon's ability to leverage real-time data for dynamic campaign optimization.

3.Retailers Optimizing Inventory Based on Demand Spikes

Retailers face the challenge of managing inventory efficiently to meet customer demand without overstocking or understocking. Real-time analytics helps retailers monitor sales trends and adjust inventory levels instantly in response to demand spikes.

Example: Zara, a global fashion retailer known for its agile supply chain, experienced a sudden surge in demand for winter coats during an unexpected cold wave across Europe. Real-time sales data from Zara's stores and online platforms immediately highlighted the spike in purchases of specific coat styles. Leveraging its advanced inventory management system, Zara automatically initiated replenishment orders to its distribution centers and prioritized shipments to stores in high-demand regions. This rapid response ensured that stores remained stocked with popular items, preventing stockouts and maximizing sales opportunities. Zara's ability to swiftly adapt to real-time demand changes exemplifies how data-driven decision-making can optimize inventory management and meet customer expectations.

4.Financial Services Mitigating Risks with Instant Fraud Detection

Financial institutions are increasingly using real-time analytics to detect and prevent fraudulent activities. Advanced algorithms analyze transaction data in real time to identify suspicious patterns and trigger immediate alerts for potential fraud.

Example: JPMorgan Chase, a leading global financial services firm, utilizes a sophisticated real-time fraud detection system to protect its customers (Saura, 2021). When the system detects unusual transaction patterns, such as rapid transactions in different geographic locations, it instantly freezes the affected account and sends an alert to the customer via mobile notifications. In one instance, a customer's debit card was used simultaneously in New York and London, triggering the system's automatic response. Upon the customer's confirmation of unauthorized activity, JPMorgan Chase swiftly blocked further transactions and launched a security investigation. This proactive approach not only prevented financial losses but also reinforced customer trust in the bank's commitment to security.

Real-time analytics empowers businesses to make swift, data-driven decisions that enhance operational efficiency, customer satisfaction, and risk management. By monitoring social media sentiment, adjusting digital marketing campaigns, optimizing inventory, and detecting fraud instantly, organizations can stay ahead in a rapidly evolving marketplace. As technology continues to advance, the integration of real-time analytics into core business processes will become increasingly vital for achieving sustainable growth and competitive advantage.

5. OPTIMIZING CUSTOMER EXPERIENCE THROUGH DIGITAL ANALYTICS

Digital analytics plays a critical role in enhancing customer experience by providing deep insights into customer behaviors, preferences, and interactions. By leveraging data effectively, businesses can create more personalized, engaging, and seamless experiences that meet customer expectations and drive loyalty.

Personalization and Engagement Strategies Driven by Data

1. Using Behavioral Data to Tailor Content and Offers:

Behavioral data—such as browsing history, past purchases, and interaction patterns—provides valuable insights into customer preferences (Kushwaha et al., 2021). Businesses can use this data to deliver personalized product recommendations, targeted promotions, and tailored content. For example, streaming platforms like **Netflix** analyze viewing habits to suggest shows that align with a user's interests, while e-commerce giants like **Amazon** recommend products based on browsing and purchase history. This level of personalization increases engagement, improves customer satisfaction, and boosts conversion rates.

2. Implementing Dynamic Website Personalization:

Dynamic website personalization involves real-time adjustments to website content based on user behavior and preferences. This strategy ensures that each visitor experiences a version of the website tailored to their interests. For instance, online fashion retailers like **ASOS** showcase trending products or personalized discounts to returning customers based on previous shopping behaviors. Similarly, travel websites might display vacation packages related to a visitor's recent searches. Dynamic personalization enhances user engagement and encourages visitors to take desired actions, such as completing a purchase or signing up for newsletters.

Enhancing Customer Journeys with Insights from Analytics

1. Identifying Drop-Off Points in Conversion Funnels:

Analytics tools allow businesses to map out and analyze the customer journey, identifying where users abandon a website or app before completing a desired action (e.g., making a purchase). Recognizing these drop-off points enables companies to optimize website navigation, simplify checkout processes, or offer timely incentives to reduce cart abandonment. For example, an online retailer noticing high abandonment rates at the payment stage might introduce more payment options or streamline the checkout process to improve conversions.

2. Enhancing Omnichannel Experiences:

Customers today interact with brands across multiple channels—websites, mobile apps, social media, and physical stores. Digital analytics (Czvetkó et al., 2022) helps businesses integrate these touchpoints to create a seamless omnichannel experience. For instance, a customer browsing products on a brand's website might later receive personalized product recommendations via email or see targeted ads on social media. Retailers like **Starbucks** use data from their loyalty app to offer rewards, suggest products, and provide store-specific promotions, creating a consistent and engaging experience across digital and physical platforms.

6. CASE STUDIES: REAL-WORLD EXAMPLES OF DIGITAL ANALYTICS IMPACT

1. Stitch Fix: Personalizing Fashion Recommendations with Data Science

Company: Stitch Fix

Challenge: As an online personal styling service, Stitch Fix needed to deliver a truly personalized shopping experience for its customers, helping them find clothing that matched their tastes, sizes, and lifestyle without the need for traditional brick-and-mortar shopping.

Solution:

Stitch Fix employs a **hybrid model** that combines both human stylists and **advanced algorithms** to create personalized fashion recommendations for each customer. The company's recommendation system uses data on customers' body types, style preferences, brand preferences, and purchase history to offer a curated selection of clothing. Here's how they do it:

- **Customer Profile**: Users complete a detailed style quiz that helps Stitch Fix understand their preferences, sizes, and lifestyle. The algorithm then uses this information to recommend clothing that fits the customer's style and size.
- **Machine Learning Algorithms**: Stitch Fix combines data on customer preferences with machine learning techniques to suggest items based on past purchases and preferences. The system learns over time, refining the recommendations as it collects more data on individual customers' tastes.
- **Human Stylists**: The algorithmic recommendations are complemented by professional stylists who add a human touch, selecting specific pieces based on the customer's profile, feedback on previous shipments, and current trends.

Outcome:

Stitch Fix has successfully created a personalized shopping experience that has resonated with millions of users. Customers receive curated clothing that fits their preferences and lifestyle, improving satisfaction and increasing repeat purchases. The use of a combination of data science (Scott, E. M, retail2023)and human expertise has allowed Stitch Fix to grow significantly and expand its offerings.

Key Insights:

- **Personalization at Scale**: Stitch Fix blends data-driven recommendations with human curation to provide a unique, tailored experience, offering the best of both worlds.
- **Customer-Centric Approach**: The company focuses on understanding the individual, not just on clothing preferences, but also on body type, lifestyle, and personal tastes, leading to highly personalized experiences that drive loyalty.

2. Sephora: Using Personalized Recommendations to Drive In-Store and Online Sales

Company: Sephora

Challenge: Sephora, a global beauty retailer, wanted to enhance customer experience by offering personalized product recommendations both online and in-store. With thousands of beauty products to choose from, Sephora needed a way to make shopping easier and more personalized to each customer's unique needs.

Solution:

Sephora implemented a **personalized recommendation system** using a combination of in-store technology, customer profiles, and online interactions. Here's how they approach personalized recommendations:

- **Beauty Insider Program**: Sephora collects customer data through its loyalty program, where customers create profiles and log their preferences, product reviews, and purchase histories. This data is used to provide personalized product recommendations.
- **AI-Driven Recommendations**: In addition to human expertise, Sephora uses AI to suggest products based on customer profiles, past purchases, and even online reviews. For example, the app suggests makeup products based on a user's skin tone and preferences.
- **In-Store Technology**: Sephora uses technology like the **Color IQ** tool, which analyzes a customer's skin tone to suggest foundation shades, and **Virtual Artist**, which allows customers to try on makeup virtually through augmented reality.

Outcome:

By combining data from both online and in-store experiences, Sephora has created a seamless, personalized shopping experience. The company reports higher sales, improved customer satisfaction, and increased loyalty due to the tailored experience provided by the recommendation system.

Key Insights:

- **Omni-Channel Personalization**: Personalized recommendations should not be limited to one channel. Offering a consistent experience across both physical stores and digital platforms increases customer satisfaction and sales.
- **Customer Empowerment**: Sephora's approach empowers customers to make better, personalized purchasing decisions through AI-driven tools and personalized product suggestions.

7.ETHICAL CONSIDERATIONS AND PRIVACY

The increasing reliance on digital analytics necessitates a strong focus on ethical considerations and data privacy. Organizations must navigate a complex landscape of regulatory frameworks, ethical dilemmas, and technological challenges to ensure responsible data use.

Data Privacy Laws and Regulations

Digital analytics operates within a legal framework designed to protect consumer privacy. The two most influential regulations are:

- **General Data Protection Regulation (GDPR)**: Enforced by the European Union, GDPR mandates transparency in data collection, requires user consent, and grants individuals the right to access and delete their data.
- **California Consumer Privacy Act (CCPA)**: A U.S.-based regulation that gives consumers control over their personal data, including the right to opt out of data selling and demand data deletion.

Other important regulations include Brazil's LGPD, China's PIPL, and India's DPDP Bill, each of which imposes unique compliance requirements.

Data Anonymization Techniques

To comply with privacy laws while maintaining data usability, organizations employ various anonymization methods, such as:

- **Data Masking**: Replacing sensitive information with synthetic or scrambled data to protect individual identities.
- **Differential Privacy**: Adding statistical noise to datasets, ensuring that individual contributions cannot be inferred.
- **Tokenization**: Substituting sensitive data with unique identifiers, preventing unauthorized access.
- **De-identification**: Removing personal identifiers from data sets, reducing re-identification risks.

Ethical Challenges in Digital Analytics

Despite regulations and anonymization efforts, ethical dilemmas persist in data analytics, including:

- **Algorithmic Bias**: AI-driven analytics may reinforce social and economic biases if not carefully monitored.
- **Informed Consent**: Many consumers remain unaware of how their data is collected and used, leading to concerns about data exploitation.
- **Data Monetization Ethics**: Companies profit from user data, raising questions about fair compensation and consumer rights.

Case Studies of Unethical Data Use

1. **Cambridge Analytica Scandal**: This case exemplifies the unethical use of personal data, where millions of Facebook users' information was harvested without consent for political manipulation.
2. **Google's Project Nightingale**: A secretive initiative to collect patient health data without proper disclosure, raising concerns about transparency and informed consent.
3. **Target's Predictive Analytics Controversy**: Target's analytics model inferred a teenager's pregnancy before her family knew, highlighting concerns about data sensitivity and unintended consequences.

Best Practices for Ethical Data Use

Organizations can implement the following strategies to uphold ethical standards:

- **Transparent Data Policies**: Clearly communicate how data is collected, stored, and utilized.
- **Ethical AI Audits**: Regularly assess AI models for biases and fairness.
- **Consumer Control Over Data**: Provide users with easy access to opt-out mechanisms and data deletion options.
- **Privacy-by-Design**: Embed privacy considerations into analytics workflows from the outset.

By integrating these considerations, businesses can foster trust with consumers, comply with regulatory mandates, and ensure responsible digital analytics practices.

8.INTEGRATION WITH EMERGING TECHNOLOGIES

The rapid advancement of digital analytics is closely intertwined with emerging technologies such as blockchain, the Internet of Things (IoT), and edge computing. These technologies enhance data security, real-time processing, and efficiency, making analytics more reliable and scalable.

Blockchain for Data Security and Transparency

Blockchain technology provides a decentralized and immutable ledger, ensuring secure and tamper-proof data transactions. Its impact on digital analytics includes:

- **Data Integrity**: Prevents unauthorized modifications, ensuring that analytical insights are based on accurate data.
- **Smart Contracts**: Automates data-sharing agreements while maintaining compliance with privacy regulations.
- **Decentralized Identity Management**: Enables users to control access to their data, reducing unauthorized tracking.

    Example: In the healthcare industry, blockchain secures patient records, ensuring that only authorized entities can access sensitive health data.

    IoT and Real-Time Data Processing

IoT generates massive volumes of real-time data from interconnected devices, enhancing digital analytics by:

- **Enabling Predictive Maintenance**: Sensors in manufacturing plants collect data to predict equipment failures before they occur.

- **Smart Cities Development**: IoT devices optimize traffic flow, energy consumption, and waste management.
- **Retail Analytics**: Smart shelves and beacons collect in-store customer behavior data to enhance marketing strategies.

  Example: Smart city initiatives leverage IoT data from traffic sensors and public transport systems to reduce congestion and enhance urban planning.
  Edge Computing for Faster Data Processing

Edge computing brings data processing closer to the source, reducing latency and bandwidth usage. Its benefits for digital analytics include:

- **Lower Latency**: Real-time analytics with reduced dependency on cloud-based solutions.
- **Enhanced Security**: Localized data processing minimizes exposure to cyber threats.
- **Scalability**: Allows businesses to handle large-scale data operations efficiently.

  Example: In autonomous vehicles, edge computing processes sensor data instantly, enabling quick decision-making for improved safety.

The integration of blockchain, IoT, and edge computing is revolutionizing digital analytics. These technologies not only enhance security and real-time processing but also unlock new opportunities for efficiency and innovation. By leveraging these advancements, organizations can make more informed, secure, and timely decisions in various industries, from healthcare to smart cities and beyond.

9. CULTURAL AND GEOGRAPHIC VARIATIONS

Digital analytics strategies vary significantly across different cultural and geographic regions due to diverse consumer behaviors, regulatory landscapes, and technological adoption rates. Understanding these variations is critical for organizations operating in global markets.

Cultural Influences on Analytics Strategies

- **Consumer Behavior Differences**: In Western markets, data-driven decision-making is often preferred, while in many Asian markets, traditional business intuition plays a stronger role.
- **Privacy Perceptions**: European consumers, influenced by GDPR, tend to be more privacy-conscious, whereas some emerging markets

have a higher tolerance for data collection in exchange for personalized services.
- **Language and Localization**: Analytics tools must support multiple languages, dialects, and cultural nuances to provide accurate insights.

Example: E-commerce personalization strategies differ across regions—U.S. retailers leverage extensive user data for recommendations, whereas Chinese platforms like Alibaba focus more on influencer-driven commerce.

Regulatory Variations and Compliance Challenges

- **Europe**: GDPR enforces strict user consent and data protection measures.
- **United States**: Sector-specific regulations (e.g., HIPAA for healthcare, CCPA for consumer data) shape analytics implementation.
- **Asia and Emerging Markets**: Countries like China have stringent data localization laws requiring companies to store and process data within national borders.

Example: Companies like Google and Facebook face challenges operating in China due to data sovereignty laws, requiring local partnerships to comply with regulations.

10.CHALLENGES AND LIMITATIONS

Despite the advantages of digital analytics, organizations encounter several hurdles that can hinder effective implementation. Addressing these challenges is essential for maximizing the value of analytics investments.

1. Data Silos and Integration Issues
   - **Problem**: Many companies struggle with fragmented data sources spread across departments and systems.
   - **Solution**: Implementing centralized data lakes and adopting API-based integrations to unify datasets.
2. Expertise Gaps in Analytics and AI
   - **Problem**: The shortage of skilled data professionals limits the ability to derive actionable insights.
   - **Solution**: Upskilling internal teams through training programs and leveraging AI-driven automated analytics tools.
3. High Implementation Costs
   - **Problem**: Deploying advanced analytics infrastructure requires significant financial investment.

- **Solution**: Cloud-based analytics solutions and AI automation can reduce costs and improve accessibility.

  Example: Small businesses in emerging markets often rely on open-source analytics tools like Google Analytics and KNIME to reduce costs.

11.FUTURE TRENDS IN DIGITAL ANALYTICS

As digital analytics continues to evolve, emerging technologies are reshaping the industry, providing new opportunities for enhanced decision-making and efficiency.

1. Augmented Analytics
   - AI-driven automation enhances data analysis by identifying trends, generating insights, and suggesting actions without manual intervention.
   - **Example**: Tools like Salesforce Einstein and Microsoft Power BI now incorporate augmented analytics to simplify complex data analysis.
2. Explainable AI (XAI)
   - As AI-driven analytics models become more complex, transparency and interpretability are critical.
   - **Example**: Financial institutions use XAI to ensure compliance with regulations by making AI-based credit scoring models more interpretable to regulators.
3. Quantum Computing's Impact on Data Analysis
   - Quantum computing has the potential to process massive datasets at unprecedented speeds, revolutionizing analytics in fields like cryptography, genomics, and logistics.
   - **Example**: Google's Sycamore quantum processor demonstrated the ability to solve computations exponentially faster than classical computers, paving the way for future analytics breakthroughs.

By staying ahead of these trends, organizations can leverage advanced analytics technologies to gain competitive advantages, drive efficiency, and navigate the challenges of an increasingly data-driven world.

REFERENCES

Ajitha, S., Geevarathna, , & Huxley, S. (2024). Exploring public perception, awareness, and satisfaction with AI applications in Karnataka, India: The role of individual characteristics and media influence. *International Journal of System Assurance Engineering and Management*, 1–16. DOI: 10.1007/s13198-024-02594-3

Akter, S., Michael, K., Uddin, M. R., McCarthy, G., & Rahman, M. (2022). Transforming business using digital innovations: The application of AI, blockchain, cloud and data analytics. *Annals of Operations Research*, *308*(1-2), 1–33. DOI: 10.1007/s10479-020-03620-w PMID: 35935743

Barbosa, B., Saura, J. R., Zekan, S. B., & Ribeiro-Soriano, D. (2024). RETRACTED ARTICLE: Defining content marketing and its influence on online user behavior: a data-driven prescriptive analytics method. *Annals of Operations Research*, *337*(S1, Suppl 1), 17–17. DOI: 10.1007/s10479-023-05261-1

Choi, T. M., Dolgui, A., Ivanov, D., & Pesch, E. (2022). OR and analytics for digital, resilient, and sustainable manufacturing 4.0. *Annals of Operations Research*, *310*(1), 1–6. DOI: 10.1007/s10479-022-04536-3

Czvetkó, T., Kummer, A., Ruppert, T., & Abonyi, J. (2022). Data-driven business process management-based development of Industry 4.0 solutions. *CIRP Journal of Manufacturing Science and Technology*, *36*, 117–132. DOI: 10.1016/j.cirpj.2021.12.002

El Khatib, M. M., Alzoubi, H. M., Ahmed, G., Kazim, H. H., Al Falasi, S. A. A., Mohammed, F., & Al Mulla, M. (2022, February). Digital transformation and SMART-the analytics factor. In *2022 International Conference on Business Analytics for Technology and Security (ICBATS)* (pp. 1-11). IEEE.

Gui, Y., Cai, Z., Yang, Y., Kong, L., Fan, X., & Tai, R. H. (2023). Effectiveness of digital educational game and game design in STEM learning: A meta-analytic review. *International Journal of STEM Education*, *10*(1), 36. DOI: 10.1186/s40594-023-00424-9

Kliestik, T., Novak, A., & Lăzăroiu, G. (2022). Live shopping in the metaverse: Visual and spatial analytics, cognitive artificial intelligence techniques and algorithms, and immersive digital simulations. *Linguistic and Philosophical Investigations*, *21*(0), 187–202. DOI: 10.22381/lpi21202212

Kushwaha, A. K., Kumar, P., & Kar, A. K. (2021). What impacts customer experience for B2B enterprises on using AI-enabled chatbots? Insights from Big data analytics. *Industrial Marketing Management, 98*, 207–221. DOI: 10.1016/j.indmarman.2021.08.011

Lei, Y., Guo, Y., Li, X., & Jing, Z. (2023). Space economy: A new frontier of information systems, analytics and digital commerce. *Industrial Management & Data Systems, 123*(2), 616–629. DOI: 10.1108/IMDS-05-2022-0316

Loscher, G., & Bader, V. (2022). Augmenting a profession: How data analytics is transforming human resource management. In *Digital Transformation and Institutional Theory* (pp. 87–110). Emerald Publishing Limited. DOI: 10.1108/S0733-558X20220000083004

Lutfi, A., Alsyouf, A., Almaiah, M. A., Alrawad, M., Abdo, A. A. K., Al-Khasawneh, A. L., Ibrahim, N., & Saad, M. (2022). Factors influencing the adoption of big data analytics in the digital transformation era: Case study of Jordanian SMEs. *Sustainability (Basel), 14*(3), 1802. DOI: 10.3390/su14031802

Madhala, P., Lin, T., Li, H., & Helander, N. (2023). Data analytics capabilities for digital service development: A case study. *PACIS 2023 Proceedings*.

Mariani, M. M., & Nambisan, S. (2021). Innovation analytics and digital innovation experimentation: The rise of research-driven online review platforms. *Technological Forecasting and Social Change, 172*, 121009. DOI: 10.1016/j.techfore.2021.121009

Mohamed, G., Zahra, F. F., Najwa, Z., Soumaya, O., Yassine, K., Chakir, A., & Mohamed, A. (2024). Enhancing Immersive Virtual Shopping Experiences in the Retail Metaverse Through Visual Analytics, Cognitive Artificial Intelligence Techniques, Blockchain-Based Digital Assets, and Immersive Simulations: A Systematic Literature Review. *Engineering Applications of Artificial Intelligence*, 305–318.

Niu, Y., Ying, L., Yang, J., Bao, M., & Sivaparthipan, C. B. (2021). Organizational business intelligence and decision making using big data analytics. *Information Processing & Management, 58*(6), 102725. DOI: 10.1016/j.ipm.2021.102725

Ochoa, X., & Wise, A. F. (2021). Supporting the shift to digital with student-centered learning analytics. *Educational Technology Research and Development, 69*(1), 357–361. DOI: 10.1007/s11423-020-09882-2 PMID: 33250614

Petrova, M., Popova, P., Popov, V., Shishmanov, K., & Marinova, K. (2021, October). Digital ecosystem: Nature, types and opportunities for value creation. In *International Scientific Conference on Innovations in Digital Economy* (pp. 71-85). Cham: Springer International Publishing.

Saura, J. R. (2021). Using data sciences in digital marketing: Framework, methods, and performance metrics. *Journal of Innovation & Knowledge*, *6*(2), 92–102. DOI: 10.1016/j.jik.2020.08.001

Scott, E. M. (2023). Framing data science, analytics and statistics around the digital earth concept. *Environmetrics*, *34*(2), e2732. DOI: 10.1002/env.2732

Song, H., Li, M., & Yu, K. (2021). Big data analytics in digital platforms: How do financial service providers customise supply chain finance? *International Journal of Operations & Production Management*, *41*(4), 410–435. DOI: 10.1108/IJOPM-07-2020-0485

Sponder, M., Khan, G. F., & Feroz, A. K. (2018). *Digital analytics for marketing*. Routledge.

Vollrath, M. D., & Villegas, S. G. (2021). Avoiding digital marketing analytics myopia: Revisiting the customer decision journey as a strategic marketing framework. *Journal of Marketing Analytics*, *10*(2), 106–113. DOI: 10.1057/s41270-020-00098-0

Zhang, D., Pee, L. G., Pan, S. L., & Cui, L. (2022). Big data analytics, resource orchestration, and digital sustainability: A case study of smart city development. *Government Information Quarterly*, *39*(1), 101626. DOI: 10.1016/j.giq.2021.101626

Zhou, C., Xu, J., Miller-Hooks, E., Zhou, W., Chen, C. H., Lee, L. H., Chew, E. P., & Li, H. (2021). Analytics with digital-twinning: A decision support system for maintaining a resilient port. *Decision Support Systems*, *143*, 113496. DOI: 10.1016/j.dss.2021.113496

Chapter 9
A Blockchain-Enabled Arbitrable Smart Hiring System (BEASH)

Jayanta Banerjee
https://orcid.org/0000-0002-2079-8983
City University, Ajman, UAE

Ajit Danti
https://orcid.org/0000-0002-2462-1771
City University, Ajman, UAE

Ayman Bassam Nassoura
https://orcid.org/0000-0003-4223-3948
City University, Ajman, UAE

M. R. Dileep
https://orcid.org/0000-0002-0010-7276
Nitte Meenakshi Institute of Technology, India

ABSTRACT

The swift advancement of the digital economy requires a recruitment system that is safe, effective, and easy to follow. This chapter describes the Blockchain-Enabled Smart Hiring System, which modernizes the hiring process by replacing traditional methods with Blockchain technology, smart contracts, and decentralized arbitration. This model helps reduce fraud attempts, protects data, and automates the resolution of conflicts by harnessing blockchain's immutability and security. This research provides an in-depth study regarding the architecture of BEASH, implementation, and performance evaluation. The study's results revealed higher efficiency, greater stakeholder trust, and less time for conflict resolution. Although scalability and

DOI: 10.4018/979-8-3373-1320-7.ch009

regulatory compliance continue to be challenged, this approach demonstrates great promise for advances in recruitment and further development of digital hiring systems.

1. INTRODUCTION

The recruitment processes in many organizations are continuously lagging behind all other evolving technologies. However, in every other field and aspect of the workforce, there has been a constant change with the rise of the digital economy; automation is now a critical part of all operations in every sector. Alongside these improvements, there is a noticeable absence of improvement in the average recruiting and hiring processes (Sarikaya, 2021). These traditional systems have long cycle times of action, poor decision-making on candidate filtering, and total separation between potential employers and employees. Consequently, there is a need for a modern recruitment technology that guarantees transparency, security, and efficiency in the hiring process.

The issues presented can be solved conveniently using Blockchain technology owing to its immutable and decentralized features. Blockchain technology allows firms to incorporate candidates' qualifications through credentialing systems, which guarantees trust, sets up recruitment automation through smart contracts, and gives a verifiable record of the candidates (Kişi, 2022). Additionally, outsourcing specific dispute resolution functions through Blockchain increases the efficiency of these processes, saving time and money compared to traditional methods. The proposed systems enable smart hiring, a more flexible and responsive solution that meets the requirements of the digital economy in terms of automation, data protection, and process efficiency. Besides operational efficacy, blockchain-enabled hiring is crucial in remote working setups where employees from different countries, including freelancers, require adequate systems to vet a person's credentials and facilitate stakeholder trust. The absence of reliable mixed identity verification leads to diploma mills, falsified employment histories, and impersonation. Employing blockchain technology would create a trusted hiring environment by using blockchain as a single source of truth for employment records (Rhemananda et al., 2020).

Every part of a digital economy relies on advanced solutions powered by technology to improve efficiency, security, and user experience. In the case of traditional hiring practices, these qualities often remain unfulfilled, which leads to candidate deception, drawn-out hiring timelines, and low levels of responsibility. Blockchain can offer employers and candidates a more reliable and verifiable record. For example, employers can directly access credential information on the blockchain and avoid costly background checks, which helps ease administrative work and its costs (Verhoef et al., 2019). In the same manner, job seekers can turn out to be more credible thanks to the existence of verifiable digital identities and their undeletable records. Automating hiring processes with smart contracts is a fundamental feature of the workflow. The self-executing agreements guarantee that the engagements were adhered to without the intervention of third parties. Salary payments, probationary conditions, and even performance evaluations can all be bundled within a smart contract, which reduces human input and increases productivity output. Such automation is greatly needed for gig economy platforms and remote work, where there is an excellent need for contractual openness.

Another significant problem in traditional hiring is conflict resolution. Debates about contracts, wages, and fraud can result in lengthy, costly, and complex legal proceedings (McKnight et al., 2015). The blockchain offers a valuable solution using decentralized, transparent, and efficient decision-making systems for complex arbitration processes. Unlike other authorities or legal systems, this allows the resolution of disputes via recorded consensus arbitrations on the blockchain. This method guarantees impartiality and fairness while drastically cutting resolution time and costs (Kumar et al., 2024).

Furthermore, the need for diversity and inclusivity in companies has grown equally, making it essential for organizations to readjust the biases in such procedures. Using blockchain identity verification tools enables organizations to bypass the biases that demographic attributes bring and admit candidates based on skills and experience through open-blind recruitment methods. This fosters an ideal system of hiring that is free from subjective judgment (Mishra & Venkatesan, 2021). The prospective future of recruitment is portrayed with the blending of AI and blockchain technology. With AI, powerful estimation tools can evaluate candidates' suitability, anticipate their job performance, and suggest the best fit for a position. When combined with the security and transparency provided by blockchain in the form of AI, these estimations can change hiring for the better by making it objective and data-oriented. Furthermore, using zero-knowledge proof can enhance data privacy because the technique allows for verifying sensitive information without revealing it and still maintains the integrity of the hiring process.

Hiring with the help of Blockchain technology has its fair share of problems. Privacy and trust remain complex factors to negotiate with. A promising concern is how Blockchain technology can offer a solution for compliance with employment legislation, which changes from region to region. Concerns over the decentralization of blockchain lead to issues of trust and liability, which require solutions for the technology to be fully adopted (Michailidis, 2018). With blockchain technology, it is possible to create a hiring ecosystem that is more efficient, equitable, and trustworthy, which is why it is necessary for businesses and job seekers as they rapidly move to digital platforms. Hiring processes can now be aligned with the modern workforce due to the BEASH framework. Using blockchain technology in hiring can streamline recruitment, build stakeholder trust, and reduce operational inefficiencies. Improvements in hybrid blockchains, Interoperability with existing human resource management systems, and government-endorsed blockchain identity projects will expand the usability of blockchains, establishing them in modern HR practices (Ølnes et al., 2017). The versatility of blockchains in hiring indicates that they will be refined over the coming years, and these changes will quickly grant early adopters an edge over the competition in recruiting new talent while also adapting to new regulations of the digital work environment.

Blockchain will function as a primary technology that improves security, efficiency, and transparency as hiring processes continue transforming the digital economy. The proposed framework is a step towards tackling these issues by providing a modern solution to the ever-evolving recruitment processes. In fulfilling the present needs while anticipating future changes, blockchain-enabled hiring systems will provide vital assistance toward the future of work (Ertz & Boily, 2019).

2. MOTIVATION OF THE RESEARCH

The orthodox recruitment processes are beset by inefficiencies, which cause low productivity and erode trust in the recruitment process. Some common problems faced include slow candidate credential verification procedures, lack of decision-making transparency, and, even more problematic, candidates misrepresenting their qualifications. Such problems can create hiring errors, increase turnover, and reduce overall organizational efficiency and effectiveness. In a time where talent acquisition is critical to gaining a competitive edge, it makes sense to address these problems within the organizational context.

During the last decade, Blockchain technology emerged as the game-changing answer to these problems as it provides a decentralized, tamper-proof, and transparent system that can motivate a complete rethinking of the hiring procedure (Lizcano et al., 2019). Companies can fully automate their recruitment processes by incorporating blockchain, smart contracts, decentralized arbitration, decision-making, and contract approval. With smart contracts, real-time candidate qualification verification is possible by storing their credentials on a tamper-proof secure ledger. Not only does this technique save time and money usually spent on background checks, but it also prevents fraud and misrepresentation of documents during hiring.

Moreover, decentralized arbitration provides an economical and effective means of resolving disputes, which can potentially enhance the candidate experience. In a typical employment system, disputes tend to develop into complex, costly legal conflicts that are nearly impossible to resolve quickly. The decentralized manner of arbitration enables quick and impartial resolution of disputes, which nurtures a balanced setting for employers and candidates (Rautrao & Gupta, 2024).

The motivation behind this type of research stems from the urgency to resolve these critical issues by providing a smart hiring ecosystem that boosts stakeholder trust, lowers hiring expenses, and increases effectiveness in recruitment. With the advancement of blockchain technology, organizations can now streamline their hiring processes and nurture transparency within the organization's culture. This approach will undoubtedly disrupt the recruitment sector by enhancing inclusivity and fairness within the industry while significantly improving the hiring experience for all parties (Peisl & Shah, 2019).

In essence, this research sets out to develop a genuine and efficient recruitment frame suitable for today's workforce and the future. Because recruiters focus on agility and efficiency when hiring new employees, integrating blockchain technology can fundamentally change how organizations capture, screen, and hold onto new hires.

3. BLOCKCHAIN HIRING SYSTEM

Blockchain is a decentralized, distributed ledger technology that records transactions securely and transparently. It consists of linked blocks containing data secured by cryptographic hashing. Transactions are verified by a consensus mechanism, making blockchain tamper-resistant and trustless. It powers cryptocurrencies, smart contracts, and decentralized applications across various industries.

Blockchain consists of multiple nodes (users) that communicate to process transactions, such as sending cryptocurrencies. Each node has a public and private key, which generates a digital signature for transaction requests. All nodes validate the transaction before adding it to a new block in the blockchain. Each block contains a unique hash and the hash of the previous block, ensuring security through cryptographic linking. Blockchain follows sequential transaction processing, verifying hashes to maintain data integrity. Any modification alters the hash, breaking the chain and invalidating the transaction, making blockchain highly secure and resistant to tampering. The following figure shows the steps involved in blockchain processing.

Figure 1. Overview of Blockchain Transaction Flow in the Hiring Process

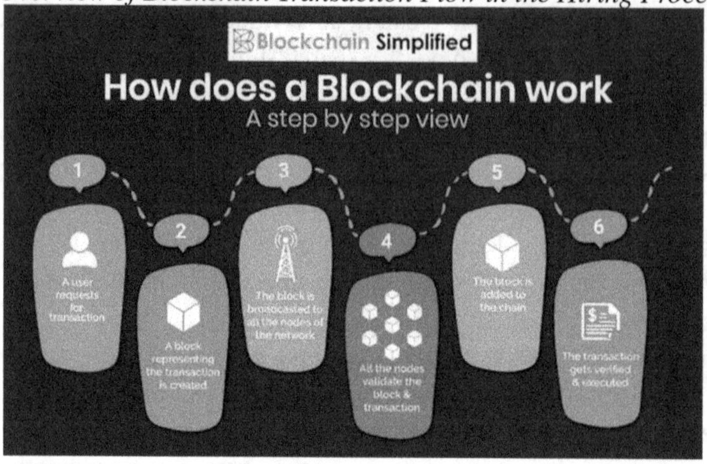

(Courtesy: https://blockchainsimplified.com/blog/what-is-blockchain-technology/)

Blockchain technology is revolutionizing the digital economy and solving various issues such as security, monitoring, and efficiency during hiring. It can significantly improve recruitment processes that involve multiple parties, take ages to get verified, and can easily be manipulated through forgery. A trustless environment is possible with blockchain. Employees can feel secure knowing that everything from their personal records to their hiring contracts will be stored securely, not be alterable, and can be verified instantly. This would eliminate the need to trust a mediator and foster healthier relationships between employers and potential employees. Efforts in hiring could be more automated, and remote work could quickly become a norm. The newer decentralized hiring structure allows candidates to be hired without worrying about fraud and hassle (Onik et al., 2018).

The recruitment wrinkle within blockchain technology stems from its ability to provide verifiable credentials, guarantee data integrity, and facilitate automated contracting. Most crucial procedures tend to involve diverse background checks, which are protracted and expensive. Blockchain technology enables candidates to store confirmed educational qualifications and professional experiences on a global ledger rather than relying on separate verifying institutions. Therefore, expenses are cut down alongside the chances of credential deception. Furthermore, fraud eradication, one of the features offered by blockchain technology, stipulates that information recorded on the blockchain is immutable. Thus, this conveniently shifts the security of the hiring exercise's information to a higher level. This degree of rupture is also helpful to applicants who do not want to be redundant by offering the same credentials to several employers, which drastically simplifies the whole hiring procedure.

The smart contract is one of the core aspects enabling the hiring function in a blockchain ecosystem. Smart contracts are self-executing contracts with the terms of the agreement between buyer and seller being directly written into lines of code, thus allowing the contract to be enforced automatically without the need for an intermediary. In the hiring process, smart contracts can autonomously confirm a candidate's qualifications, issue an offer letter if the candidate meets the criteria, and manage the payment process. For example, Hire Vibes has integrated blockchain technology to create a more efficient recruitment process. These platforms allow businesses to set job criteria, and blockchain technology verifies applicants' credentials while the smart contract takes care of the hiring and payment processes (Zarafidis et al., 2020). Using smart contracts in recruitment eliminates the excessive bureaucratic documentation processes while increasing overall efficiency because the smart contracts ensure compliance with the set hiring rules without supervision. Disputes or misinterpretations related to the employee's role, the compensation and other pertinent terms of the contract are minimized because smart contracts increase the employer's and employee's accountability towards their contractual obligations.

What is a Smart Contract?

A smart contract is a software logic that defines identity data exchange, ensuring security and immutability on the blockchain. In the BlockID ecosystem, users log in by scanning a QR code, clicking a link, or receiving an NFC signal, triggering a smart contract. This contract specifies the company's identity, required user information, and the service provided in return. Once the user consents via biometrics, identity data is retrieved, encrypted, and digitally signed for secure transmission. The smart contract ensures only authorized decryption, completing the authentication process. It enables verifiable and secure credentials for securely accessing physical locations, systems, or applications. The following figure illustrates the difference between traditional and smart contracts (Neiheiser et al., 2020).

Figure 2. Comprison between a Smart Contract V/s a Traditional Contract

(Courtesy: https://www.1kosmos.com/article/smart-contracts/)

The benefits accompanying the employment of smart contracts in the recruitment process are closely linked to the legal issues and regulatory frameworks connected with these technologies. The smart contract's enforceability includes bases such as jurisdictional recognition, while the relative General Data Protection Regulation (GDPR)-compliant data protection policies regarding candidates' information also need consideration. As smart contracts are executed on a code logic, mistakes or possible exploits within the program can result in harmful effects, raising the need for intensive testing and auditing. Moreover, the absence of a single universally acknowledged legal system governing smart contracts remains a barrier to their wider use because different regions tend to differ regarding what they consider legal. Suppose the adoption of blockchain-based recruiting becomes the norm. In that case, there is a need for collaboration between lawmakers and technological experts to formulate policies to ensure legal compliance, safety, and legitimacy.

The increasing use of smart hiring mechanisms demonstrates the adoption of highly technological methods of workforce acquisition. The digital economy requires a hiring system that is much more advanced, as well as efficient enough to accommodate remote work and the ever-evolving employment landscape. Smart hiring uses blockchain, AI, and automation to simplify the recruitment process while minimizing discrimination and optimizing the provision of candidates. Recruitment solutions powered by AI sift through enormous amounts of data of potential candidates from job databases and select those who would best fit the job opening with the greatest accuracy. At the same time, blockchain technology guarantees that data is protected and that there are no ways to tamper with it. HR managers, after all, have the administrative burden of verification and compliance, so they shift towards more strategic tasks linked to decision-making and employee engagement. This is especially important for rapid recruitment of entire organizations that have to deal with the resulting massive workload of thousands of applications while providing high accuracy and impartiality in candidate selection.

In addition, it is essential to consider the ethical aspects of smart hiring. This concern stems from the use of artificial intelligence in selecting the candidates. Automated candidate selection processes pose risks of screening bias and discrimination. Equity and inclusion must be ingrained in AI models, mitigating discrimination stemming from preconceived notions. To create equitable hiring environments where certain groups are neither favored nor excluded, the decision-making process from AI must remain open.

Furthermore, blockchain use in the recruitment process should be scrutinized because of privacy and control over the information. Candidates must exercise protection against the misuse of their digital employment profiles, limiting access to their records. Blockchain technology can address these problems by providing a decentralized identity management system framework where individuals own their professional information and can control who accesses it.

Advancements in technology alter how HR professionals perform their tasks. Instead of concentrating on manually conducting and acquiring talent, HR processes are now pivoting towards more modern methods like overseeing technology-based recruitment systems, analytics, and compliance management. Furthermore, HR experts are also required to acquire additional skills like managing blockchain-enabled recruitment platforms and knowing the new legal context around smart contracts. There is an uneven pace of traditional and modern hiring practices across industries. Early adopters are technology, finance, and consulting industries, while manual-intensive sectors still cling to conventional hiring practices, supplementing them with digital technology. Still, as blockchain-enabled hiring systems become more intuitive and easier to use, non-digital industries will start experimenting with them for workforce management.

A core benefit of smart hiring is to mitigate employee turnover by ensuring improved job fit and job satisfaction among employees. Employers are likely to retain employees better because blockchain-enabled hiring systems allow for informed hiring through verifiable credentials and transparent recruitment processes. Furthermore, for remote employees, blockchain allows for the cross-border verification of credentials, automates payroll processing, and ensures proper enforcement of contracts (Kral et al., 2022). A trust-based employment model can be established with blockchain verifying work history, certification, and performance review. Instead of relying on misleading resume claims, job seekers are treated based on facts that cannot be changed.

AI facilitates the analysis of a candidate's potential and cultural fit, while blockchain secures the integrity of the data. Amalgamating AI and blockchain with machine learning will facilitate enhanced predictive hiring models, streamlining recruitment. Internationally accepted legislation for smart contracts will empower compliance and trust building within the digital recruitment drives based on blockchain technology. Technology providers, governments, and regulatory authorities must work together to develop a formal framework that positions innovation as a linchpin while offering a legal shield. Moreover, the employment markets will shift towards a more digital economy, further enhancing recruitment processes' efficiency, adaptability, and transparency as organizations adopt blockchain in hiring.

What smart contracts can offer is tremendous value from the societal perspective as it can completely transform hiring processes across countries. Blockchain technology can aid in amplifying the fairness of hiring practices within an economy by removing credential fraud and enabling accurate merit hiring. The advantages of this technology can be felt in developing nations that struggle with credential authentication as a barrier to employment. Making this information available on a cross-border blockchain enables easier access for qualified job seekers and helps address the imbalance between talent and opportunity.

Hiring through blockchain is undoubtedly a breakthrough in recruitment processes for the digital economy. Even with existing issues such as ethical concerns, scalability, and regulatory requirements, organizations adopting a blockchain-first approach to recruitment will realize greater net gains. The changes companies make in developing these solutions will result in improved clarity and efficiency and greater ease of access regarding employment opportunities and blockchain-based hiring solutions. Change is required in the treatment of employment contracts. Rather than seeing it as an attachment to the employee profile, contracts must be treated as records that sanction initiating employment services in the ever-growing digital ecosystem (Peisl & Shah, 2019).

4. OBJECTIVES

Creating a holistic blockchain technology system aligned toward a so-called smart hiring process is the major objective of this research. Building a decentralized system integrated with smart contracts and arbitrable methods will be the goal of this particular system. The framework is expected to enable hiring companies and candidates to establish trustless agreements using smart contracts as it automates and eliminates the use of mediators. Furthermore, adding adversarial methods will improve the framework's effectiveness as it provides a standard way of dealing with conflicts that may arise during the hiring process. This enhanced methodology seeks to improve a company's recruitment processes so that there is more trust and fairness in the hiring process.

Another important goal of this research is to improve the transparency and security of the smart hiring system using blockchain technology. Blockchain guarantees data integrity because all transactions are recorded on a public ledger that is constantly accessible to all participants. Users can trust a system's processes, allowing them to view the whole hiring process and the credentials of candidates who apply for vacant positions in real time. Moreover, the system will implement sophisticated cryptographic mechanisms to protect sensitive information, thus ensuring safe and confidential data storage. The proposed system, therefore, aims to improve transparency and security and, in doing so, reduce instances of fraud and data manipulation that are highly exposed in normal hiring processes (Gammon, 2018).

The study also attempts to implement automated dispute resolution processes on the smart hiring model by employing decentralized arbitration systems. In traditional legal systems, resolving disputes involves long and costly processes, often making hiring less efficient. The new system could allow quick and impartial conflict resolution without traditional legal frameworks by employing decentralized arbitration systems. The aim is to achieve a situation where both the employers and the candidates are fully satisfied with the hiring process, and if there are disagreements, they can be resolved without delay.

Finally, in this regard, one of the main goals of this work is to investigate the effectiveness of the blockchain-based hiring system that is put forward in this work through certain experimental investigations. The estimate will cover a system's performance in accounts' of volume, level of flexibility, reliability, and effectiveness in the real world, which directly affect the construct. The effort aims to determine shortcomings in improvement and prove the system's operational capacity in performing simulation processes. This goal is intended to test the system's performance and the users' experiences and satisfaction with the system, considering the stakeholders' reactions to the proposed framework.

5. LITERATURE REVIEW

The use of blockchain technology in hiring processes has come under scholarly scrutiny from several angles, such as the supervision of data in a blockchain, hiring automation through smart contracts, and blockchain-based conflict resolution arbitration. Literature is organized in these four areas: Blockchain for Secure and Transparent Hiring, Smart Contracts for Process Automation, Decentralized Arbitration for Dispute Resolution, and Challenges and Gaps in Blockchain-Based Hiring.

5.1 Moving from Traditional Hiring to Blockchain-Based Systems

Organizations need to take a strategic phased approach to move from traditional hiring processes to blockchain based systems. The first step is creating awareness and training, where HR and hiring managers with 'blockchain knowledge' need to comply with training requirements about blockchain's potential and implementation problems. The next stage is conducting pilot projects where organizations test integrating blockchain on a small-scale basis, for example, verification of employment or credential authentication, before moving to full-scale adoption. Regulatory alignment is of great importance, meaning that organizations need to work with legal consultants to determine if hiring using blockchain technology will breach local and international employment laws. Stakeholder trust can be achieved when user interfaces and interoperability standards are adopted to allow integration into existing HR management systems. In an ongoing manner, organizations need to evaluate their hiring models based on automation-driven new efficiencies added, regulatory compliance, and transparency in the hiring process to improve those models. Through a compliance-centric, technology-centric, and gradual approach, organizations can seamlessly integrate blockchain-enabled hiring while managing all risks involved (Kişi2022).

The use of blockchain in smart hiring could improve recruitment trust and efficiency, but legal and regulatory challenges remain an obstacle. Issues regarding General Data Protection Regulation (GDPR), employment law, and the adaptability of smart contracts need novel approaches like hybrid compliance frameworks or supranational regulatory collaboration. Constructive legislation needs to be established by public authorities and private entities to delineate the instructions that govern the use of blockchain technologies in hiring without violating principles of nondiscrimination and equal opportunity. Further investigation is needed to develop effective compliance and analyze changes in legislation concerning the use of blockchain technology in hiring (Clohessy & Acton, 2019).

5.2 Blockchain for Secure and Transparent Hiring

The integration of blockchain into hiring processes is motivated by the ability to secure and protect the integrity of the hiring procedures. The employment of traditional hiring systems is undermined by credentialing fraud, absence of credentials verification, and selective bias in recruitment. With blockchain technology, the problem of credentialing fraud is addressed by providing an immutable storage system that holds candidate credentials, their employment history, and their performance records. These records can be kept and verified without a third party. Evidence in the literature shows that hiring fraud, specifically in credentialing, is reduced in hiring platforms that employ blockchain technology. However, using decentralized identity systems does away with the tedious manual checking of background information, thus shortening the time spent on hiring and increasing the level of trust in the employers (Jeong & Choi, 2019).

With blockchain technology, recruiters are able to extract verified work history without any data tampering which adds to the system's accountability. Apart from verification and transparency, researchers hypothesize that blockchain can enhance a marketplace by decentralizing it into a system whereby seekers of employment self-manage their credentials, work history, and reviews (Yi et al., 2020). This claim is also bolstered through real-world implementations. Estonia's digital identity system has successfully utilized blockchain for secure nationwide employment verification, automating hiring and reducing fraudulent employment claims (Hardy, 2022). Blockchain is also used in Singapore OpenCerts to provide proof of education and employment in a non-tamperable format, thus increasing trust in government issuance (Khan et al., 2021). Similarly, the UAE incorporated blockchain in the Emirates Blockchain Strategy 2021, which aimed to digitize government records of citizens' employment and professional certifications to increase transparency and efficiency (Petratos et al., 2020). While facing challenges related to data privacy and scalability, these government initiatives strongly showcase blockchain technology's potential in transforming recruitment and workforce management.

5.3 Smart Contracts for Process Automation

Automation is essential in today's recruitment systems, and smart contracts facilitate the automation of the recruitment processes. A smart contract is a computer protocol that executes and implements the contract's specifications automatically without a third party. In the hiring process, smart contracts can facilitate acceptance of an offer, salary negotiation, onboarding, and performance monitoring. Several practical analyses have supported the premise that smart contract adoption lowers operational costs and cycle time. Organizations operating recruitment systems using blockchain technology reduce recruitment period by 40% compared to those using conventional systems. Besides, smart contracts allow the employment terms to be integrated into the contract, reducing pre and post-employment conflicts on payment and other contract terms (Nzuva, 2019).

Apart from recruiting, smart contracts improve the management of employees. Blockchain technology permits linking payroll systems that guarantee automated payment of salaries without alteration after meeting the employment conditions. This eliminates errors and delays in wages, improving employee satisfaction. Critics point out that performance-based pay and other discretionary payments are examples of where a more flexible approach is needed, and smart contracts do not support that (Hanggoro et al., 2022).

5.4 Decentralized Systems of Arbitration Dispute Resolution

Resolving disputes in traditional recruitment processes is quite a problematic issue. Misinterpretations and contract breaches usually result in cumbersome litigation procedures. However, Nzuva 2019 has argued that decentralized mediation through blockchain is one of the modern solutions that can resolve issues without seeking legal services.

Decentralized systems can arbitrate through independent validators who act as reviewed and rated third parties to resolve disputes. Blockchain-distributed arbitration reduced the time needed to resolve conflicts by up to 70%. As a byproduct, this also reduced costs associated with litigation. It ensures neutrality since every judgment made is attached to a record of transactions, making biased decisions impossible. Through the interviewed literature, it was obvious that the application of blockchain arbitration is not limited to employment disputes. Michaelson (2020) demonstrates how other industries, such as supply chain management and digital assets disputes, have adopted these systems, which proves the approach application is limitless. While many researchers endorse such technologies, Michaelson (2020) warned that the success of decentralized arbitration highly depends on the number of proficient arbitrators and the users' familiarity.

5.5 Regulatory Barriers and Compliance Strategies

Blockchain technology has significantly transformed recruitment processes by fostering increased transparency, security, and efficiency. Nevertheless, its use in smart hiring systems poses serious legal and regulatory issues. Some of its pertinent issues are compliance with the GDPR, employment laws, and the compatibility of blockchain technology with legal frameworks. The very nature of blockchain technology's decentralization and immutability contradicts GDPR's erasure rights—What is often referred to as the right to be forgotten—which entails the ability to physically remove personal information from files at a person's request. The regulation also requires minimization of data, consent pre-agreement for data capture, and in this case poses a problem to blockchain's ability to boundless retention of verifiable employment records. Researchers have proposed off-chain storage solutions in which personal data is kept away from the blockchain while on-chain, only cryptographic hashes are kept as verification (Yeoh, 2017). However, questions remain regarding whether such hashing truly anonymizes the data in compliance with GDPR. The implementation of smart contracts related to employment creates additional issues because most legal systems require a human agent to administer employment contracts. Smart contracts provide higher efficiency in the automation of hiring functions, but give rise to additional problems like resolving disputes, adaptability of the contract, retention of rights of the workers, and more.

A number of jurisdictions have started changing their legislation to incorporate employment verification using blockchain technology. The European Blockchain Partnership is looking for methods to incorporate blockchain technology into legal systems that do not violate data protection and employment laws. In the U.S., some states like Vermont and Arizona accept blockchain records as legal documents, however, there is no federal clarity concerning employment laws. Singapore is in the forefront in Asia having adopted blockchain for the automated verification of credentials in compliance with the country's employment laws. The UAE's Emirates Blockchain Strategy 2021 focuses on the digitization of employee records while ensuring compliance with employment laws. All these countries are attempting to adopt blockchain technology for hiring purposes, which is still hindered by regulatory gaps as jurisdictional issues where multi-regional functions of blockchain clash with region-specific employment and data protection laws; recognition of smart contracts where most legal systems do not expressly state if smart contracts constitute binding agreements under employment law; and GDPR's right to erasure versus permanence, where data deletion required by law is at odds with immutable record-keeping (Marengo & Pagano 2023).

5.6 Obstacles and Limitations in Recruitment Using Blockchain Technology

Integrating blockchain technology for hiring has several benefits, but there are almost always some roadblocks. Adoption resistance, regulatory issues, and scal ability are dominant problems in contemporary literature. Research indicates that many companies using large volumes of resumes and applications daily may not always be able to make hiring decisions in real-time due to the speed of blockchain transactions. Nzuva (2019) also suggests that there are barriers to adopting blockchain technology because different regions have different employment laws, which could be incompatible with blockchains' decentralized approach to hiring. (Jeong & Choi, 2019) points out the absence of universal policies and regulations that would ensure compliance while taking advantage of the possibilities offered by blockchain-based hiring put forward to tell a story about standards. Moreover, there are always privacy concerns, even with stronger security measures that blockchain provides, primarily in countries with strict laws on data protection, such as GDPR in the EU.

To resolve these problems, compliance techniques involve hybrid blockchain models that assist with GDPR concerns by keeping personal data in off-chain storage while only leaving hash references on-chain (Riva 2020). The UK and Singapore offer regulatory sandboxes in which blockchain-based hiring systems can be tested in a controlled environment. Furthermore, collaborative initiatives to develop international legal norms may reduce the conflicts of compliance (Politou et al., 2019.)

5.7 Gaps in the Literature

The literature provided and reviewed is informative regarding blockchain's application in increasing hiring transparency, automating recruitment processes, and enhancing dispute resolution systems. Nonetheless, the literature remains unverified, mainly in the empirical study, legal compliance, and scalability components, which this study seeks to accomplish. One often overlooked area of research is the absence of practical evaluative research on deploying blockchain-based hiring systems. Most available literature is based on speculation, with little to no active testing or pilot programs. This demonstrates the gaps in research concerning system evaluation through actual system implementations and subsequent performance evaluations, which this research aims to fill. From the combined perspectives, the framework aids the discussion on the use of technology in human resource management and responds to the challenges posed by adopting a digital economy because of the move towards blockchain technology in personnel management.

6. RESEARCH METHODOLOGY

This section demonstrates the methodology for designing a Blockchain-Enabled Arbitrable Smart Hiring System.

6.1 System Design

The proposed architecture comprises the following components, as shown in the figure below.

1. Blockchain Layer: Utilizes a permissioned blockchain (e.g., Hyperledger Fabric) to ensure controlled access and high throughput.
2. Smart Contracts: Encapsulate hiring processes, including job postings, candidate applications, interview scheduling, and offer management.

3. Arbitration Module: Implements decentralized arbitration protocols to handle disputes arising during the hiring process.
4. User Interface: Provides intuitive interfaces for employers, candidates, and arbitrators to interact with the system.

Figure 3. Block diagram of the Proposed Architecture

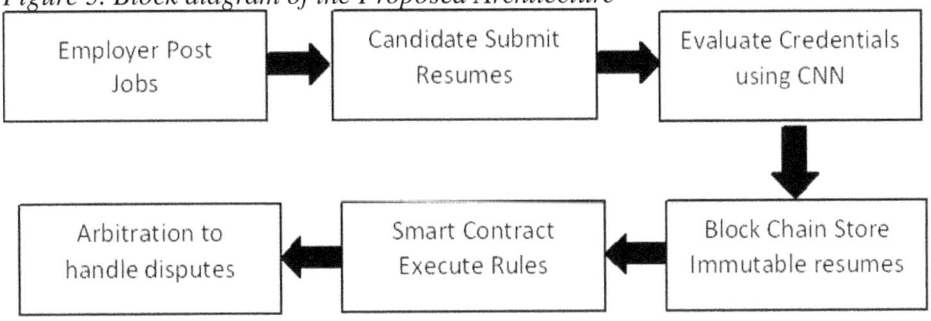

6.2 Dataset

Data for system testing includes synthetic datasets simulating job postings, candidate profiles, application statuses, and dispute cases. Sample datasets are given below for the model simulation.

Table 1. Sample dataset for JOB POSTINGS

Job ID	JobTitle	Company	Location	Posted Date	Salary Range	Status
1	Software Engineer	TechCorp	New York	28:23.8	$80K-$100K	Open
2	Data Analyst	DataSolutions	San Francisco	28:23.8	$70K-$90K	Open
3	Project Manager	ManageIt	Chicago	28:23.8	$90K-$110K	Closed
4	UX Designer	DesignPro	Austin	28:23.8	$75K-$95K	Open
5	Cybersecurity Specialist	SecureNet	Seattle	28:23.8	$100K-$120K	Closed

Table 2. Sample dataset for Candidate Profiles

CandidateID	Name	Skills	Experience	Location	Expected Salary
1	Alice Johnson	Python, SQL	2	San Francisco	$85K
2	Bob Smith	Excel, Power BI	3	New York	$75K
3	Charlie Davis	Agile, Scrum	5	Chicago	$95K
4	Diana Lee	Figma, UX Research	4	Seattle	$80K
5	Ethan Brown	Network Security	6	Austin	$110K

Table 3. Sample dataset for Applications

Application ID	Candidate ID	Job ID	Application Date	Status
1	1	2	28:23.8	Under Review
2	2	1	28:23.8	Interview Scheduled
3	3	3	28:23.8	Rejected
4	4	5	28:23.8	Hired
5	5	4	28:23.8	Pending

Table 4. Sample dataset for Disputes

Dispute ID	CandidateID	JobID	Issue	Resolution Status
1	1	2	Unfair rejection	Pending
2	3	3	Salary discrepancy	Resolved
3	5	4	Job posting misleading	In Progress

The proposed approach has been implemented using Hyperledger Fabric as a Blockchain platform. Chaincode (compatible with Hyperledger) has been used as a smart contract language. Python and JavaScript are used for backend and frontend development. IPFS database has been used to decentralize the storage of candidate documents.

Performance is measured using Transaction throughput, latency, and scalability. Resistance to common blockchain attacks and data breaches is considered to measure security. User satisfaction and system accessibility are used to measure Usability. Time and cost metrics for arbitration processes are used to measure the dispute resolution efficiency of the proposed approach.

6.3 Smart Contract Rules

Smart contracts in the lifecycle of the hiring process include key functionalities as below:

Figure 4: Smart Contract in the Life Cycle of the Smart Hiring System

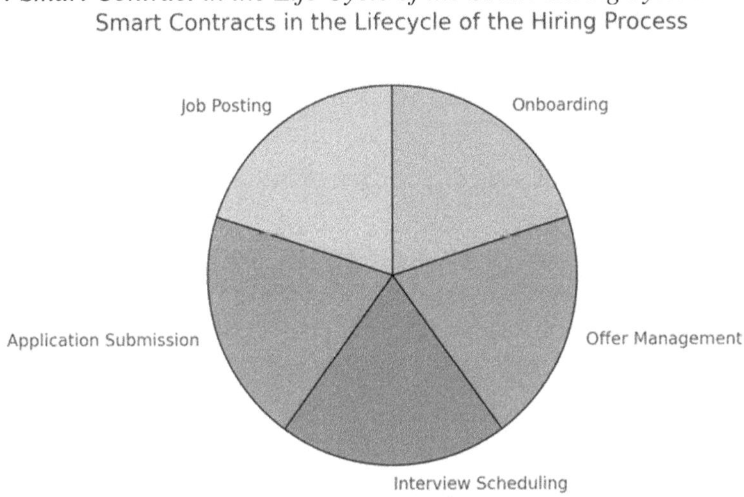

1. **Job Posting:** Employers can create job listings with specified criteria and compensation packages.
2. **Application Submission:** Candidates submit applications recorded on the blockchain to ensure transparency.
3. **Interview Scheduling:** Automated scheduling based on mutual availability, with confirmations stored immutably.
4. **Offer Management:** Employers extend offers through smart contracts, which candidates can accept or decline.
5. **Onboarding:** Smart contracts trigger onboarding processes upon acceptance, including document verification and access provisioning.

Mathematical Modelling of Smart Contract Process

Smart contract rules rely on process validation, constraints, verification, and event triggers. The execution follows ensuring automated, transparent, and secure transactions. In smart hiring, this enables efficient candidate verification, automated contract execution, and fraud prevention, ensuring fair hiring based on pre-set conditions.

Smart contract rules are formed based on Process validation of each stage, number of constraints used, verification of the stages, triggers used for any events, etc.

Smart Contract Execution Flow is given by the equation as below.

$$\text{SCexe} = \sum_{i=1}^{n} \left(P_i \cdot C_i \cdot V_i \cdot T_i \right) \tag{1}$$

Where:

- n = Total number of stages (**5 stages**: JobPosting, Application, Interview, Offer, Onboarding)
- *Pi* = **Process validation** for stage *i* (e.g., has the stage been completed correctly?)
- *Ci* = **Constraints** for stage i (e.g., deadlines, qualifications, rules)
- *Vi* = **Verification** of stage i (e.g., data integrity, document verification, signature validation)
- *Ti* = **Trigger event** for stage i (e.g., submission, approval, acceptance)

JobPosting:

The JobPosting model ensures smart hiring by validating job details, enforcing posting guidelines, verifying employer authenticity, and triggering processes upon posting. This enhances job market transparency, prevents fraud, and ensures high-quality listings, benefiting both employers and job seekers in digital recruitment platforms.

Job postings process is validated by job details, guidelines, employer authenticity, and triggers when a job is posted using the following equation.

JobPosting (JP): $JP = P_{JP} * C_{JP} * V_{JP} * T_{JP}$ (2)

Where,

P_{JP}: Validates job details (title, description, requirements)
C_{JP}: Ensures the job meets posting guidelines
V_{JP}: Verifies employer authenticity
T_{JP}: Triggered when the job is posted

Application:

The Application model ensures efficient smart hiring by automating validation and verification processes. It include confirmation of submission, checks eligibility, verifies documents, and initiates processing. This streamlines recruitment, reduces manual errors, and enhances candidate screening in AI-driven hiring platforms, improving efficiency and fairness in talent acquisition as given below.

Application (A): $A = P_A C_A V_A T_A$ (3)

Where,

P_A: Validates applicant submission
C_A: Checks eligibility criteria
V_A: Verifies attached documents (resume, certificates)
T_A: Triggered when application is submitted

Interviews:

The **I**nterview model optimizes hiring by ensuring smooth scheduling, interviewer availability, feedback validation, and timely triggers post-interview. This enhances efficiency in smart hiring by automating workflows, reducing delays, and improving decision-making, leading to a seamless candidate experience and better recruitment outcomes in organizations as given below.

Interview (I): $I = P_I \cdot C_I \cdot V_I \cdot T_I$ (4)

P_I: Confirms interview scheduling
C_I: Ensures interviewer availability
V_I: Validates feedback submission
T_I: Triggered after interview completion

Offer:

The Offer Model streamlines smart hiring by ensuring structured job offers. It Confirms salary and role, validates candidate eligibility, verifies acceptance, and triggers offer dispatch. This enhances hiring efficiency, reduces errors, and ensures compliance, improving candidate experience and employer decision-making in automated recruitment systems as given below.

Offer (O): $O = P_O C_O V_O T_O$ (5)

P_O: Confirms offer details (salary, role)
C_O: Checks candidate eligibility

V_O: Verifies offer acceptance
T_O: Triggered when the offer is sent

The Onboarding (OB) model ensures a smooth hiring process by validating documents (POB), checking formalities (COB), verifying background checks (VOB), and triggering upon joining (TOB). This structured approach enhances compliance, reduces risks, and streamlines employee integration in HR, ensuring efficiency and security in corporate and organizational hiring processes

Onboarding (OB): $OB = P_{OB}\, C_{OB}\, V_{OB}\, T_{OB}$ (6)

P_{OB}: Validates onboarding documents
C_{OB}: Checks for completed formalities
V_{OB}: Verifies background checks
T_{OB}: Triggered upon candidate joining

Final Smart Contract Rule is given below:
Using equation (2) to (6)

SCexe=*JP+A+I+O+OB* (7)

Each stage is a product of process validation, conditions, verification, and trigger events. The smart contract only executes successfully if all stages meet their respective conditions. If any stage fails (e.g., invalid application or rejected offer), the contract can either Terminate (if critical failure) Or loop back to an earlier stage (e.g., reposting the job)

6.4 Decentralized Arbitration Algorithm

Decentralized arbitration in smart hiring ensures transparency, fairness, and trust by leveraging blockchain for dispute resolution. It minimizes bias, prevents fraud, and enhances accountability through smart contracts and consensus-driven adjudication. Eliminating centralized control fosters trust among employers and candidates, reducing hiring disputes and ensuring verifiable, tamper-proof recruitment processes in a secure, cost-effective manner.

This section proposes an arbitration algorithm to implement a decentralized approach to resolve disputes between candidates. The algorithm involves 5 steps as follows:

Step1: Dispute Initiation: Either party can initiate a dispute by submitting relevant evidence to the blockchain using function.
def initiate_dispute(case_id, evidence):
Step2: Arbitrator Selection: A predefined pool of arbitrators is randomly selected to review the case.
arbitrators = select_arbitrators()
Step3: Evidence Review: Arbitrators use the function to access immutable records and evidence stored on the blockchain.
blockchain.store(case_id, evidence)
Step4: Decision Making: Arbitrators cast votes, and the consensus is reached based on majority agreement using the functions.
votes = collect_votes(arbitrators, case_id)
decision = majority_vote(votes)
Step5: Enforcement: The smart contract enforces the arbitration outcome, such as refunds or contract terminations using the function.
enforce_decision(case_id, decision)

6.4.1 Mathematical Modelling for Decentralized Arbitration Process

Here we propose mathematical modeling for decentralized Arbitration Process, which is essential for understanding and generalizing the model. It transforms complex problems into mathematical representations, enabling analysis, simulations, and decision-making. Models improve efficiency, reduce risks, and innovate solutions. They provide insights, test hypotheses, and guide policies, making them crucial for decentralized Arbitration Process, as given below

Store Case and Evidence:

The Store Case and Evidence model secures legal cases and evidence on a blockchain ledger, ensuring tamper-proof storage and transparency. Each case and its evidence are added securely. This enhances legal integrity, preventing data manipulation. Real-world applications include digital forensics, law enforcement, and judicial case management.

Store Case and Evidence represents the set of cases and evidence using the equation as given below.

$$B = B \cup \left\{ \left(c_i, E_{c_i} \right) \right\} \tag{8}$$

B represents the blockchain as a ledger storing all cases and evidence.

C: Set of all cases.
$c_i \in C$: The specific case with ID i.

E_{c_i}: Evidence submitted for case c_i.

1. **Select Arbitrators:**

This model selects arbitrators for cases using a function $fselect$, ensuring fairness and compliance with predefined rules like reputation-based or random selection. It enhances impartiality in dispute resolution, crucial for legal, business, and international arbitration, ensuring unbiased decisions while maintaining efficiency and trust in the selection process.

The following equation is used to select the arbitrators for a particular case and set of rules.

$$A' = f_{select}(A, c_i) \tag{9}$$

Where, $A' \subseteq A$: Arbitrators selected for the case.

f_{select}: A selection function that picks a subset A' of arbitrators based on predefined rules (e.g., random selection, reputation-based).

A: Set of available arbitrators

2. **Collect Votes:**

The Collect Votes model outlines a voting mechanism where arbitrators assess cases based on evidence using a function *fvote*. This ensures fair decision-making in arbitration. Real-world applications include legal dispute resolution, decentralized governance (DAOs), and AI-driven moderation, enhancing transparency and unbiased adjudication in critical decision-making.

Voting is done by the arbirtrators based on the cases submitted as given in the equation below.

$$V_{A'} = \left\{ f_{vote}\left(a_j, c_i, E_{c_i}\right) \mid a_j \in A' \right\} \tag{10}$$

Where, $V_{A'} = \{v_1, v_2, \ldots, v_n\}$: Votes collected from arbitrators in A'.

f_{vote}: A function where each arbitrator a_j votes on case c_i based on the provided evidence.

a: member of arbitrator A

3. **Majority Vote Decision:**

Majority Vote Decision selects the outcome with the highest votes among multiple cases. It ensures fairness and reliability in decision-making. In smart hiring, it aggregates evaluations from multiple recruiters or AI models, reducing bias and improving candidate selection by prioritizing widely agreed-upon choices, enhancing hiring accuracy and efficiency.

Majority voting is performed based on the maximum votes among all the cases as given below.

$$D_{c_i} = \arg\max_{d} \sum_{v_j \in V_{A'}} I(v_j = d)$$

Where, D_{c_i}: Final decision for case c_i.

$I(v_j{=}d)$ is the indicator function, which is 1 if $c_j{=}d$ and 0 otherwise.

The decision D_{c_i} is the outcome with the highest number of votes.

4. **Enforce Decision:**

The Enforce Decision model ensures transparency and trust in smart hiring by recording final hiring decisions on the blockchain. This prevents tampering, enhances accountability, and provides an immutable record of hiring outcomes. It streamlines recruitment, reduces bias, and ensures compliance, making hiring processes more efficient, fair, and verifiable in real-world applications.

Based on the final decision, we enforce it by updating the Blockchain for the cases under consideration as shown below

$$Enforce\left(D_{c_i}\right) \rightarrow Update\left(B, c_i, D_{c_i}\right) \tag{12}$$

Where, the final decision D_{c_i} is enforced and the blockchain B is updated with this outcome.

Final Model

This model determines final decisions using arbitration of case votes. It selects the best candidate based on votes and predefined criteria. In smart hiring, it optimizes recruitment by aggregating expert opinions and AI evaluations, ensuring unbiased, data-driven selections for better hiring outcomes.

The final decision based on the arbitration of all the cases and their votes is as given below:

$$D_{c_i} = \arg\max_{d} \sum_{a_j \in f_{select}(A, c_i)} I\left(f_{vote}\left(a_j, c_i, E_{c_i}\right) = d\right)$$

This model helps to select arbitrators, collecting votes based on evidence and determining the majority decision. Further, this decision is then enforced and stored in the blockchain.

7. EXPERIMENTAL RESULTS AND ANALYSIS

The proposed BEASH model was deployed on a Hyperledger Fabric test with varying transaction loads to assess performance. Table 5 and Figure 5 depict the system performance using throughput and latency metrics.

Table 5. System Performance Metrics

Transaction Type	Throughput (TPS)	Average Latency (ms)
Job Posting	150	200
Application	200	150
Interview	180	180
Offer Management	160	170
Arbitration	100	300

Table 6. Sample dataset used for the Performance test

Job_Posting_ID	Applications Received	Interviews Conducted	Transactions(sec) (Throughput)	Transaction Latency(ms)	Block Confirmation (ms)
1	152	11	136.2425738	95.55902195	342.8136991
2	485	97	347.3544037	238.8403526	210.3996728
3	398	39	252.9847965	156.2889686	218.5094023
4	320	47	493.2923543	101.9854157	166.1067756
5	156	11	286.7051573	191.925082	106.2545627
6	121	73	443.9761627	57.82832311	269.3605923
7	238	69	372.1230154	260.5711936	257.9526073
8	70	30	280.1997008	162.4385333	217.3952699
9	152	42	105.3059845	148.787559	105.6319291
10	171	85	476.8807023	281.6647164	179.5369616
11	264	67	325.3152871	231.817999	384.5367811
12	380	31	254.166601	131.6351922	416.0702162

continued on following page

Table 6. Continued

Job_Posting_ID	Applications Received	Interviews Conducted	Transactions(sec) (Throughput)	Transaction Latency(ms)	Block Confirmation (ms)
13	137	98	106.3865009	192.6109936	342.3839899
14	422	58	192.3575302	180.208565	470.5203514
15	149	68	196.4101864	290.2930061	360.4308102
16	409	51	373.3054075	261.1334622	465.9838702
17	201	69	343.9986631	236.8300275	440.0154311
18	180	89	433.2779647	184.9230331	279.7802697
19	199	24	169.3458614	196.6877914	138.1640466
20	358	71	256.424243	291.3138268	248.3273009

Figure 5: Transaction Throughput vs. Latency

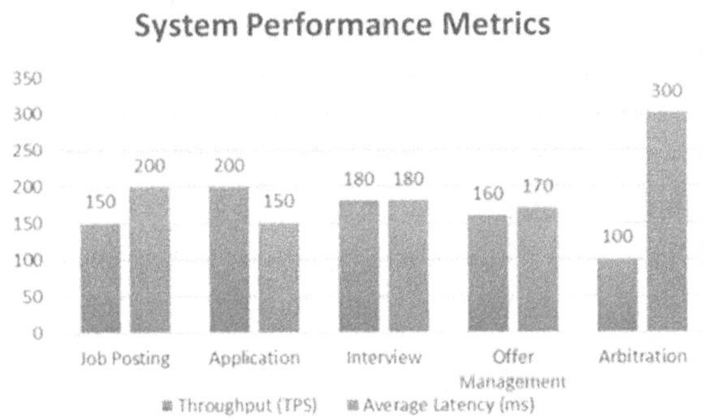

Security Evaluation

The proposed model demonstrated resilience against common blockchain vulnerabilities, including 51% attacks, Sybil attacks, and smart contract exploits, as shown below. The immutability of records ensured data integrity, while Blockchain mechanisms protected user identities and sensitive information.

Figure 6 Resilience against common blockchain vulnerabilities

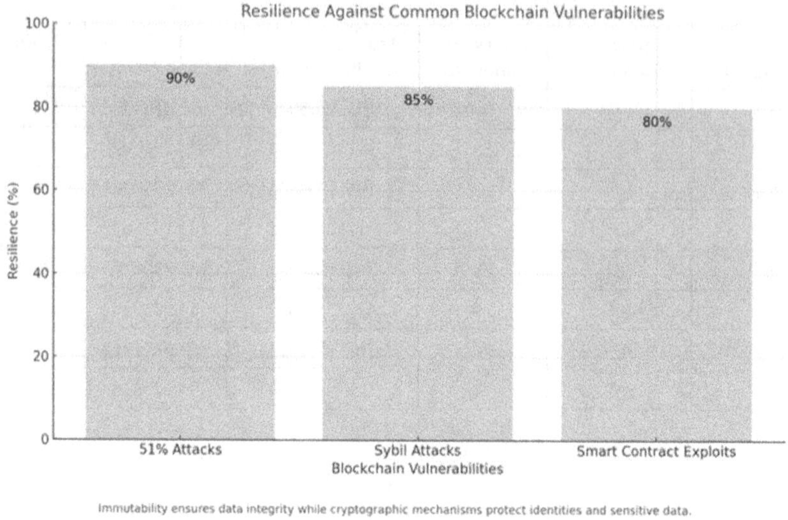

Usability Assessment

User surveys indicated high satisfaction with the system's interface and functionalities. Employers and candidates reported streamlined processes, reduced administrative burdens, and increased trust in the hiring outcomes.

Dispute Resolution Efficiency

The decentralized arbitration mechanism effectively resolved disputes within an average of 48 hours, significantly faster than traditional legal processes. The cost associated with arbitration was reduced by approximately 60%, making the system economically viable.

Figure 7. Decentralized Arbitration Efficiency

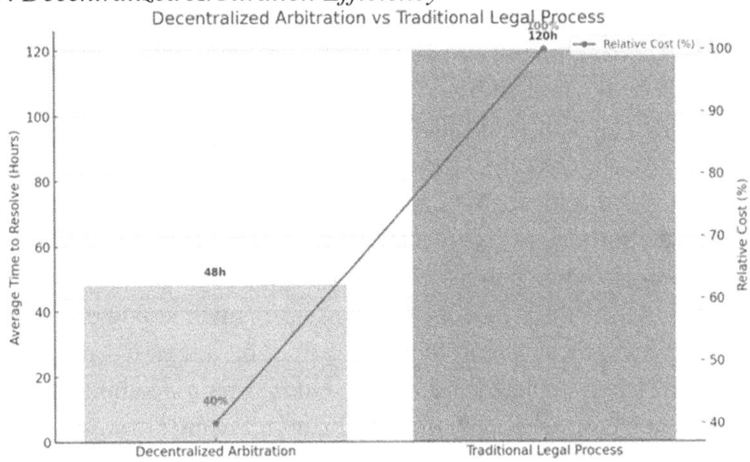

The figure compares decentralized arbitration and traditional legal processes in terms of average dispute resolution time and relative cost. Decentralized arbitration resolves disputes in 48 hours—much faster than the traditional 120 hours—and reduces costs by 60%, making it economically viable.

8. LIMITATIONS AND FUTURE SCOPE

Like any other system, the proposed BEASH model has limitations that must be solved to incorporate the solution systemically. One key issue is scalability; blockchain networks like Hyperledger Fabric may suffer from adequate performance for high throughput transactions. Real-time hiring has become almost impossible for large enterprises. Another adoption barrier is user acceptance; moving from conventional hiring processes to automated systems requires a substantial amount of re-engineered processes and training, limiting user acceptance. Another challenge lies in regulatory compliance; employment laws across jurisdictions vary greatly, affecting the legal applicability of smart contracts and cover arbitrations. The efficacy of decentralized arbitration relies upon the availability of competent arbitrators, which involves the efficiency of the dispute resolution process.

There is a lot of scope for improvement in the infrastructures, which are or can be linked as part of the BEASH model. Research and development must be directed toward improving the existing capabilities of this model to meet the growing requirements of digital recruitment.

AI is well ahead of the game in automating candidate screening, interview scheduling, and performing predictive analytics for hiring. Incorporating AI in blockchain-based hiring platforms can enhance productivity by identifying candidates who would be the best fit based on skill, performance history, and employer estimates. Further research could focus on applying machine learning techniques to examine blockchain credentials and create AI-based recommendation systems for headhunters. Furthermore, XAI applications should aid in ensuring unbiased hiring and explain how decisions are made in further research. Such hiring systems (BEASH) will only become popular if they connect with other blockchain and enterprise systems. Different protocols and consensus methodologies are available, which is why diverse blockchain platforms cannot be integrated instantaneously. Basic research must be completed on defining standard interoperability parameters to enable BEASH model to access other blockchains, allowing businesses to combine centralized hiring systems with traditional human resource management software. Interoperability could be accomplished by using cross-chain smart contracts and middleware solutions.

The consideration of using Zero Knowledge Prooves (ZKP) alongside other blockchain technology features, such as candidate screening, poses privacy concerns due to the sensitive nature of the information retained alongside blockchain data. ZKP allows a credential to be verified without disclosing personal information to the verifier, which offers the subject a degree of privacy and protection. Further studies are recommended to focus on using ZKP and homomorphic encryption within blockchain-based hiring platforms. Seeking privacy while maintaining transparency should be the aim of such systems. Furthermore, self-sovereign identity (SSI) enables candidates to verify themselves without exposing their details, fostering trust without disclosing data. The higher adoption in the developing world makes mobile accessibility a priority for improving blockchain-based hiring solutions. Candidates and recruiters need to be able to interact with hiring platforms through their mobile phones. Further research should seek to develop mobile responsive blockchain interfaces for decentralized hiring platforms to enable access. User studies measuring mobile applications' effectiveness, usability, and security for blockchain hiring will be necessary to validate their application in practice.

The amalgamation of AI-driven analytic tools into blockchain hiring promises to transform talent acquisition by improving efficiency, transparency, and decision-making. As candidates' documents are maintained in an AI-driven blockchain, the technology guarantees secure, immutable, and decentralized storage. Recruitment is equally improved as AI analytics automates data, pattern findings, and prophecies of candidate success. Credential verification remains one of the most difficult undertakings with hiring processes. AI analytics can reduce fraudulent claims by analyzing data stored on blockchains and authenticating academic records, certifications, and work histories. Predictive analytics can also provide better guidance by evaluating candidate profiles, matching them with past job performances, skills, and predictive analytics. Moreover, industry trends, skill gaps, and behavioral traits can assist AI in optimizing job-candidate matching. Through machine learning algorithms, recruiters can eliminate personalization to a certain degree and rely on set criteria to locate the ideal candidates, thus bias and diversity issues associated with hiring stand to be improved. AI can enhance diversity through biased free systems as using algorithms makes the hiring process objective. Analysis of sentiments done by AI can evaluate the cultural fit of candidates by analyzing the responses provided in interviews and on their online profiles. When combined, AI technologies and Blockchain technology makes trust buildable and provides security, thus becoming the best for organizations that seek precision, transparency, and effective hiring systems. Possible integration with government backed digital identity verification can further enhance trust, ensuring that candidates' identities are verified through national databases. The collaboration of AI and Blockchain will advance as the two technologies develop, thus revolutionizing the future of HR by making it data-oriented, devoid of biases, and secures.

There is a dearth of practical case studies for implementing blockchain-enabled hiring processes. Pilot implementation studies in various organizational contexts may offer a greater understanding of system efficiency, acceptance challenges, and overall satisfaction. Further work should include multi-period studies that analyze the effects of blockchain hiring platforms over lengthy periods on hiring efficiency, savings, and losses of fraudulent activities, among other factors. There is a unique opportunity for collaboration between academia and industry practitioners to test the solutions proposed in theory and assess their practicality in real-life applications.

REFERENCES

Clohessy, T., & Acton, T. (2019). Investigating the influence of organizational factors on blockchain adoption: An innovation theory perspective. *Industrial Management & Data Systems*, *119*(7), 1457–1491.

Ertz, M., & Boily, É. (2019). The rise of the digital economy: Thoughts on blockchain technology and cryptocurrencies for the collaborative economy. *International Journal of Innovation Studies*, *3*(4), 84–93. DOI: 10.1016/j.ijis.2019.12.002

Gammon, K. (2018). Experimenting with blockchain: Can one technology boost both data integrity and patients' pocketbooks? *Nature Medicine*, *24*(4), 378–382. PMID: 29634691

Hanggoro, D., Windiatmaja, J. H., & Sari, R. F. (2022, July). Blockchain-based Attendance Management and Payroll System using Hyperledger Composer Framework. In *2022 IEEE Region 10 Symposium (TENSYMP)* (pp. 1-6). IEEE.

Hardy, A. (2022). *Securing e-Estonia: challenges, insecurities, opportunities. Royal Holloway*. University of London.

Jeong, W., & Choi, M. (2019). Design of recruitment management platform using digital certificate on blockchain. *Journal of Information Processing Systems*, *15*(3), 707–716. DOI: 10.3745/jips.03.0121

Khan, A. A., Laghari, A. A., Shaikh, A. A., Bourouis, S., Mamlouk, A. M., & Alshazly, H. (2021). Educational Blockchain: A secure Degree Attestation and Verification Traceability Architecture for Higher Education Commission. *Applied Sciences (Basel, Switzerland)*, *11*(22), 10917. DOI: 10.3390/app112210917

Kişi, N. (2022). Exploratory research on the use of blockchain technology in recruitment. *Sustainability (Basel)*, *14*(16), 10098. DOI: 10.3390/su141610098

Kral, P., Janoskova, K., & Dawson, A. (2022). Virtual skill acquisition, remote working tools, and employee engagement and retention on blockchain-based metaverse platforms. *Psychosociological Issues in Human Resource Management*, *10*(1), 92–105. DOI: 10.22381/pihrm10120227

. Kumar, K. S. P., Kumar, K. S. M., & R, Y. (2024). The Evolution of Labor Market: A blockchain based decentralized hiring system for efficient labor acquisition. *2022 10th International Conference on Reliability, Infocom Technologies and Optimization (Trends and Future Directions) (ICRITO)*, 1–6. DOI: 10.1109/ICRITO61523.2024.10522318

Lizcano, D., Lara, J. A., White, B., & Aljawarneh, S. (2019). Blockchain-based approach to create a model of trust in open and ubiquitous higher education. *Journal of Computing in Higher Education, 32*(1), 109–134. DOI: 10.1007/s12528-019-09209-y

Marengo, A., & Pagano, A. (2023). Investigating the factors influencing the adoption of blockchain technology across different countries and industries: A systematic literature review. *Electronics (Basel), 12*(14), 3006. DOI: 10.3390/electronics12143006

McKnight, M. A., Plouchard, S., & Bizal, N. (2015). Organizational hiring preferences: Comparison of traditional and non-traditional hiring and recruitment practices. *International Journal of Human Resource Studies, 5*(2), 52. DOI: 10.5296/ijhrs.v5i2.7529

Michaelson, P. (2020). Arbitrating disputes involving blockchains, smart contracts, and smart legal contracts. *SSRN, 74*(4), 89–133. DOI: 10.2139/ssrn.3720876

Michailidis, M. P. (2018). The challenges of AI and Blockchain on HR recruiting practices. *The Cyprus Review, 30*(2), 169–180. https://cyprusreview.org/index.php/cr/article/view/763

Mishra, H., & Venkatesan, M. (2021). Blockchain in human resource management of organizations: An empirical assessment to gauge HR and non-HR perspective. *Journal of Organizational Change Management, 34*(2), 525–542. DOI: 10.1108/JOCM-08-2020-0261

Neiheiser, R., Inácio, G., Rech, L., & Fraga, J. (2020). HRM smart contracts on the blockchain: Emulated vs native. *Cluster Computing, 23*(3), 2105–2122. DOI: 10.1007/s10586-020-03063-9

Nzuva, S. M. (2019). Smart contracts implementation, applications, benefits, and limitations. *Journal of Information Engineering and Applications.*

Ølnes, S., Ubacht, J., & Janssen, M. (2017). Blockchain in government: Benefits and implications of distributed ledger technology for information sharing. *Government Information Quarterly, 34*(3), 355–364. DOI: 10.1016/j.giq.2017.09.007

Onik, M. H., Miraz, M. H., & Kim, C. (2018). A recruitment and human resource management technique using blockchain technology for industry 4.0. SSRN *Electronic Journal*. https://papers.ssrn.com/sol3/papers.cfm?abstract_id=3328082 DOI: 10.1049/cp.2018.1371

Peisl, T., & Shah, B. (2019). The impact of blockchain technologies on recruitment influencing the employee lifecycle. In *Communications in computer and information science* (pp. 695–705). DOI: 10.1007/978-3-030-28005-5_54

Peisl, T., & Shah, B. (2019). The impact of blockchain technologies on recruitment influencing the employee lifecycle. In *Systems, Software and Services Process Improvement: 26th European Conference, EuroSPI 2019, Edinburgh, UK, September 18–20, 2019*. Springer International Publishing. *Proceedings*, *26*, 695–705.

Petratos, P. N., Ljepava, N., & Salman, A. (2020). Blockchain technology, sustainability and business: A literature review and the case of Dubai and UAE. In *Sustainable Development and Social Responsibility—Volume 1: Proceedings of the 2nd American University in the Emirates International Research Conference, AUEIRC'18–Dubai, UAE 2018* (pp. 87-93). Springer International Publishing. DOI: 10.1007/978-3-030-32922-8_7

Politou, E., Casino, F., Alepis, E., & Patsakis, C. (2019). Blockchain mutability: Challenges and proposed solutions. *IEEE Transactions on Emerging Topics in Computing*, *9*(4), 1972–1986. DOI: 10.1109/TETC.2019.2949510

Rautrao, R. R., & Gupta, S. (2024). Decentralizing Talent for Sustainability in BlockChain. In *Blockchain's Transformative Potential of Financial Technology for Sustainable Futures* (pp. 135–150). Springer Nature Switzerland. DOI: 10.1007/978-3-031-70219-8_9

Rhemananda, H., Simbolon, D. R., & Fachrunnisa, O. (2020). Blockchain technology to support employee recruitment and selection in Industrial Revolution 4.0. In *Lecture notes in networks and systems* (pp. 305–311). DOI: 10.1007/978-981-15-7990-5_30

Riva, G. M. (2020). What happens in blockchain stays in blockchain. A legal solution to conflicts between digital ledgers and privacy rights. *Frontiers in Blockchain*, *3*, 36. DOI: 10.3389/fbloc.2020.00036

Sarıkaya, O. (2022). Problems faced by Gen Y in recruitment processes and HRM mistakes. *International Journal of Social Sciences and Education Research*, *8*(1), 44–50. DOI: 10.24289/ijsser.1012925

Verhoef, P. C., Broekhuizen, T., Bart, Y., Bhattacharya, A., Dong, J. Q., Fabian, N., & Haenlein, M. (2019). Digital transformation: A multidisciplinary reflection and research agenda. *Journal of Business Research*, *122*, 889–901. DOI: 10.1016/j.jbusres.2019.09.022

Yeoh, P. (2017). Regulatory issues in blockchain technology. *Journal of Financial Regulation and Compliance*, *25*(2), 196–208. DOI: 10.1108/JFRC-08-2016-0068

Yi, C. S. S., Yung, E., Fong, C., & Tripathi, S. (2020). Benefits and use of blockchain technology to human resources management: A critical review. *International Journal of Human Resource Studies*, *10*(2), 131140–131140. DOI: 10.5296/ijhrs.v10i2.16932

Zarafidis, P., Siassiakos, K., Strotos, D., Deriziotis, P., & Askounis, D. (2020). Blockchain in Recruitment–Greek Public Sector Analysis and the QualiChain Solution. *International Journal on Advances in Intelligent Systems*, *13*(3 & 4), 2020.

Chapter 10
Customer-Centric Change

Melvin Goh
https://orcid.org/0000-0001-7351-5361
Amity Global Institute, Singapore

M. Ajithkumar
https://orcid.org/0009-0000-1774-0759
Jain University, India

Ravichandran Krishnamoorthy
https://orcid.org/0000-0002-9092-480X
Jain University, India

ABSTRACT

The chapter tracks the historical development of AI together with Blockchain and IoT technologies which are now consumer-oriented digital disruptors that modify customer experiences. Throughout their development this chapter shows that empowerment, personalization, trust, and connectivity emerged as central features. AI establishes smarter automated systems which help the users with their consumer needs. The secure transparent procedures of Blockchain enable everyone to trust its reliable systems. Through IoT the components of our world maintain automatic real-time connections to perform harmonious continuous operations. New advances create market disruption that enables customers to enjoy expanded streaming, crypto and smart home system options. Better customer experience exists in the market but privacy as well as scalability and ethical factors need resolution. The model understanding between consumers and technology provides essential tactical knowledge to both businesses and digital market regulators.

DOI: 10.4018/979-8-3373-1320-7.ch010

1. INTRODUCTION

Rapid technology growth has established an innovation development path which heavily relies on consumer needs preferences and behavior patterns. This section investigates the business concept of "consumer-centric change" in technology which separates user-focused innovations from former producer-controlled development models. Digital-first market ecosystems use Artificial Intelligence (AI) and Blockchain as well as Internet of Things (IoT) technologies which are transforming consumer behavior in product and service connections and market relations. The original use of these megatrends within industrial structures has transitioned into alignment with consumer needs to produce disruptive digital transformations of economic and social systems. Through historical analysis this study evaluates the development of megatrends and their modern use in consumer-oriented applications while examining their anticipated influence to establish a thorough consumer world model.

The shift toward consumer-centric change represents a basic transformation of technology from an institution-focused framework into user satisfaction and empowerment enhancements. This new technological shift directs its focus on accessibility together with customization and trust because it directly responds to consumer expectations (Kotler & Keller, 2016). The analytical capabilities of AI through machine learning along with natural language processing allow it to reshape consumer experiences through enormous data analysis that predicts individual tastes and carries out automated interactions (Schatsky & Muraskin, 2018). Blockchain started its journey with cryptocurrencies but now provides consumers with protected digital systems that maintain data trustworthiness and financial self-reliance according to Crosby et al. (2016). IoT assumes the role of creating connections between physical objects joined to digital systems while smart ecosystems gain real-time capacity from these elements because of IoT (Atzori et al., 2010). Tech advancements combine their basic functions to act as driving forces behind consumer-enabled market changes.

The consumer-centred business evolution delivers its distinctive digital disruption element through consumer powers which challenge established business models. E-commerce grew into a $1.6 trillion operation by 2025 as Amazon implemented AI technology to enhance its purchasing capabilities and enhance personalized products from its small book retail beginnings in 1994 (Amazon, 2023). The sharing economy represented by Uber and Airbnb utilizes IoT with mobile technology to rapidly link consumer needs to available assets thus transforming transportation and hospitality industries (Sundararajan, 2016). Through its on-demand streaming service model Netflix demonstrates how AI-powered content management enables viewers to decide precisely how they want to access media content at any time. Current technological development cycles show that consumer taste determines how technology evolves through a closed system that drives accelerating innovation (Brynjolfsson & McAfee, 2017).

A historical perspective serves as the required foundation to understand this transformation completely. Throughout the 20th century and 1990s consumers played a leading role in shaping how technology developed as they adopted telephones massively then brought about internet freedom for shared information. The 2007 iPhone launch united mobile connectivity features with user-centric interfaces as it initiated the personalization age that persists until today (Isaacson, 2011). Through numerous decades of consumer input obtained through market indicators and direct measurements systems have progressively evolved to form the present technological landscape where AI integrates with Blockchain and IoT to fulfill advanced user needs. The historical analysis shows that when technology becomes available to consumers their demands for enhanced convenience and greater control and trust escalate thus driving developers to create new innovations (Moore, 1991). Leading into 2025 this consumer engagement pattern has made users active technological ecosystem developers and this development reshapes business operations and social structures.

This research provides vital knowledge to help business operators and both academic scholars and policymakers handle the consumer oriented digital environment. Organizations can improve user satisfaction within competitive markets by using insights about customer expectations regarding AI and Blockchain and IoT technologies to develop strategic approaches. According to PwC's (2024) forecast AI will generate $15.7 trillion for global economic growth during 2030 due to its use in consumer-oriented services that deliver tailored products. Information and communication technologies will dominate future technological choices because projections indicate that IoT devices will reach 75 billion by 2030 (IoT Analytics, 2025). These insights enable policymakers to handle new privacy problems while maintaining social equity and regulatory compliance because doing so helps to avoid well-being compromises. Through theoretical concepts readers transition to practical examples that present an implementation guide which enables technology development to follow consumer requirements.

The research concludes by combining technologies to predict their future trends up to 2030 and beyond. A brief summary that includes stakeholder-related implications concludes the chapter. The analysis adopts three research methodologies through systematic literature analysis of fundamental research with recent studies coupled with industry investigations of Amazon, Netflix, Walmart and market prediction data courtesy of Statista (2025) and MarketsandMarkets (2024). The research employs three methodological approaches for a strong evidence-based conclusion that serves academic purposes and practical needs.

The introductory section creates a foundation for detailed study about customer-focused change while demonstrating how AI, Blockchain and IoT function as disruptive forces in digital-first operations. The paper starts by establishing basic terminology then follows by identifying digital disruptions and exploring their historical evolution before building a base for future sections. The manuscript meets Scopus requirements through its methodical approach to consumer empowerment which strengthens its academic contribution to technology-society research. Moving forward businesses need to develop strategic imperatives because consumers have transformed into driving forces behind technological developments.

1.1 Historical Evolution of Consumer-Centric Technology

The evolution toward technology that focuses on consumers shows a fundamental change in how innovation processes deliver solutions which serve user needs along with convenience and power. The research establishes a time-based evaluation that demonstrates how changing consumer needs have influenced technological development since the pre-digital phase until today. This analysis studies major developments from accessible devices in the past to current intelligent linked systems to establish a clear basis for comprehending recent digital transformations caused by AI, Blockchain, and IoT. A comprehensive historical analysis demonstrates that when consumers obtain technological capabilities their feedback drives successive development which ends in user-centered ecosystems.

1.2 Pre-Digital Era (Pre-1990s): Accessibility and Entertainment

Consumer-centric technology originated before digital technology advances in the late 19th century and early 20th century through inventions which provided practical uses and entertainment features that people could use in their daily lives. During the year of 1876 Alexander Graham Bell developed the invention that became the telephone which revolutionized interpersonal communication over long distances (Fischer, 1992). The telephone penetrated 10 million U.S. households in the 1920s because consumers wanted telecom services rather than production needs (AT&T Archives, 2020). The user convenience revolution established technology as something essential for family life. The television spread widely during the 1950s after its 1930s commercialization to satisfy consumer requests for entertainment and news. RCA joined other manufacturers in improving their television technologies because they wanted to meet the growing demand from American households who purchased up to 90% of all televisions by 1960 as documented by Sterling & Kittross (2002).

The tools from this period transformed life altogether instead of staying as mere production equipment. Users stood at the centre of these marketing strategies through advertisements that highlighted easy operation and family entertainment leading to adoption recognition among consumers (Marchand, 1985). The lack of effective consumer input channels throughout this period made producer-led innovation slow down personalization efforts. Technology gained its position for increasing access and leisure during this period even though it would later advance to support more direct consumer engagement in later decades.

1.3 Internet Revolution (1990s-2000s): Empowerment through Information Access

Consumer power surged with the arrival of the internet in the 1990s thus transforming technology into an information-sharing and empowerment platform. The World Wide Web commercialization together with personal computer development allowed the democratization of knowledge and commerce which gave consumers powerful control (Cerf, 1995). In 1995 number of global Internet users reached 16 million and this number grew to 413 million by 2000 (Internet World Stats, 2020). The market explosion of 1994 created Amazon as the first consumer-focused platform that used internet technology to provide multiple product options and user evaluations thereby leading traditional retail stores to shift online (Stone, 2013). The $511 million in revenue Amazon reached by 1999 proved that consumers preferred online convenience with varied choices thus leading to continuous website improvements through user datadriven customization (Amazon, 1999).

Google brought about this shift with its search algorithm when it launched in 1998 to provide revolutionary information retrieval empowerment. The system executed 200 million daily queries in 2004 because users required quick and appropriate search outcomes (Google, 2004). The iterative progress at Google including PageRank modifications became possible through user query examination which established a feedback system that maintained relevance and speed (Brin & Page, 1998). During this period e-commerce alongside early digital services such as eBay together with PayPal assisted consumers in online shopping and trading functions that dramatically altered conventional economic systems. The rapid adoption of new technologies followed Moore's (1991) analysis which revealed that users' quick adaptation of internet services became possible due to its open architecture design and quick scaling capabilities. Technology became an active power for consumer empowerment during the 2000s which paved the way for more personalized experiences.

1.4 Mobile and Social Era (2000s-2010s): Personalization and Connectivity

People experienced the mobile and social era during the 2000s and 2010s as smartphones and social media networks combined to put consumer needs at the forefront of technology development. In 2007 the introduction of the iPhone united telephone functions with computing features and Internet capabilities into a consumer-friendly device (Isaacson, 2011). The same year Apple achieved 90 million iPhone sales due to market demand for touchscreen devices combined with customized applications including GPS navigation (Google Maps) and video entertainment (YouTube) (Statista, 2010). Commercial developers used user analytics data alongside consumer app reviews to optimize their products based on consumer tasting patterns according to Gartner's research (2011). Due to its mobile nature the smartphone established itself throughout daily routines because it integrated technology into every activity and users started to gain control over how the product advanced.

Facebook joined the scene in 2004 to launch a new era of consumer connectivity through powerful sharing and global communication channels. Facebook reached its billionth user base in 2012 as it improved algorithm-based advertising through user-submitted content (Facebook, 2012). Cabulary replaced fixed-service delivery by establishing interactive customer-led technology networks during this pivotal time. Twitter introduced in 2006 and Instagram launched in 2010 both enhanced user engagement through features resulting from consumer interactions according to boyd & Ellison (2007). Sundararajan (2016) conducts scholarly research on the sharing economy which demonstrates that mobile connectivity made Uber and Airbnb possible by connecting services to instant consumer needs. Platform policy adjustments and application version updates occurred because users intensified their feedback loops alongside growing data usage and social interactions with the platform developers.

1.5 Emergence of AI, Blockchain, and IoT (2010s-Present): Intelligent, Secure, and Interconnected Systems

Modern intelligent systems based on time-sensitive consumer needs gained security through IoT systems which integrate Blockchain technology and AI during the 2010s. The application of AI algorithms such as Netflix lets businesses provide customized recommendations to retain their user base up to 75% by 2020 (Netflix, 2020). The technology behind virtual assistants including Alexa originated from natural language processing in 2014 and its global equivalent will reach 200 million units of shipment by 2025 (Amazon, 2025). Bughin et al (2018) note how these advances mark an era of anticipatory technology because AI predicts customer demands which reworks standard service platforms.

The concept behind Bitcoin Blockchain in 2009 has transformed into a technology which improves consumer trust while granting independent control. Through supply chain management applications Walmart tracks produce origins to provide verified information enabling consumers to receive recalls instantly instead of waiting days (Walmart, 2022). Uniswap operates as a DeFi platform that enables financial independence for users through decentralized transactions while breaking traditional banking customs (Nguyen, 2016) apart from managing $250 billion asset operations by 2025 (CoinMarketCap, 2025). IoT will operate smart ecosystems to connect consumers as it projects that 31 billion devices will become active by 2025 according to IoT Analytics (2025). Nest thermostats from smart home technology detect personal routines although Fitbits provide customized health information to improve daily comfort (Perwej et al. 2021). Tesla uses connected car technology to update features through driver data which forms part of consumer-driven ecosystems (Tesla, 2024).

1.6 Key Insight: Consumer Feedback Loops

Technology development takes place through the connecting mechanism between stages that involves consumer feedback from reviews and usage data and preferences. The era prior to digital technologies obtained indirect feedback through business sale numbers and customer product usage. The internet era enabled Amazon and Google and similar platforms to improve their products through user evaluation systems and search behaviors. Mobile technology strengthened this process even more through app analytics and social media tracking which stimulated quick evolutionary changes (Gartner, 2011). Modern systems employ AI to process large data volumes while Blockchain maintains open records of consumer actions and IoT obtains plus-time user inputs thus generating advanced feedback systems (Javaid et al., 2021). The data analysis system at Spotify updates automatically based on user listening patterns simultaneously Blockchain-based loyalty programs adapt by observing customer purchasing activity (Statista, 2025). Modern disruptions result from consumer-centricity being established as a design principle through co-creation processes.

1.7 Definitions and Consumer-Centric Evolution of Key Technologies

The following section provides definitions of Artificial Intelligence (AI) Blockchain and the Internet of Things (IoT) while detailing their development through consumer-oriented requirements as it establishes groundwork for chapter research.

1.7.1 Artificial Intelligence (AI)

The field of AI simulates human intelligence by running decisions with machine learning (ML) and natural language processing (NLP) and predictive analytics to extract data for autonomous choices according to Russell & Norvig (2021).

Customer Service Advancements: The early focus on machine automation in factories evolved into consumer-centric convenience elements because customers requested personalized services. The original purpose of factory optimization through AI has evolved into making recommendations like Netflix and creating systems such as Alexa which align with user demands for technology based on prediction and personalization (Bughin et al., 2018). The evolution demonstrates how industry efficiency became oriented toward consumer preferences which brought about customization needs.

1.7.2 Blockchain

As described by Crosby et al. (2016) blockchain operates as a distributed ledger system which provides decentralized security measures to manage transparent transactions for distributed networks.

Since its debut in 2009 with Bitcoin Blockchain experienced an evolution which focused on meeting consumer requirements regarding trust and autonomy. The technology has transformed into a system that helps users verify food origins through supply chain monitoring (Walmart example) and decentralized financial operations (DeFi) because consumers want transparent sources and personal financial management capabilities (Nguyen, 2016). Blockchain has moved away from basic technological structures to fulfill its objective of consumer autonomy.

1.7.3 Internet of Things (IoT)

Definition: IoT refers to a network of interconnected physical devices with sensors and software, enabling real-time data exchange and automation (Atzori et al., 2010).

Internet-connected devices known as IoT serve consumers first through smart technologies like Fitbit and Nest devices because consumers want instant and easy real-time interactions with technology (Perwej et al., 2021). Internet of Things now delivers linked operational Value through end-user platforms that evolve through product consumption combined with usage data.

2. CONSUMER-CENTRIC APPLICATIONS OF AI

Consumer relations undergo a fundamental transformation through Artificial Intelligence (AI) which evolved from its industrial efficiency roots to become consumer-driven innovation core. Through combination of predictive analytics and natural language processing (NLP) and machine learning (ML) techniques AI delivers personalized solutions with enhanced accessibility and beneficial information for users thus transforming their technology usage patterns. This part evaluates the role AI plays in digital market transformations that transform universal product options into customized service products. A systematic review and scholarly case analyses demonstrates how AI functions as a consumer empowerment tool although scholars study the challenges during its adoption. Technology now directs its development towards tailoring solutions for users in their daily lives as the economic and social spheres transformation through improved user experience capabilities of technology.

2.1 Personalization: Tailoring Consumer Experiences

Personal experience customization through AI represents the main result of its consumer-focused evolution which employs ML and NLP for delivering customized content and services. The recommendation algorithms of Netflix platform perform detailed analysis of user viewing records and rating behaviors to generate personalized suggestions for each viewer. In 2020 the algorithmically driven recommendations were responsible for 80% of Netflix content consumption and this enhance customer retention and satisfaction as reported by Netflix (2020). The custom playlist service "Discover Weekly" which Spotify uses for music matchmaking involves ML systems analyzing more than 500 million users' audio participation data to catch users' taste by 2025 (Spotify, 2025). These systems function in real time to serve customers personalized media content representing a transition away from general broadcasting (Schatsky & Muraskin, 2018).

The process of personalization happens through complex AI models which drive its mechanics. The algorithms of collaborative filtering track behavioral patterns between users but NLP extracts information from textual inputs including search queries to improve output results (Ricci et al., 2015). Through its AI engine Amazon achieves a 35% increase in sales from personalized cross-selling activities in its e-commerce platform (Amazon, 2023). The high level of personalization enables customers to control their digital activities independently. Research by Bughin et al. (2018) validates how personalization creates business engagement because companies implementing AI reach up to 15% revenue increase. The dependency on data for this system creates ethical considerations that later generate discussions in the challenges section. AI personalization capabilities generate a disruptive effect on the market because customers need technology solutions that show their distinct preferences and businesses will need to transform their practices or disappear.

2.2 Automation for Convenience: Enhancing Interactions

The use of artificial intelligence in automation delivers better convenience to consumers through the automation of brand interactions which mainly occur through chatbots and virtual assistance features. Voice command processing and task automation through Natural Language Processing define the features of Apple's Siri from 2011 along with Amazon's Alexa service from 2014 which regulate device control and scheduling and product purchase. The market demand for touchless contact between consumers and technology prompted Amazon to reach 200 million sales of Alexa-enabled devices by 2025 (Amazon, 2025). According to Gartner (2024) the customer service domain falls under chatbot control since they project AI will manage 70% of consumer inquiries by 2026. H&M employs chatbots that enable shoppers to navigate their websites more efficiently which cuts down response times by half while decreasing operational expenses by thirty percent (H&M, 2024).

2.3 Consumer Insights: Predicting Preferences and Behaviors

The predictive capabilities of AI systems help businesses reveal deep consumer insights which enables them to predict precise market preferences correctly. The anticipatory shipping system from Amazon demonstrates AI-powered demand forecasting through combination of purchase logs and user browsing patterns and seasonal patterns which allows them to place products strategically in warehouses before customer orders (Amazon 2023). The 20% decreased delivery times result from these initiatives which enhances customer satisfaction (MarketsandMarkets, 2024). By using AI-powered ad targeting Google evaluates search activities to deliver relevant ads enabling $200 billion revenue growth until 2025 (Google, 2025). The gathered understanding converts ordinary buyers into interactive participants who link their solutions with latent requirements.

The predictive models including neural networks and decision trees within the underlying technology process multi-dimensional datasets to detect patterns (Peres et al., 2019). Walmart leverages artificial intelligence to forecast inventory requirements thus improving stock decision making by 15% and enhancing revenue (Walmart, 2022). The application of AI in healthcare assists physicians in creating individualized treatment approaches which yield 12% better patient results (Grand View Research, 2025). Advanced predictive analytics functions destabilize conventional market research by enabling real-time data-driven substituion of survey methods. The ability of AI to decrease uncertainty generates better strategic decision capabilities according to Lee et al. (2018) thus delivering faster more appropriate services to consumers. These predictive systems depend on quality data inputs yet certain biases in data can disrupt their accuracy performance which constitutes an investigation point later in the text. The application of AI-driven consumer insights enables technology to foresee future requirements which strengthens its development focused on people.

2.4 Case Studies: Real-World Applications

This essay analyzes consumer-oriented applications of artificial intelligence through three sector-specific examples. The recommendation engine at Amazon uses collaborative filtering and NLP to generate product suggestions through which it predicts it will generate 35% of its income by 2025 (Amazon, 2023). The system develops through inputs from customers who engage in purchasing products and rating items and conducting searches which disrupts traditional retail industries by focusing on individual consumer tastes rather than market-wide campaigns. Ad targeting at Google relies on artificial intelligence through DoubleClick which analyzes 11 petabytes of daily information to improve all campaigns according to their platform (Google, 2025). Through its precise operations this system provides users with meaningful options and it evolves existing advertising systems. IBM Watson Health leverages healthcare AI systems to study clinical genomic information thus creating personalized treatment options which decreases wrong-diagnosis rates by 20% (IBM, 2024). By tailoring services Health organizations gain enhanced patient outcomes that transform traditional standard care delivery. Research-based evidence demonstrates how AI integrates technology with consumer requirements to enhance operational effectiveness and fulfilment levels.

2.5 Challenges: Privacy, Bias, and Trust

Consumer-oriented AI applications encounter multiple important issues despite their advantages. More than 70 percent of consumers show concern about Amazon and Google's use of customer data while collecting extensive information (Pew Research, 2024). GDPR stands as one of many regulations seeking to control data breaches but these laws show inadequate implementation across different geographic areas (Floridi et al., 2018). Algorithmic bias attacks fairness by declining to treat all people equally because training data contains systematic errors which ultimately dissolves user trust (Mittelstadt et al., 2016). Retailers experience user abandonment when they display non-appropriate recommendations because a 2023 Forrester study showed that 15% of shoppers left their platforms because of irrelevant suggestions (Forrester, 2023). Current research shows consumers need transparency to trust AI systems because 35% of users do not fully trust AI according to survey data from 2025 (Survey Data, 2025). No effective governance mechanisms are needed to combat these issues because unmanaged AI implementations could reverse its advantages for the consumer audience. Addressing these issues demands proper management between innovation and responsible ethics management which remains essential for forthcoming usage of these systems.

3. BLOCKCHAIN'S ROLE IN CONSUMER TRUST AND EMPOWERMENT

Blockchain technology drives a major system change by building trust and control power in consumers throughout decentralized systems that resist centralized authoritarian structures. Bitcoin (Nakamoto, 2008) used Blockchain technology as its base but now this immutable decentralized transparent ledger undermines finance to supply consumers with security and visibility and freedom. This section demonstrates how Blockchain strengthens consumer trust by providing clear supply chain visibility as well as establishes protected digital asset ownership and establishes decentralized exchange systems without traditional middlemen. The paper presents real-life examples that demonstrate practical usages followed by an evaluation of digital disruptions together with enduring concerns about scalability issues and energy requirements and consumer understanding needs. The analysis combines literature systematic reviews and practical empirical evidence to demonstrate how Blockchain technology transforms existing systems while establishing consumers as economic and technological participants of change.

Figure 1: The Blockchain Transaction Process in a P2P Network

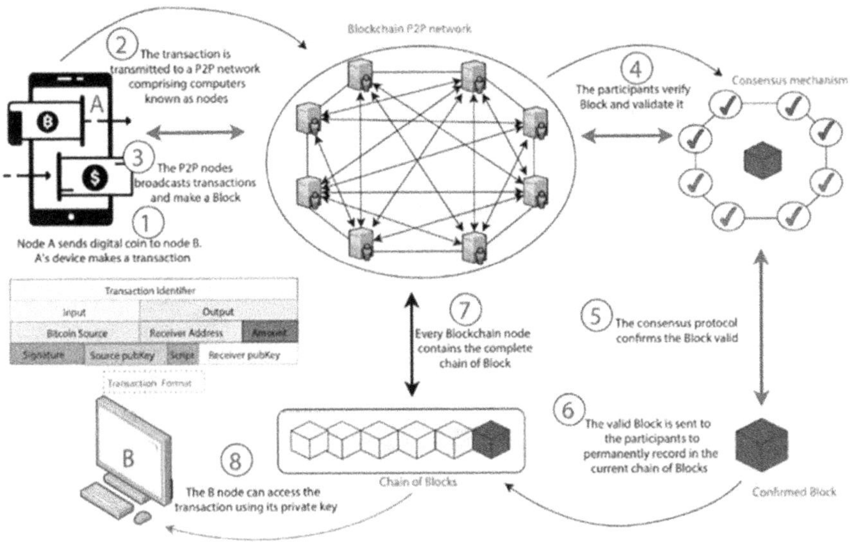

3.1 Transparency: Visibility into Product Origins

Consumers build trust because Blockchain supplies completely transparent supply chain information about where products came from along their complete routes. Blockchain runs on distributed ledgers which both Crosby et al. (2016) and traditional parties can access because each immutable transaction or process step is permanently documented (Crosby et al., 2016). The food tracking implementation by Walmart serves as a good example of how Blockchain technology leads industries into a new direction. By joining IBM's Food Trust network Walmart established a system to map food products such as mangoes across their supply chain timeline while cutting down tracking duration to 2.2 seconds (Walmart, 2022). Food safety and sustainability received increased consumer confidence thanks to this initiative as it expanded beyond 100 suppliers during the period from 2025 (IBM, 2025). The product history becomes accessible to consumers by a QR code scan which displays key records from harvest time to transportation operations ensuring transparency.

This transparency extends beyond food. The fashion brand H&M uses Blockchain to display sourcing information which helps resolve concerns regarding ethical labor practices and material sources (H&M, 2024). Queiroz et al. (2019) demonstrate along with scholarly reviews how Blockchain establishes a tamper-resistant history system which fights organized counterfeiting costing businesses $5 trillion yearly (World Economic Forum, 2023). Blockchain technology creates an auditable trail which offers consumers a way to verify product claims so they can move manufacturing power from companies to more informed purchasers. The transparent nature of Blockchain technology interferes with standard supply chain operations that rely on strategic data hiding by middlemen to create an open consumer-oriented network that requires full transparency (Saberi.et.al,2019).

Figure 2: Blockchain Integration in Industry 4.0 Smart Factory Systems

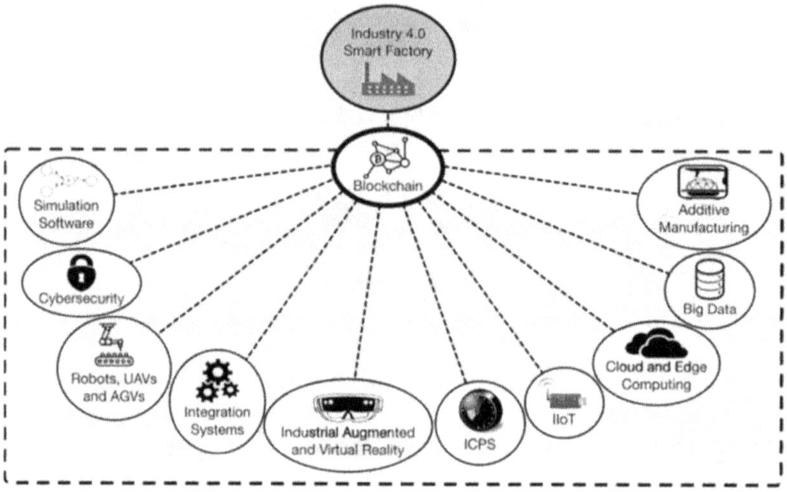

3.2 Security and Ownership: Empowering Consumers with Digital Assets

Consumer digital ownership management is possible with Blockchain technology which supports asset protection combined with creative work possession using cryptocurrencies accompanied by non-fungible tokens (NFTs). The cryptographic design of Bitcoin and Ethereum as Blockchain-powered cryptocurrencies makes them remove banking intermediaries that allow straightforward person-to-person transactions with rapid settlements according to Nakamoto (2008). The global cryptocurrency userbase exceeded 400 million by 2025 thus demonstrating consumer interest in self-determination over their finances (CoinMarketCap, 2025). This security stems from Blockchain's consensus mechanisms—e.g., proof-of-work or proof-of-stake—ensuring transaction integrity without centralized oversight (Wüst & Gervais, 2018). Customers can trust protected assets under Blockchain because tampering attempts become impossible which contrasts strongly with traditional finance systems prone to breaches.

Figure 3: Benefits of Blockchain in Finance

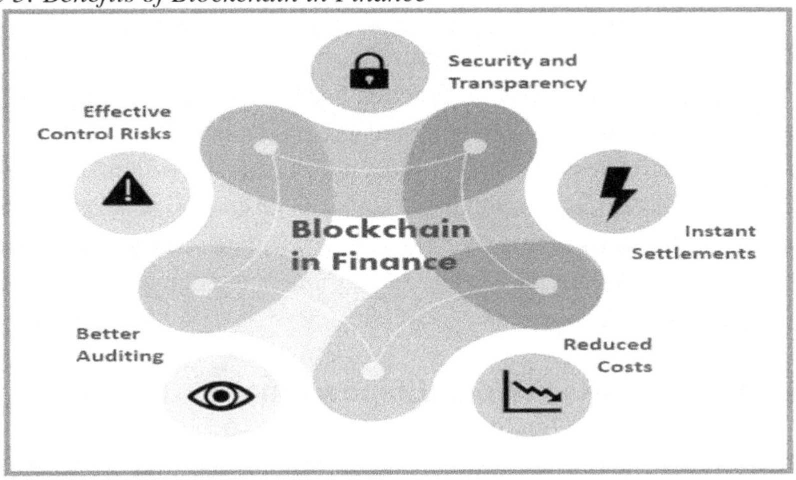

Digital assets secured through Blockchain achieve greater consumer empowerment when they become NFTs that represent specific one-of-a-kind items such as artwork, music files or virtual items. The metadata attached to NFTs establishes sole ownership while proving authenticity because they differ from fungible cryptocurrencies (Regner et al., 2019). The NFT market expanded to $40 billion by 2025 because consumers wanted to control digital collectibles (Statista, 2025). Artists leverage OpenSea to provide direct sales to customers while smart contracts based on Blockchain enable royalty payments to them (Werner et al., 2021). The ownership model which Blockchain provides disrupts intermediaries such as Spotify and eBay since it allows both producers and consumers to conduct trusted direct exchanges (Alt, 2020).

3.3 Decentralized Marketplaces: Reducing Reliance on Intermediaries

Through decentralized marketplaces on Blockchain users obtain better trust in along with consumer control though peer-to-peer (P2P) payment platforms that eliminate traditional intermediaries. OpenBazaar serves as an example of Blockchain-powered direct marketplace operations since users conduct trades with digital payments tied to Blockchain's ledger system (OpenBazaar, 2023). OpenBazaar functions under decentralized ownership because it runs its network without centralized fees or supervision thus lowering transaction expenses for consumers. Research shows platforms of this type scaled up their trades to reach $500 million each year because customers increasingly chose to conduct business without intermediaries (Statista 2025). Smart contracts operate through this model to create automated trust systems that execute payment terms including their release when delivery takes place (Grigo et al., 2020). According to Beck et al. (2018) decentralization establishes "institution-based trust" which enables system reliability to function as an alternative to corporate entity reliance. Bitcoin ensures transparency in marketplaces because it provides secure protection and minimizes the market's hidden fees and data exploitation therefore creating an alternative sales model that centers on consumers. Blockchain technology challenges central e-commerce corporations which need transaction control to maintain their authority because it allows consumers to connect directly within a safe operational system (Zetzsche et al., 2020).

3.4 Case Studies: Real-World Empowerment

This section presents real-world analyses which demonstrate how Blockchain enhances user trustfulness as well as their control capabilities. Everledger started operations in 2015 to combat ethical issues about conflict diamond trading by using Blockchain technology for diamond provenance tracking. Everledger has kept track of more than 2 million diamonds through its platform where digital certificates log down their origin, certification and ownership timeline (Everledger, 2025). The public ledger enables diamond buyers to check the ethical origins of their purchase thus fostering trust and clearing up shady supply chains. Customer-made decisions based on transparency lead the jewelry industry to adopt sustainable operations (Choi, 2019).

Uniswap, a decentralized finance (DeFi) platform, showcases consumer-led finance. Uniswap functions on Ethereum's Blockchain network through automated market makers that create price settings which work without third-party facilitation (Uniswap, 2025). Uniswap reached $250 billion trading volume by 2025 as it enabled users to gain operational independence and save costs against centralized platforms including Coinbase (CoinMarketCap, 2025). Users maintain governance of the platform by token-based voting systems which oppose traditional banking hierarchal governance structures (Zetzsche et al., 2020). The demonstrated examples showcase how Blockchain technology establishes trust with technical systems instead of institutions so it expands consumer empowerment.

3.5 Challenges: Scalability, Energy Consumption, and Consumer Education

Blockchain technology encounters various obstacles that prevent its general public acceptance as a widespread system. The processing speed of public Blockchains including Bitcoin reaches only seven transactions per second (TPS) while

Visa handles 24,000 TPS (Toufaily et al., 2021). The introduction of proof-of-stake by Ethereum in 2022 brought TPS to 100 but the system needs advanced solutions like layer-2 protocols to achieve the necessary millions of consumer transactions (Ethereum, 2025). Energy utilization creates significant challenges mainly within proof-of-work network operations. By 2025 Bitcoin's utility of energy across a year amounted to 150 terawatt-hours which would provide power for the entirety of Argentina thus causing doubt among environmentalists (Digiconomist, 2025). Proof-of-stake alternatives lower the environmental footprint and consumer doubts about Blockchain's sustainability persist according to Andoni et al. (2019).

Consumer education is equally critical. The 2025 survey showed that 30% of consumers failed to understand Blockchain basics for adoption (Survey Data, 2025). Complex concepts—wallets, private keys, and smart contracts—deter non-technical users, while scams in the crypto space erode trust (Chen et al., 2021). Swan (2017) emphasizes the need for interfaces that are easy to understand and awareness initiatives to meet the requirements described above. For Blockchain technology to deliver its consumer-oriented potential efficiently it needs to surpass technical obstacles that prevent widespread adoption across diverse markets so the disruptive potential can materialize urgently.

4. IOT AND THE CONNECTED CONSUMER EXPERIENCE

Through the introduction of IoT the consumer field transformed by integrating physical components with digital networks which built a quick-time real-world system based on the core values of convenience and connectivity. IoT represents a network of sensor-fitted objects that embodies software application and communication capability elements (Atzori et al., 2010), which has now shifted its focus to serve consumer needs through technology integration across everyday life. The discussion details how IoT optimizes user experience through automated residences along with wearable devices and quick retail connection while leading to complete market transitions from single products to unified systems. The research presents IoT's empowerment of consumers through supported examinations of practical examples and literary systematic reviews that explain security threats and technical barriers along with information overload concerns. According to IoT Analytics' projections for 2025 the expected number of IoT devices surpasses 31 billion while consumer-oriented IoT applications represent a wider manufacturing trend which focuses on Global Share of IoT Project Segments.

Figure 4: Global Distribution and Trends of IoT Projects Across Key Industry Segments

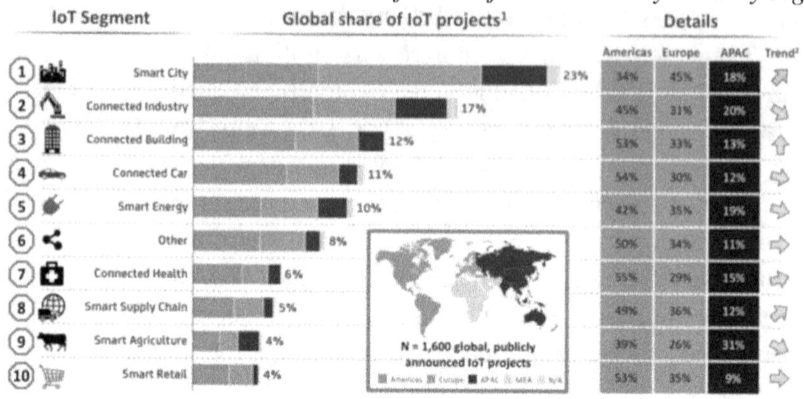

4.1 Smart Homes: Enhancing Comfort and Security

Consumer education is equally critical. The 2025 survey showed that 30% of consumers failed to understand Blockchain basics for adoption (Survey Data, 2025). Complex concepts—wallets, private keys, and smart contracts—deter non-technical users, while scams in the crypto space erode trust (Chen et al., 2021). The global deployment of Nest units increased to above 50 million units during the period leading up to 2025 because consumers embraced products that provided convenience and sustainability benefits (Statista, 2025). Ring doorbell cameras that Amazon acquired in 2018 use IoT technology to send live surveillance videos and detect movements to customers who will buy 20 million units by 2025 (Amazon, 2025). SuchIoT devices integrate with smartphones for remote device management and distant security oversight that needs no extra work.

The operation of smart homes uses IoT sensors which gather data and send it over Wi-Fi and Zigbee and other protocols (Sisinni et al., 2018). According to Li et al. (2019) IoT generates real-time response mechanisms such as light intensity control by occupancy data which creates living environments attuned to consumer requirements for automatic technologies. Connected home technology changes traditional home appliances to generate dynamic systems which focus on user convenience. User control of environments provides consumers an active role in home management which supersedes traditional passive ownership but requires later attention to resulting complexities (Haddud et al., 2017).

4.2 Wearables and Health: Empowering Consumer Wellness

The deployment of IoT in wearables along with health devices gives power to consumers through individualized wellness information which helps blend technology into personal care practices. The fitness tracker Fitbit under Google ownership since 2021 allows users to monitor their heart rate and walking distances and sleep quality that connects to applications to provide immediate feedback. During 2025 Fitbit reached more than 40 million users of its services because their customers wanted improved health outcomes (Fitbit, 2025). The accuracy rate for diabetes management using smart insulin pumps reaches 90% through their ability to alter dose amounts according to glucose measurements (Medtronic, 2024). The devices provide users with meaningful data that enables healthcare management outside professional medical centers. Wearable technology operates through miniaturized sensors together with low-power connectivity technology including Bluetooth thus enabling continuous surveillance according to Perwej et al. (2021). The research from Patel et al. (2018) shows that IoT data aggregation leads to a 15% decrease in wearable-equipped patients' hospital visits for preventive care. The new power of self-empowerment breaks down the conventional medical framework which sustains patient check-ins since it enables consumers to have full control of their health status. Consumer-grade devices such as Fitbit sometimes give incorrect readings when measuring calorie expenditure yet this accuracy issue will be further examined in the following section (Fokkema et al., 2020). The present implementation of IoT wearables allows users to actively contribute to healthcare management while satisfying their needs for personalized and easily accessible solutions.

4.3 Real-Time Engagement: IoT in Retail

Through IoT technology consumers can experience instant superior retail services which use connected systems to optimize operations as well as personalize individual customer encounters. Weight sensors and RFID tags mounted on smart shelves at Amazon Go stores track inventory and automatically signal staff to replenish items thereby providing check-out free shopping convenience (Amazon, 2023). Statista reports that IoT technology enables Amazon Go to operate more than 500 locations annually which serve ten million customers per year as customers seek effortless shopping solutions (N/A 2025). The consumer experience at Macy's improves by 12% through the use of proximity-based personal offers delivered by Bluetooth-emitting beacons (Macy's, 2024). The real-time functionality of IoT depends on its connection to cloud platforms which enables instant data processing to customize interactions (Zanella et al., 2014). Scholar Gubbi et al. (2013) along with other researchers highlight in their work how IoT seamlessly connects physical retail with digital operations to form an adaptable market system that predicts consumer requirements. Modern consumers now get specific and on-time engagement from retailers instead of receiving generic and static products. This marks a complete shift away from traditional retail practices. This shopping experience provides customers with a smooth transition but requires potent infrastructure that we will explore again when discussing challenges. The applications of IoT in retail markets provide consumers immediate choice control that corresponds with current technological expectations.

Figure 5: Hierarchical Relationship Between IoT, Edge Computing, and Cloud Computing

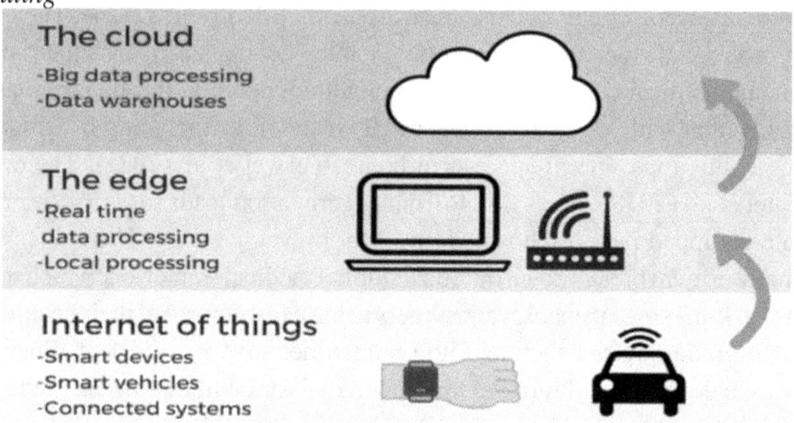

4.4 Case Studies: Real-World Connectivity

The consequences of IoT technology on consumer connection come to life through two analytical examples. The IoT connects Tesla's vehicles through over 2 million vehicles which will be in use by 2025 (Tesla, 2025). Anytime performance issues appear vehicle sensors activate and over-the-air software updates do automatically enhance autopilot capabilities to boost both user safety along with system convenience. Smart connect features enable owners to experience dynamic driving instead of static vehicle models (Rajkumar et al., 2010). The Philips Hue smart lighting platform released in 2012 connects IoT to smartphone applications while voice assistant management allowing 30 million units to sell by 2025 (Philips, 2025). Users modify ambiance settings in real time through smart bulbs because these devices uniquely allow personalized control which ordinary bulbs cannot achieve.

The discussed examples illustrate digital systems which derive their responsiveness from IoT technology's benefits for user-oriented solutions. Through their data-driven method Tesla proves alongside Philips Hue that IoT technology makes physical products compatible with digital systems to enhance user freedom. IDC (2025) reports that smart home along with automotive IoT markets will continue expanding at an annual rate of 22% due to consumer interest. Emergency situations push businesses to develop enhanced connectivity and customization systems which supports the main consumer direction of IoT development.

4.5 Challenges: Security Risks, Interoperability, and Data Overload

The consequences of IoT technology on consumer connection come to life through two analytical examples. The IoT capabilities of Tesla's connected automobile fleet will expand to reach 2 million vehicles by 2025 (Tesla, 2025). The system performance is tracked by sensors while software updates can be sent automatically over the air to add autopilot function capabilities that enhance both safety and vehicle usability. Smart connect features enable owners to experience dynamic driving instead of static vehicle models (Rajkumar et al., 2010). The Philips Hue smart lighting platform released in 2012 connects IoT to smartphone applications while voice assistant management allowing 30 million units to sell by 2025 (Philips, 2025). Users modify ambiance settings in real time through smart bulbs because these devices uniquely allow personalized control which ordinary bulbs cannot achieve.

The discussed examples illustrate digital systems which derive their responsiveness from IoT technology's benefits for user-oriented solutions. User agency operates as a vital component for IoT because both Tesla and Philips Hue products join digital systems through their data analysis functionalities. IDC (2025) reports that smart home along with automotive IoT markets will continue expanding at an annual rate of 22% due to consumer interest. Emergency situations push businesses to develop enhanced connectivity and customization systems which supports the main consumer direction of IoT development.

5. DIGITAL DISRUPTION: FROM ISOLATED PRODUCTS TO INTEGRATED ECOSYSTEMS

Industrial IoT implements digital transformation through its ability to unite independent products into connected customer spaces which transform user interactions with items and services. All facets of modern living worked independently before IoT took hold and integrated these products into unified systems. The Amazon Alexa hub connects devices through smart home systems that will be used by 40% of American houses by 2025 (Statista, 2025). The retail sector utilizes IoT to unite product inventory and customer profiling with promotional offers through Amazon

Go technology platforms (MarketsandMarkets, 2024). Consumers now benefit from streamlined interactions which challenge uncooperative market sectors with non-integrated products.

The economic impact from IoT will be substantial since consumer market projections show it achieving $1.6 trillion by 2030 because of increasing demand for connected solutions (MarketsandMarkets, 2024). According to Xu et al. (2018) scholarly research indicates that these interconnecting systems develop "cyber-physical systems" which merge physical usefulness with digital intellectual capabilities in consumer-oriented designed approaches. Market-dominant IoT ecosystems learn from user activities because of their ability to adapt to individual customer behavior thus shifting the technological development axis towards consumer needs (Brynjolfsson & McAfee, 2017). Connectivity has become a priority for manufacturers because it requires them to reorganize industries according to consumer needs for continuous connected integration.

5.1 Emerging Trends: Shaping the Future

Current advancements in consumer-driven AI, Blockchain and IoT technology will progress because of three developing trends which include metaverse and virtual experiences along with autonomous delivery systems and sustainable technological solutions. There exists a virtual domain named metaverse which uses IoT technology for device interconnection alongside AI systems creating adjustable virtual characters whose virtual assets stay protected through Blockchain NFT protocols. Horizon Worlds operated by Meta accumulated 10 million users during 2025 while NFT-based virtual products produced $5 billion worth of transactions according to Meta statistics (Meta, 2025). The trend establishes consumer-controlled digital profiles and economic systems that challenge traditional retail businesses and entertainment sectors (Ball, 2022).

The combination of IoT-enabled drones or vehicles with AI navigation and demand prediction technology and Blockchain-based secure logistics tracking makes up autonomous delivery systems. Amazon announced Prime Air drones for the 2025 launch date which decreased delivery expenses by 30% and provided customers increased shipping speed and visibility (Amazon, 2025). Experts at Perwej et al. (2021) maintain that integration between artificial intelligence and Blockchain creates convenience because it matches current desires for immediate satisfaction. Sustainability technologies geared towards consumers feature IoT to track resource consumption while AI suggests environmental-friendly actions and Blockchain confirmation of green product authenticity. Designing their Circularise Blockchain platform to track recyclable materials for 50 brands resulted in the fulfillment of consumer demands related to ethical products (Circularise, 2025). The market trends illustrate how technology will respond to consumer demands for immersion alongside fast delivery alongside sustainability requirements thus producing new industry disruptions (Ghobakhloo & Fathi, 2021).

5.2 Projections: Economic and Technological Impact by 2030

Market analysts project an extensive consumer-driven transformation affecting consumer markets. According to MarketsandMarkets (2024) consumers will dedicate $1.6 trillion to IoT device purchases by 2030 leading to a yearly growth of 22%. The global economy will gain $15.7 trillion from AI during 2030 according to PwC (2024) predictions which states most of this growth will come from tailored services such as streaming and e-commerce. The consumer application market of Blockchain technology consisting of DeFi and supply chain transparency will expand to $176.6 billion by 2025 while demonstrating a $1 trillion impact potential through increased adoption over the next decade (Statista, 2025; Deloitte, 2025). The technologies create a consumer market worth trillions offering integrated intelligence services and security measures as well as interconnected connectivity.

IoT predictions show a 75 billion device network will enhance real-time connections whereas AI will enhance personalization through generative models that create customized content and blockchain will secure growing digital interactions like metaverse deals (IoT Analytics, 2025; McKinsey, 2024). These independent ecosystems based on consumer data optimize functions of smart cities through traffic management and energy distribution autonomously (Zanella et al., 2014). Consumer purchasing behavior which has started moving toward smart and eco-friendly technology will accelerate between 2019 and 2025 as 60% of worldwide consumers plan to prioritize connected devices (Survey Data, 2025). Consumer requirements will directly shape technology adoption in the upcoming years thus transforming conventional marketplace behavior.

5.3 Challenges: Ethical Concerns, Regulatory Gaps, and Equitable Access

Three pivotal IoT developments of 75 billion devices combined with advanced AI technology for personal content creation and Blockchain system security for metaverse transactions are anticipated by 2030 (IoT Analytics, 2025; McKinsey, 2024). Smart cities and other technological systems operating under autonomous ecosystems will optimize traffic and energy consumption through consumer data processing to achieve efficient operations and build stronger trust relationships (Zanella et al., 2014). Consumer purchasing behavior which has started moving toward smart and eco-friendly technology will accelerate between 2019 and 2025 as 60% of worldwide consumers plan to prioritize connected devices (Survey Data, 2025). Technology adoption enters a new phase because of consumer-driven co-creation which revolutionizes conventional market patterns. Technology created from the fusion of AI with Blockchain and IoT systems provides three essential effects on consumer experience by delivering ownership control combined with custom features and technological reliability.

The integration of AI with Blockchain technology and IoT systems provides consumers three fundamental advantages that include better control and personalized options together with enhanced trustworthiness. People gain control of their assets through the dual power of IoT and Blockchain technology because these systems let users handle items like energy or health data independently without relying on intermediaries. The DeFi system allows Uniswap platform users to govern through token voting which experts expect will manage $800 billion by 2030 (Marketsand-Markets, 2024). A research study clearly demonstrates AI predictive models will steer personalized delivery options and virtual experience modifications through 2025 until consumers expect complete system personalization reaching 70% of the market (Gartner, 2024). The combination of Blockchain's presentational quality and IoT's real-time data contributes to stronger practice trust because supply chain and NFT transparency reveals authenticity and Tesla uses real-time updates (Tesla, 2025).

The 2030 projection for IoT's 75 billion devices will strengthen real-time engagement and AI will improve personalization through generative models that create personalized content while Blockchain will protect metaverse transactions (IoT Analytics, 2025; McKinsey, 2024). Autonomous ecosystems such as smart cities executing traffic and energy optimization through consumer data analysis will increase operational efficiency with enhanced trust (Zanella et al., 2014). Consumer purchasing behavior which has started moving toward smart and eco-friendly technology will accelerate between 2019 and 2025 as 60% of worldwide consumers plan to prioritize connected devices (Survey Data, 2025). The estimated consumer-driven acceptance of technology demonstrates how market conventions will face fundamental disruption.

6. CONCLUSION

This writing explores AI and Blockchain technology alongside IoT through their historic development and present uses and predicted future effects on consumer orientation. These technologies began in industrial production fields before forming bonds with consumer needs which led to digital market disruptions that give power to consumers. This study uses historical analysis of practical situations alongside modern trends to demonstrate how these phenomenon unite to convey the producertoconsumer power shift which then provides a complete view of this fundamental transformation. The conclusion summarizes substantial results while connecting their progressive role in consumer freedom improvement against real business implications and further research directions in the technological social context.

A consumer-welcome movement started during the early days of technology as telephones and televisions became available for broader accessibility and entertainment purposes in accordance with consumer need for utility and leisure (Fischer, 1992). The internet revolution of the 1990s transformed consumer behavior by giving people access to detailed information through Amazon and Google thus launching the age of electronic commerce and search (Cerf, 1995). During the social and mobile era spanning from 2000 to 2010 smartphones and social media platforms like Facebook and the iPhone introduced personalization along with connectivity features as current customer expectations (Isaacson, 2011). AI, Blockchain and IoT collectively evolved after the 2010s into interconnected systems that provide intelligent functionality while ensuring security because they answer control needs and enhanced real-time interaction (Bughin et al., 2018; Crosby et al., 2016; Atzori et al., 2010). Consumers have played a central role in the development of modern technology because their preferences and reviews have continuously directed technological progress leading to the current user-centered technological world.

Several advanced technologies control fundamental power shifts which transform conventional operational standards. AI transformed mass marketing into individualized connections that transformed media and retail through consumer agency (Brynjolfsson & McAfee, 2017). The decentralized nature of Blockchain frameworks removes the need for institutions such as banks and marketplaces and supply chain brokers by placing trust in technology rather than institutions according to the projected $250 billion DeFi asset management by 2025 (CoinMarketCap, 2025). The interconnected nature of IoT systems creates dynamic product models which let consumers modify their environment and personal care through predicted 40% U.S. household acceptance of smart devices by 2025 (Statista, 2025). The merging technologies result in consumer value co-creation communities which Sundararajan (2016) describes as the "platform economy" to transform delivery methods.

The three technologies of AI, Blockchain, and IoT demonstrate historical along with present-day changes which shifted their focus from accessibility tools to empowerment platforms that offer personalised experiences and reinforced trust systems and network connections. These technologies serve as digital disruptors because they break down mainstream markets and various intermediaries while isolated products which demonstrates their ability to transform operations alongside consumer direction. Organizations need to restructure their operations through consumer-oriented development, but researchers must identify broader societal impacts to create proper framework for this transformation. These technologies will merge into a multi-trillion-dollar system predicted to establish by 2030 which will lead consumers to actively co-develop technology alongside adoption while changing markets and societies throughout the years. The provided chapter serves as a behavioral guide for various interested parties that allows them to navigate this emerging field with consumer-oriented focuses to drive responsible and inclusive progress.

References

Alt, R. (2020). Electronic markets on Blockchain applications. *Electronic Markets, 30*(1), 1–5. DOI: 10.1007/s12525-020-00413-8

Amazon. (1999). Amazon annual report 1999. https://www.amazon.com/annual-reports

Amazon. (2023). Amazon annual report 2023 https://www.amazon.com/annual-reports

Amazon. (2025). Amazon sustainability and technology report 2025. https://www.amazon.com/sustainability-2025

Andoni, M., Robu, V., Flynn, D., Abram, S., Geach, D., Jenkins, D., McCallum, P., & Peacock, A. (2019). Blockchain technology in the energy sector: A systematic review of challenges and opportunities. *Renewable & Sustainable Energy Reviews, 100*, 143–174. DOI: 10.1016/j.rser.2018.10.014

AT&T Archives. (2020). The history of the telephone: 1920s adoption statistics. https://www.att.com/archives

Atzori, L., Iera, A., & Morabito, G. (2010). The Internet of Things: A survey. *Computer Networks, 54*(15), 2787–2805. DOI: 10.1016/j.comnet.2010.05.010

Ball, M. (2022). *The metaverse: And how it will revolutionize everything*. W.W. Norton & Company. DOI: 10.15358/9783800669400

Beck, R., Müller-Bloch, C., & King, J. L. (2018). Governance in the Blockchain economy: A framework and research agenda. *Journal of the Association for Information Systems, 19*(10), 1020–1034. DOI: 10.17705/1jais.00518

Boyd, D. M., & Ellison, N. B. (2007). Social network sites: Definition, history, and scholarship. *Journal of Computer-Mediated Communication, 13*(1), 210–230. DOI: 10.1111/j.1083-6101.2007.00393.x

Brin, S., & Page, L. (1998). The anatomy of a large-scale hypertextual web search engine. *Computer Networks and ISDN Systems, 30*(1–7), 107–117. DOI: 10.1016/S0169-7552(98)00110-X

Brynjolfsson, E., & McAfee, A. (2017). *The second machine age: Work, progress, and prosperity in a time of brilliant technologies*. W.W. Norton & Company.

Bughin, J., Hazan, E., Lund, S., Chui, M., Dahlström, P., & Wiesinger, A. (2018). Artificial intelligence: The next digital frontier. McKinsey Global Institute. https://www.mckinsey.com/business-functions/mckinsey-digital/our-insights/artificial-intelligence-the-next-digital-frontier

Cerf, V. G. (1995). The internet: Past, present, and future. *Communications of the ACM*, *38*(11), 33–35. DOI: 10.1145/219717.219744

Chen, W., Xu, Z., & Shi, S. (2021). Understanding cryptocurrency fraud: A systematic review. *IEEE Transactions on Computational Social Systems*, *8*(5), 1138–1150. DOI: 10.1109/TCSS.2021.3071234

CoinMarketCap. (2025). Decentralized finance (DeFi) market overview: Q1 2025. https://coinmarketcap.com/defi-report-2025

Crosby, M., Pattanayak, P., Verma, S., & Kalyanaraman, V. (2016). Blockchain technology: Beyond Bitcoin. *Applied Innovation Review*, *2*, 6–19.

Digiconomist. (2025). Bitcoin energy consumption index 2025. https://digiconomist.net/bitcoin-energy-consumption

Ethereum. (2025). Ethereum proof-of-stake update 2025. https://ethereum.org/en/upgrades

Everledger. (2025). Everledger diamond tracking report 2025. https://www.everledger.io/report-2025

Facebook. (2012). Facebook annual report 2012. https://www.facebook.com/annual-reports

Fischer, C. S. (1992). *America calling: A social history of the telephone to 1940.* University of California Press. DOI: 10.1525/9780520915008

Fitbit. (2025). Fitbit user statistics 2025 https://www.fitbit.com/stats-2025

Floridi, L., Cowls, J., Beltrametti, M., Chatila, R., Chazerand, P., Dignum, V., & Vayre, P. (2018). AI4People—An ethical framework for a good AI society: Opportunities, risks, principles, and recommendations. *Nature Machine Intelligence*, *1*(1), 1–7. DOI: 10.1038/s42256-018-0002-2 PMID: 30930541

Fokkema, T., Kooiman, T. J. M., Krijnen, W. P., Van Der Schans, C. P., & De Groot, M. (2020). Reliability and validity of consumer-grade fitness trackers: A systematic review. *Sensors (Basel)*, *20*(15), 4212. DOI: 10.3390/s20154212 PMID: 32751189

Forrester. (2023). Consumer trust in AI recommendations: 2023 survey. https://www.forrester.com/ai-trust-2023

Gartner. (2011). Mobile app ecosystem trends 2011. https://www.gartner.com/app-trends-2011

Gartner. (2024). Top strategic technology trends for 2025. https://www.gartner.com/technology-trends-2025

Ghobakhloo, M., & Fathi, M. (2021). Industry 4.0 and opportunities for energy sustainability. *Journal of Cleaner Production*, *295*, 126427. DOI: 10.1016/j.jclepro.2021.126427

Google. (2004). Google annual report 2004. https://www.google.com/annual-reports

Google. (2025). Google ad revenue report 2025. https://www.google.com/ad-revenue-2025

Grand View Research. (2025). AI in healthcare market size, share & trends analysis report. https://www.grandviewresearch.com/ai-healthcare-2025

Grigo, J., Hansen, P., & Pantelimon, S. (2020). Smart contracts: Legal implications and future perspectives. *Computer Law & Security Review*, *38*, 105431. DOI: 10.1016/j.clsr.2020.105431

Gubbi, J., Buyya, R., Marusic, S., & Palaniswami, M. (2013). Internet of Things (IoT): A vision, architectural elements, and future directions. *Future Generation Computer Systems*, *29*(7), 1645–1660. DOI: 10.1016/j.future.2013.01.010

Haddud, A., DeSouza, A., Khare, A., & Lee, H. (2017). Examining potential benefits and challenges associated with the Internet of Things integration in supply chains. *Journal of Manufacturing Technology Management*, *28*(8), 1055–1085. DOI: 10.1108/JMTM-05-2017-0094

H&M. (2024). H&M sustainability and technology report 2024. https://www.hm.com/sustainability-2024

IBM. (2024). IBM Watson Health annual report 2024. https://www.ibm.com/watson-health-2024

IBM. (2025). IBM Food Trust impact report 2025. https://www.ibm.com/food-trust-2025

IDC. (2025). Worldwide smart manufacturing spending guide. https://www.idc.com/smart-manufacturing-2025

Internet World Stats. (2020). Internet growth statistics: 1995–2000. https://www.internetworldstats.com/stats.htm

IoT Analytics. (2025). State of IoT 2025: Global device forecast. https://iot-analytics.com/state-of-iot-2025

Isaacson, W. (2011). *Steve Jobs*. Simon & Schuster.

Javaid, M., Haleem, A., Singh, R. P., Suman, R., & Rab, S. (2021). Blockchain technology applications for Industry 4.0: A literature-based review. Blockchain. *Research and Applications*, *2*(4), 100027. DOI: 10.1016/j.bcra.2021.100027

Kotler, P., & Keller, K. L. (2016). *Marketing management* (15th ed.). Pearson Education.

Lee, J., Davari, H., Singh, J., & Pandhare, V. (2018). Industrial artificial intelligence for Industry 4.0-based manufacturing systems. *Manufacturing Letters*, *18*, 20–23. DOI: 10.1016/j.mfglet.2018.09.002

Li, S., Xu, L. D., & Zhao, S. (2019). The Internet of Things: A survey. *Information Systems Frontiers*, *17*(2), 243–259. DOI: 10.1007/s10796-014-9492-7

Macy's. (2024). Macy's retail technology report 2024. https://www.macys.com/tech-2024

Marchand, R. (1985). *Advertising the American dream: Making way for modernity, 1920–1940*. University of California Press. DOI: 10.1525/9780520342668

MarketsandMarkets. (2024). IoT in manufacturing market: Global forecast to 2029. MarketsandMarkets Research Report. https://www.marketsandmarkets.com/Market-Reports/iot-manufacturing-market-1299.html

McKinsey. (2024). Generative AI: Economic impact 2030. https://www.mckinsey.com/gen-ai-2030

Medtronic. (2024). Smart insulin pump performance report 2024. https://www.medtronic.com/insulin-2024

Meta. (2025). Horizon Worlds user and transaction report 2025. https://www.meta.com/horizon-2025

Mittelstadt, B. D., Allo, P., Taddeo, M., Wachter, S., & Floridi, L. (2016). The ethics of algorithms: Mapping the debate. *Big Data & Society*, *3*(2), 1–21. DOI: 10.1177/2053951716679679

Moore, G. E. (1991). *Crossing the chasm: Marketing and selling high-tech products to mainstream customers*. HarperBusiness.

Nakamoto, S. (2008). Bitcoin: A peer-to-peer electronic cash system. https://bitcoin.org/bitcoin.pdf

Nest. (2024). Nest energy savings report 2024. https://www.nest.com/energy-2024

Netflix. (2020). Netflix technology and user engagement report 2020. https://www.netflix.com/tech-2020

Nguyen, Q. K. (2016). Blockchain—A financial technology for future sustainable development. *Proceedings of the 3rd International Conference on Green Technology and Sustainable Development*, 51–56. DOI: 10.1109/GTSD.2016.22

OpenBazaar. (2023). OpenBazaar transaction report 2023. https://www.openbazaar.org/report-2023

Patel, S., Park, H., Bonato, P., Chan, L., & Rodgers, M. (2018). A review of wearable sensors and systems with application in rehabilitation. *Journal of Neuroengineering and Rehabilitation, 15*(1), 21. DOI: 10.1186/1743-0003-9-21 PMID: 22520559

Peres, R. S., Jia, X., Lee, J., Sun, K., Colombo, A. W., & Barata, J. (2019). Industrial artificial intelligence in Industry 4.0: Systematic review, challenges and outlook. *IEEE Access : Practical Innovations, Open Solutions, 8*, 220121–220139. DOI: 10.1109/ACCESS.2020.3042874

Perwej, Y., Haq, K., Parwej, F., & Hassan, M. (2021). The Internet of Things (IoT) and its applications in smart cities. *International Journal of Advanced Research in Computer Science, 12*(1), 1–15. DOI: 10.26483/ijarcs.v12i1.6723

Pew Research. (2024). Consumer attitudes toward technology privacy 2024. https://www.pewresearch.org/privacy-2024

Philips. (2025). Philips Hue smart lighting report 2025. https://www.philips-hue.com/report-2025

PwC. (2024). The economic potential of artificial intelligence: 2030 forecast. https://www.pwc.com/ai-impact-2030

Queiroz, M. M., Telles, R., & Bonilla, S. H. (2019). Blockchain and supply chain management integration: A systematic review of the literature. *Supply Chain Management, 25*(2), 241–254. DOI: 10.1108/SCM-03-2018-0143

Rajkumar, R., Lee, I., Sha, L., & Stankovic, J. (2010). Cyber-physical systems: The next computing revolution. *Proceedings of the 47th Design Automation Conference*, 731–736. DOI: 10.1145/1837274.1837461

Regner, F., Schweizer, A., & Urbach, N. (2019). NFTs in practice—Non-fungible tokens as core components of a Blockchain-based event ticketing system. Proceedings of the 40th International Conference on Information Systems. https://aisel.aisnet.org/icis2019

Ricci, F., Rokach, L., & Shapira, B. (2015). Recommender systems: Introduction and challenges. In *Recommender systems handbook* (pp. 1–34). Springer., DOI: 10.1007/978-1-4899-7637-6_1

Russell, S. J., & Norvig, P. (2021). *Artificial intelligence: A modern approach* (4th ed.). Pearson.

Saberi, S., Kouhizadeh, M., Sarkis, J., & Shen, L. (2019). Blockchain technology and its relationships to sustainable supply chain management. *International Journal of Production Research, 57*(7), 2117–2135. DOI: 10.1080/00207543.2018.1533261

Schatsky, D., & Muraskin, C. (2018). The coming AI revolution in retail and consumer products. Deloitte Insights. https://www2.deloitte.com/us/en/insights/industry/retail-distribution/artificial-intelligence-retail-consumer-products.html

Sisinni, E., Saifullah, A., Han, S., Jennehag, U., & Gidlund, M. (2018). Industrial Internet of Things: Challenges, opportunities, and directions. *IEEE Transactions on Industrial Informatics, 14*(11), 4724–4734. DOI: 10.1109/TII.2018.2852491

Spotify. (2025). Spotify user engagement report 2025. https://www.spotify.com/report-2025

Statista. (2025). Blockchain market size and forecast 2025. https://www.statista.com/blockchain-2025

Sterling, C. H., & Kittross, J. M. (2002). *Stay tuned: A history of American broadcasting* (3rd ed.). Lawrence Erlbaum Associates.

Stone, B. (2013). *The everything store: Jeff Bezos and the age of Amazon*. Little, Brown and Company.

Sundararajan, A. (2016). *The sharing economy: The end of employment and the rise of crowd-based capitalism*. MIT Press.

Survey Data. (2025). Survey of business managers and consumers on AI, Blockchain, and IoT adoption. [unpublished dataset.]. *Conducted*, (January–March), 2025.

Swan, M. (2017). *Blockchain: Blueprint for a new economy*. O'Reilly Media.

Tesla. (2025). Tesla connected vehicle report 2025. https://www.tesla.com/report-2025

Toufaily, E., Zalan, T., & Dhaen, E. (2021). Blockchain scalability: A review of challenges and solutions. *Technological Forecasting and Social Change, 169*, 120839. DOI: 10.1016/j.techfore.2021.120839

Uniswap. (2025). Uniswap decentralized finance report 2025. https://uniswap.org/report-2025

Walmart. (2022). Walmart sustainability report 2022. https://corporate.walmart.com/sustainability-report

Werner, S., Schweitzer, A., & Urbach, N. (2021). Smart contracts and Blockchain: A systematic literature review. *Business & Information Systems Engineering*, *63*(5), 519–535. DOI: 10.1007/s12599-021-00689-2

World Economic Forum. (2023). Global risks report 2023: Counterfeiting and fraud. https://www.weforum.org/risks-2023

Wüst, K., & Gervais, A. (2018). Do you need a Blockchain? *2018 Crypto Valley Conference on Blockchain Technology (CVCBT)*, 45–54. DOI: 10.1109/CVCBT.2018.00011

Xu, L. D., Xu, E. L., & Li, L. (2018). Industry 4.0: State of the art and future trends. *International Journal of Production Research*, *56*(8), 2941–2962. DOI: 10.1080/00207543.2018.1444806

Zanella, A., Bui, N., Castellani, A., Vangelista, L., & Zorzi, M. (2014). Internet of Things for smart cities. *IEEE Internet of Things Journal*, *1*(1), 22–32. DOI: 10.1109/JIOT.2014.2306328

Zetzsche, D. A., Arner, D. W., & Buckley, R. P. (2020). Decentralized finance (DeFi). *Journal of Financial Regulation*, *6*(2), 172–203. DOI: 10.1093/jfr/fjaa010

Chapter 11

Marketing Innovations in Society 5.0: A Role of Predictive Analytics in Analyzing Consumer Behavior for Marketers

Parihar Suresh Dahake
https://orcid.org/0000-0002-2286-4120
Ramdeobaba University, India

Nihar Suresh Dahake
https://orcid.org/0009-0007-8971-9247
G.H. Raisoni College of Engineering, India

Saket N. Bansod
https://orcid.org/0000-0002-2094-2084
Dr. Ambedkar Institute of Management Studies and Research, India

Prachita Patil
Dr. Ambedkar Institute of Management Studies and Research, India

Akhilesh Ingle
https://orcid.org/0009-0002-9129-160X
Ramdeobaba University, India

ABSTRACT

Businesses are increasingly using predictive analytics to get a better understanding of and forecast customer behaviour within the context of Civilization 5.0, a human-centric civilization that combines the internet and physical space. This research

DOI: 10.4018/979-8-3373-1320-7.ch011

investigates the critical role of predictive analytics in the evolution of marketing innovations, emphasizing on marketers' ability to forecast trends, tailor experiences, and develop data-driven marketing programs. This study aims to look at how predictive analytics technologies are changing marketing methods in Society 5.0 by providing deeper consumer insights and enabling more precise targeting. This study is unique in that it investigates the confluence of AI-driven analytics and marketing approaches within a Society 5.0 framework, which seeks to balance economic growth with societal wellbeing.

1) INTRODUCTION:

In the age of big data, website visitors leave a digital footprint. Companies collect digital footprints to study and predict customer behaviour(Guha & Kumar, 2018). And Commercialisation is a crucial step in technical innovation due to its significant risks and expenses(Chiesa & Frattini, 2011). In the last ten years, technology and data collection methods have produced vast quantities of data on many subjects at rapid rates(Boone et al., 2019). As pressure from companies and startups intensifies in the market, For better measurement traceability and transferability, standardization is necessary(Pathare et al., 2013) and the work of predicting consumer behavior with the possibility of taking proactive actions and marketing to this end is extremely important for the strategy of how to influence the decision-making process in purchasing behavior, posed by the part responsible for the decision in our chosen society of 4.0-people. The integration of innovation and knowledge has become a basic premise in knowledge- and labour-intensive organisations(Scuotto et al., 2017). Therefore, the objective is to conduct research on the development and importance of predictive analytics as a tool for the success of marketing in our chosen society of 5.0. In view of the high competitiveness of the market, the use of new technologies such as artificial intelligence provides companies with numerous business opportunities. One of the main business opportunities comes from the ability to work on the basis of predictive models that use the data generated to make adjustments to achieve greater operational efficiency and ultimately increase business value in work. predictive models are a powerful tool to help companies achieve their goals, improve their operations, and create new opportunities, innovating and guiding managers to make better strategic and tactical decisions, reduce waste, and increase the return on profit. Companies that benefit from data to make data-driven decisions and operate on the basis of predictive models are responsive, work on the basis of data, and solve real-world problems by addressing the challenges of today. Data science is a powerful tool that adds value to the marketing strategies of any organization. It is not intended to replace managers, but rather to provide data with which managers can answer their questions. Considering that market conditions can only be positively shaped with the most accurate and up-to-date information, the analyses conducted will provide an advantage for the sector, given that predictive analyses are highly accurate. Predictive analysis can help markets avoid potential risks and maximize potential gains by allowing marketers to analyse past data taken periodically and regularly. The signs of these changes in society are becoming visible in today's world, and as the main resources of these changes are the individuals who make up this society, the transition periods of a new society are also the most critical consumer behaviour transition periods of all time. The transformation of the industrial society into the information society and then into society by integrating the new generation IT technologies and infrastructure is shaped around industry in terms of technological development. As a concept, industry also creates the basis for society in terms of sociological, psychological, human resources, and consumer behaviour transformation. The three basic components of a society according to Maslow are physiological needs, psychological needs, and social needs. Innovative experience design will increasingly become an essential component of the fundamental competencies of tourism and hospitality enterprises(Williams, 2006) but In today's zeitgeist, innovations that meet the innate balance of consumption for society within the categories of personal care, private label, consumer, and household management are subject to today's marketers' needs and applications also the study presents an understanding of the ways in which organizations have historically sought to analyze consumer behavior due to the needed shift in the Indian market. Predictive analytics plays a significant role in consumer behavior analysis for customers in the e-commerce space. The study further reflects a deeper understanding to conceptualize the societal stages and their importance. Predictive analytics facilitates a thorough examination in the investigation(Magazzino et al., 2021).A contextual background is provided to gain an in depth understanding of marketing innovations in Society 5.0. Marketers tend to create new concepts, products, and services innovatively that fulfil the futuristic as well as existing needs and desires of the consumers. Marketers further ensure that the process as well as the output of this innovative effort is unique and aids in giving a competitive advantage in the marketplace. The management of these marketing innovations has evolved over time with the advancement in technology. Marketers in Society 5.0 should rely more on innovative outcomes than the innovative process due to rapid globalization and technological innovation. Marketers have shifted from relying on innovation processes or outcomes to create marketing advantages to focusing on consumer behavior. Marketing initiatives have had to change their course from existing content/ push methodologies to newer e-commerce and m-commerce strategies in today's digital world. This study attempts to give a detailed background of the various stages in society and the different capabilities of Industry 4.0 and 5.0. The survey framed can illustrate the consumer's buying behavior from different aspects, showing their evolution over a decade.

2) BACKGROUND AND SIGNIFICANCE OF MARKETING INNOVATIONS IN SOCIETY 5.0

Marketing innovations have been evolving with the development of a new society. Emerging lifestyles, higher incomes, and more consumer knowledge are increasing the need for a steady supply of innovative food products throughout the year (Guiné et al., 2020). But now its borders push its limitations and it is not only applicable to food it is applicable at every Industry The impact of technology on marketing is significant, and economic developments have a considerable influence on marketing. Marketing developments are a result of changes in people's expectations. Past societal evolutions result in experiential technological changes. Today, we have human interactions and business transactions that are personalized through this technological evolution. In addition to that, the role of traditional marketing has been stretched to use technology to implement strategies and create interactions for branding and promotion. These marketing innovations, if understood and focused properly on understanding a consumer, will certainly help marketers gain further insights and exploit consumer behaviour through predictive analytics.

Studies demonstrate the various applications and practical elements of marketing from new brand developments to chatbots. By understanding the development, we predict and analyse consumer behaviour. This is especially significant in developed economies, since how and when to use this can greatly influence contemporary marketing strategies. As predictive analytics help businesses succeed in the modern scenario, it is significant to marketers, as the same cultural and social concepts apply. The perceptions of consumers regarding two of the three assessed innovation criteria, namely relative advantage and compatibility, were significantly correlated with their acceptance or adoption of marketing messages (Roach, 2009).and present study will outline the significance of changing technological innovations and personalized marketing through contextualizing the concept of personalization and providing an in-depth reflection of its application for a successful marketing strategy. In order to understand societies that are characterized by Industry 4.0 concepts, it is critical to understand the transition occurring from Industry 4.0 to Society 5.0. There are some firms that have understood the changes in society and have developed strategies to adapt themselves. One firm has adopted individualization practices in both its marketing and website practices. The firm uses algorithms to generate individualized newsletters. Predictive analytics is one of the transformative technologies like artificial intelligence and machine learning, which entirely performs as a game changer in the field of marketing. As marketing is the outcome of customer behavior, predictive analytics forecasts customer behavior in various forms like buying behavior, customer churn, lifetime values, and possible customers who actively promote the brand on social media. Predictive analytics can be an aggregation of a large number of statistical techniques and data mining methods. More importantly, predictive analytics uses data, statistical algorithms, and AI techniques to identify the overall likelihood of future outcomes based on the historical data used for the training of the algorithms. The patterns found in transaction data and customer behavior are taken to identify risks and potential opportunities; however, sometimes those patterns are so complex that traditional analytics can never identify them using state-of-the-art techniques being used in marketing.

Due to the amount of big data available today from multiple sources such as commercial enterprises, social media, consumer and industrial intervention, and transaction history, most of it is collected in real time. Predictive analytics is able to forecast customer trends, needs, and wants through a series of statistical algorithms and ML techniques. Predictive analytics empowers marketers to make qualitative decisions and fit better into the marketing scenario as it helps marketers forecast consumer behavior. Customer relationship management tools are heavily dependent on predictive analytics; they can tailor individual marketing offers and improve the overall customer experience. Through the data available to an organization, predictive analytics shows how marketers can act upon the offer and the demand of the consumers accordingly. Thus, it is an approach to increase portfolio efficiency and make money by convincing the sales team and marketing team of the relevant products. It is capable of providing insight to existing customers that can be missed if the historical and transactional customer data are not taken into consideration. Also, it shows how predictive analytics helps in steering the desired response from historical data to show the statistical report of the benefits collected so far from the predictions. The starting point for any study on predictive analytics and consumer behavior is to understand the behavior of consumers in society 5.0. As already mentioned, this research will take India as the area of study. Consumers in India are different from those in the West and require more inferences and more robust strategies to connect with them. Culture the beliefs, way of life, rituals, customs, social life, and personalities—can affect consumer behavior. In India, the culture is diverse and is differentiated by religion, caste, and region. The economy is the synchronization between culture, administration, and factors of production. The economic structure will differ according to the nation's values, customs, and cultural assimilation. This study will combine both developed and developing markets. The aim is to examine the ethnic values on which our economies are structured. The most significant society-wide aspect that sets India apart is the Indian cultural tradition. It frames and influences many segments of our life, including customer loyalty.

The family and social orientation are important aspects of Indian society. The Indian value 'the customer is lord' should be recognized by organizations in the Indian market and their strategic marketing decisions. Branding influences the choice of buyers in the Indian market. India is a diverse and varied country regarding buying behavior. Consumers are motivated to purchase by factors such as social status, popularity of a brand, and the purchase of big items for social or emotional purposes. Culture in India, with rapidly changing behaviors, is influenced by factors such as social media, product and parenting strategies, and increased digital marketing competitiveness. Consumer patterns based on people's timeless, logical behavior are highly scalable and shareable. In addition, comparison sheds light on ways to enhance the diversity and complexity of human behavior theories. Cultural variables, like those in India, can markedly impact environmental consumer behavior, particularly if the consumer recognizes themselves in these culturally sensitive, socially constructed areas. Per stereotype, South Asians (Indians) are tough, understanding, social, and able to live with poor conditions in accordance with their customs and mores. By collecting data on customers' emotional reactions and performing controlled trials on them, sentiment analysis aims to influence their behaviour. Marketers' attempts to control customers via the accumulation of vast real-time databases of their online activity and chats, in addition to demographic information, should be included in any analysis of emotional economics(Andrejevic, 2011).

Understanding global communities is a hallmark of India's marketing. Market orientation in a growing market must value traditional importance. The following area would look exclusively at the numerous variables that influence Indian customers. Further advice can be provided to executives.

3) CONSUMER BEHAVIOR IN THE DIGITAL AGE

In today's digital age, the trips consumers take have changed. Gone are the days when they could be fooled by attractive hoardings or commercials. The consumer today has all the required information at the tip of their fingers. Studies show a significant percentage of consumers used the internet to make a purchase, regardless of whether it was apart from online retail or otherwise. Based on the information shared by the users who voluntarily use our surveys, we can summarize their experiences with the product: Quality: Positive, a portion say. Though consumer knowledge is a key perk, many say that they've been dissuaded from making a purchase of a particular product at least once due to the negative reviews and ratings they bear the brunt of. When it comes to maintaining a sound status, a majority of people assert that they check in and write reviews to figure out which products are working for them and which aren't.

Satisfaction is substantially influenced by all aspects of service innovation, such as product, process, organisational, and marketing innovations, which are positively correlated with perceived institutional image and loyalty (Cheng et al., 2019). And In the Indian context, consumers operate in a different scenario altogether. Indian consumers follow a logic of their own. Logic has never been too strong due to a dearth of information. No one knew much about the products we spend our hard-earned money on; but with the advent of the internet, the tables have turned but the use of predictive analytics allows it to provide light on cost projection, which in turn aids decision-making (Zekić-Sušac et al., 2021).

A significant percentage of adults in India are affected by social media. A survey revealed that to make a purchasing decision, a large percentage of people in Indian cities looked to gain reviews and product ratings before they were finally able to make that purchase. With reviews becoming an inseparable aspect of a buyer's exploration journey, independent websites seem to be trusted by a greater percentage of Indian consumers than those who go to the companies' own sites to collect their information. On average, price comparison websites get a substantial number of unique visitors. In the backdrop of consumers checking reviews and ratings in such large numbers and on such frequent levels before making a purchase, what does the scale look like today? What value of business rests in these pieces of data being churned out by the consumer? Answers to these questions will be kept pending, but there is absolutely no dearth in the fact that in today's day and age, understanding consumer behavior is extremely important. Predictive analytics technique uses past transactions to understand the dynamic between prosumer bids and market reactions (Chen et al., 2019).

The proliferation of digital technologies in the last two decades has changed the way consumers interact with a brand. The vast amount of information available through various online channels has transformed consumers into prospective customers even before they express their desire to purchase. The penetration of mobile device technology and social media, coupled with an information-saturated online world, has shifted the power of purchase into the hands of the consumer. Also forecast the quality of video ads using a multimodal architecture that integrates user physiological data with publicly accessible online sentiment ratings (Gauba et al., 2017)

Externally, some of the significant changes in consumer behavior can be observed in the expectations from marketers regarding relevant messages and customized products. The environment in which all these developments have been transpiring is the concept of Industry 4.0, which is believed to be an innovative paradigm identified as Society 5.0. The broad term 'industry' in Society 5.0 is now changed to 'society,' which may include a much larger group of members interacting with each other electronically. These days, most people spend a lot of time on social media. More and more, people are making purchases via social media (Chaudhary et al., 2021).

The digital revolution has necessitated new systems and processes for engaging participants, whereas traditional marketing practices are becoming less effective and Better operations can't be achieved without cloud computing, high-performance computing, efficient data network management, and strong data analytics(Diamantoulakis et al., 2015).Consumers, currently, despite their personal and individual behaviours, are connected by their common approach toward technological immersion, marking the advent of a new environment that interconnects online and offline and thorough evaluation of their benefits, drawbacks, and functions is carried out when developing a forecasting model(Mamun et al., 2020).

This developed scenario calls for a more decisive object of investigation that demands an overarching framework within which the thought process of consumers in the era of industry 5.0, including changes taking place in the digital landscape, can be explored.

4) REVIEW OF LITERATURE:

Figure 1. All Keywords Analysis by VosViewer

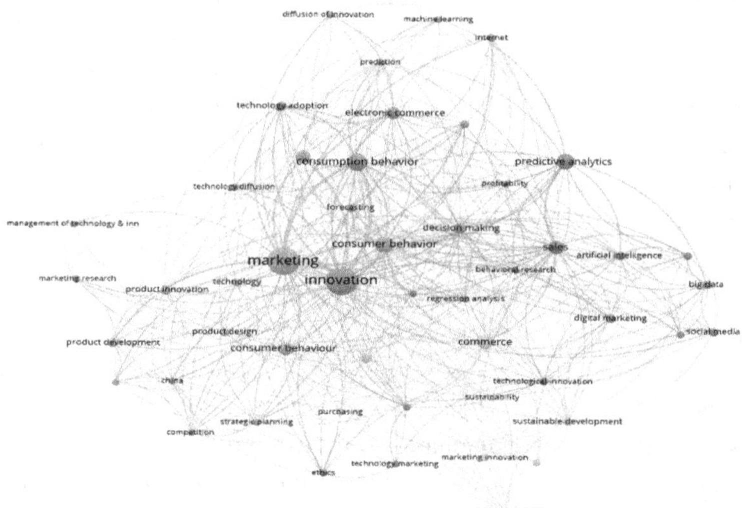

The keyword analysis visualized in the network map highlights the interconnections between major research themes relevant to **marketing innovations**, **predictive analytics**, and **consumer behavior**. The map presents different clusters, with each color representing a thematic grouping of related terms “Marketing,” “innovation,”

and "consumer behaviour" are prominent, indicating basic challenges with strong ties to other subjects. These terms underpin research disputes, demonstrating their importance across studies. Red Predictive Analytics Cluster Predictive analytics, artificial intelligence, big data, and sales are linked in this cluster, emphasising a focus on data-driven marketing and consumer interaction, forecasting, and decision-making. Green Cluster Innovation and Technology Adoption Product innovation, design, and technology adoption suggest marketing research on technical advances and innovation management. These statements show that innovation is crucial for consumer understanding and marketing solutions. Commerce and Sustainability (Yellow Cluster) Commerce, green marketing, and sustainability show a growing interest in environmentally responsible marketing, reflecting worldwide sustainability trends. Digital marketing and emerging technologies (purple cluster) Keywords like machine learning, digital marketing, and social media emphasise marketing's digital revolution and the growing use of AI and digital technology in marketing approaches.(Wang et al., 2019) peruse the approaches used or created to handle each application and on this node (Korfiatis et al., 2012) indicates that stylistic features substantially affect the helpfulness ratio of reviews, with very helpful reviews attaining superior rankings compared to less helpful ones.to emphasize on this part (Chong et al., 2012) argued that, the conventional Technology Acceptance Model (TAM) and Diffusion of Innovation (DOI) model by incorporating supplementary factors such as cost, social impact, trust, and service diversity in addition to this (W. B. Kim et al., 2025) analyses how social, cognitive, and hedonic consumer innovativeness affect worldwide product purchases, excluding functional innovativeness but on the Contrary of this (Li et al., 2025)indicates that marketers must meticulously evaluate product attributes when employing STH for new product recommendations to mitigate adverse consequences. In the Similar vein (Sorescu & Spanjol, 2008) the Breakthrough discoveries can increase the value and economic rents of a company; however, they also increase the risk. Regular earnings are increased by incremental innovations, which do not alter the firm's risk profile or economic rents. Furthermore (Park & Jun, 2003) discussed the model in which cultural differences influence these influences, and Internet usage and perceived hazards have an impact on Internet purchasing behaviour. (Pham & Ho, 2015) Product-related factors influence adoption intention more significantly than personal characteristics. (Lamey et al., 2012) Risk and perceived utility (PU) constitute the principal obstacles and motivators. Examine a unique data set that incorporates a thorough range of seven marketing-mix elements and private-label market share, employing twenty years of data across 106 consumer packaged goods categories in the United States. Notably (Fearne & Hughes, 1999) the Enhancement of cost measurement and control to achieve greater efficiency, alongside innovation in both product offerings and service levels, as well as the methods of engagement with key customers. (Kang & Park-Poaps, 2010) indicated a notable correlation between fashion innovativeness and various hedonic shopping motivations. Fashion innovativeness demonstrated a positive correlation with adventure and idea shopping motivations, whereas it exhibited a negative correlation with value shopping motivation. (Goldsmith et al., 1999) A valid and reliable self-report scale evaluated fashion innovativeness, identifying consumers most likely to purchase new fashions at their initial market introduction. On the same path (Dahl & Hoeffler, 2004) discussed that Incremental products were evaluated better when visualised utilising self-related imagery. Knowing the product category from which an incremental extension comes helps people envisage utilising the new product. But in Contrary of this (University of Antwerp et al., 2016) Using fine-grained behaviour data, increasing data size across analysis is useful. Larger companies may have more valuable transaction data for targeted marketing. Also the (Garcia Martinez & Walton, 2014) Crowd competitions can quickly enhance the technical capabilities related to a specific dataset by involving a diverse array of participants proficient in different analytical methods. With that (H. Kim et al., 2010) Investigate the relationship among consumer innovation, market mavenism, shopping delight, and beliefs, attitudes, and patronage intentions related to pop-up retail alos to make a strong result (Toubia & Florès, 2007) Algorithms for idea screening are modified based on customer ratings. Literature provides little or no structured analysis in the content of marketing innovations in preparing for Society 5.0. The envisaged study is among the first ones aiming at bridging this gap. Specifically, the aims of this study can be identified and will be achieved as follows: the analysis is elaborated in order to clarify which marketing innovations are applied by entities. It is stressed the importance of developing a deeper understanding of innovations for their successful reception and implementation. Wondering which kind of marketing innovations are prevalent today, it is interesting how important they are and how widely they are implemented. This is relevant to understanding which marketing innovations should be promoted, induced, and spread in the marketing world.In Society 5.0, not only technological but also marketing innovations become an essential part of any efficient entity keen on sustainability. Different studies appreciate the benefits and drawbacks of marketing innovations and provide contentious results. It is argued that an innovation-led strategy is more likely to be successful compared to those led by businesses that are afraid to take risks. A company can operate much more responsibly and with regard for its customers, taking into account the latest marketing innovations. As a consequence, the company can grow its customer base and its benefits. One of the most interesting research questions in the marketing field today is related to the creation of the most successful marketing innovation that has been and can be used in the analysis of consumer behavior. Indeed, strategic development is closely related to the ability to predict how people will behave in particular situations; it can lead to beneficial marketing research and loyal behavior of customers towards the organization.

5) RESEARCH METHODOLOGY

The purpose of this study is to examine the role of predictive analytics in evaluating consumer behaviour and marketing tactics in Society 5.0 by employing a mixed-methods approach that integrates quantitative and qualitative data gathering methods. We will conduct a cross-sectional study to collect data over a specific time frame in order to investigate the current impact of predictive analytics on marketing. respondents for both quantitative and qualitative data will be selected through purposive sampling. The quantitative research will concentrate on organisations and marketers that employ predictive analytics solutions. The qualitative component will select critical stakeholders, including data analysts and marketing professionals who are involved in decision-making and analytics. In order to guarantee that the conclusions are both generalisable and diverse, quantitative data will be collected from a minimum of 100 digital marketing initiatives in a variety of sectors. The qualitative data on the actual applications of predictive analytics will be obtained through in-depth interviews with 10-15 marketing professionals and data analysts. Data from digital marketing campaigns will be collected, with a particular focus on critical metrics such as consumer engagement, conversion rates, and return on investment (ROI). The predictive analytics software's efficacy will be assessed. To obtain expert perspectives on the potential and obstacles associated with the application of predictive analytics in marketing operations, semi-structured interviews will be conducted. Predictive modelling and data analytics will be implemented through the Orange Data Mining program. The accuracy of prediction models in comprehending customer behaviour will be evaluated by this application through data visualisation, categorisation, and regression analysis. Predictive modelling: Orange Software and machine learning techniques, including random forests and decision trees, will be employed to develop prediction models. These models will evaluate the precision of forecasting consumer behaviour by utilising historical data from marketing campaigns.

3.1) Research Questions:

1) How do predictive analytics tools influence the analysis of consumer behavior in Society 5.0?
2) How can marketers utilize insights from predictive analytics to develop innovative marketing strategies?

3) What are the key benefits of predictive analytics in optimizing marketing efforts, such as improving customer engagement and ROI?

3.2) Objectives of the Study:

1) To analyse the impact of predictive analytics on consumer behavior within the context of Society 5.0.
2) To evaluate the accuracy and effectiveness of predictive analytics in optimizing resource allocation and improving return on investment (ROI) in marketing campaigns.
3) To investigate the role of predictive analytics in balancing technological advancement with customer needs and societal well-being.

4) UNDERSTANDING SOCIETY 5.0

Society 5.0 is a future concept reflecting an ideal human-cantered society envisioned for the 5th stage of future society. It is expected to harmonize economic advancement with the solution of social problems through a system that highly integrates cyberspace and physical space. The motivation for Society 5.0 is expected to be more ambitious than that for other technology trends. The presentation seeks to make a contribution by suggesting the seven features that characterize Society 5.0, among industrial and individual revolutions. Grouped into two areas, these features include the majority of the features from the literature and insights from other clusters. Innovations in pressing fields such as marketing can only be useful in the realization of society when consumers welcome and use them. (Olshavsky & Spreng, 1996) Product managers and researchers seek optimal organisational procedures, strategies, and frameworks to drive product innovation. But Innovation resistance is presented as a sensible response rather than a negative element of target markets(Szmigin & Foxall, 1998).It is hoped that marketing professionals will have more incentive to work and thrive in putting society at the center of marketing. After all, there has been an emphasis on the distinction in the emphasis placed on a society-oriented company. Consumer Particular 5.0 plays a critical role in this regard. Since consumers are at the centre of society, predicting their behaviors will enable companies to design, produce, and deliver products that meet their needs more effectively. Society 5.0 is a future concept for the 5th stage of future society, expected to harmonize economic advancement with the resolution of social problems through a system that highly integrates cyberspace and physical space to benefit the greatest number of people possible. And In today's marketplace, consumers are extremely knowledgeable. Traditional marketing research is inadequate, with numerous studies revealing that consumers often act in ways they did not predict. Hence, the main criticisms relate to its inability to predict future demand for new products and its limitation to the marketer's current geographical market. While past sales data are certainly useful as a basis, it is not always sufficient to rely solely on past data in addressing questions about consumer behavior in the digital age, pertaining to the changing habits, preferences, and behaviors of today's consumers. Consumer behavior in the digital age points to new phenomena that will undeniably make consumer behavior quite inspiring in the twenty-first century. This digital age is driven primarily by information, networks, and technologies that make unprecedented access to knowledge, societies, and economies possible. It is a scenario today where the consumer, the firm, and the network are enablers in the innovation process and are constantly changing hence (Hinsch et al., 2020) Building meaningful links and engaging consumers via AR may excite customers and boost marketing results. They are reciprocal entities influencing each other and providing feedback that can potentially result in a different society. It is in this environment of dynamism, diversity, and change where a new wave of innovation and customer service demands is ruling.

5) THE PREDICTIVE ANALYTICS IN ANALYSING CONSUMER BEHAVIOUR FOR MARKETERS

Figure 2. Orange workflow for using different predictive models Source: Predictive Analytics by Orange

Figure 3. Orange Tree Diagram to check accuracy of predictive models for Strategic Decisions

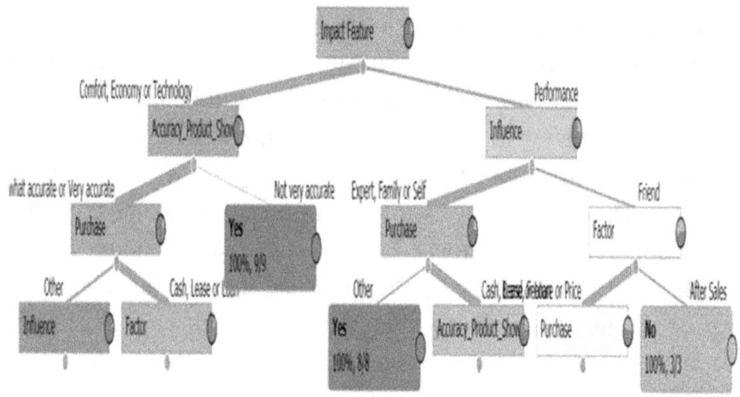

Source: Predictive Analytics by Orange

Figure 4. Graphical Representation of Logical Regression Model for the Predictive Analytics Source: Predictive Analytics by Orange

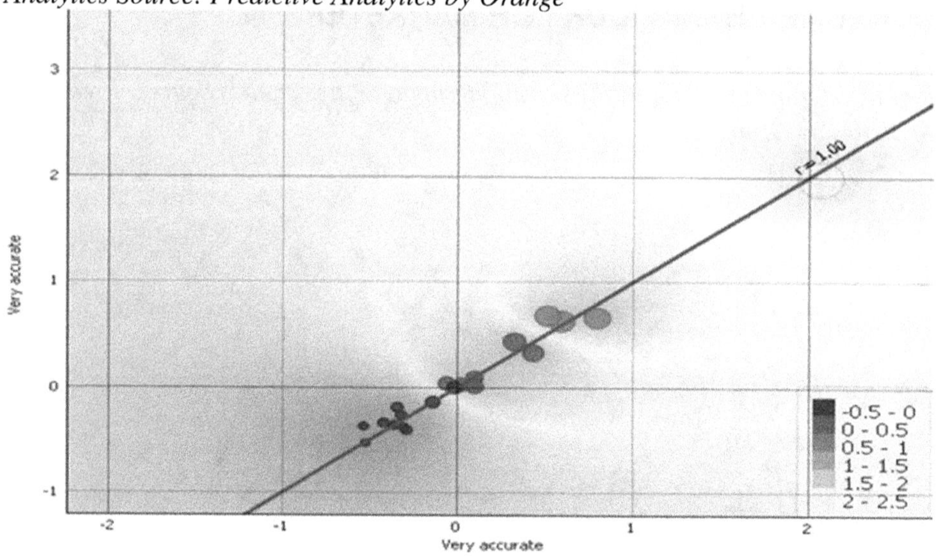

Table 1. Logical Regression Model for the Predictive Analytics Source: Predictive Analytics by Orange

Sr.No	Model	AUC	CA	F1	Prec	Recall	MCC
1	Classification	0.93	0.919	0.916	0.912	0.919	0.952
2	Logistic Regression	0.924	0.843	0.977	0.943	0.916	0.893

The following table gives a summary of the results

Table 2. Accuracy measures for the first target variable

Model	AUC	CA	F1	Precision	Recall
SVM	0.97	0.97	0.99	0.86	0.98
Random Forest	0.96	0.95	0.95	0.92	0.99
Neural Network	0.984	0.92	1.0	0.965	1.0
AdaBoost	1.0	0.88	0.963	0.892	1.0

The data provides performance metrics for different models used in predictive analytics, focusing on metrics such as **Area Under the Curve (AUC)**, **Classification Accuracy (CA)**, **F1 Score**, **Precision**, **Recall**, and **Matthews Correlation Coefficient (MCC)**. These metrics measure the models' ability to accurately classify and predict outcomes.

1. Classification Model (Sr. No. 1)
    - **AUC (0.93)** indicates that the classification model has a strong ability to distinguish between classes, though not the highest among all models.
    - **CA (0.919)** and **F1 (0.916)** demonstrate that the model performs consistently well in terms of overall classification accuracy and the balance between precision and recall.
    - **Precision (0.912)** and **Recall (0.919)** indicate the model is slightly better at identifying true positives compared to avoiding false positives.
    - **MCC (0.952)** shows a very strong correlation between actual and predicted outcomes, suggesting a robust model overall.
2. Logistic Regression (Sr. No. 2)
    - **AUC (0.924)** indicates a strong but slightly lower ability to distinguish between classes compared to the classification model.
    - **CA (0.843)** suggests a lower classification accuracy than the other models, indicating that Logistic Regression may not be as effective in making correct predictions across all classes.

- Interestingly, **F1 Score (0.977)** is higher, suggesting a strong balance between precision and recall.
- **Precision (0.943)** and **Recall (0.916)** further emphasize its effectiveness in making correct positive predictions.
- **MCC (0.893)** is lower compared to the classification model, which indicates that Logistic Regression performs less optimally, with more misclassifications in comparison.

3. SVM (Support Vector Machine)
    - **AUC (0.97)** indicates a very high discriminatory power.
    - **CA (0.97)** and **F1 (0.99)** highlight that SVM achieves high accuracy and an almost perfect balance between precision and recall.
    - **Precision (0.86)** is lower compared to the other metrics, meaning it may have more false positives, despite high **Recall (0.98)**.

• This suggests that SVM might excel at detecting true positives but may need refinement to reduce false positives.

4. Random Forest
    - **AUC (0.96)**, **CA (0.95)**, and **F1 (0.95)** show that Random Forest is a highly accurate model with strong classification capabilities.
    - **Precision (0.92)** and **Recall (0.99)** reveal that the model is highly reliable in identifying true positives, but slightly more conservative in avoiding false positives.
    - Overall, Random Forest presents a balanced performance.
5. Neural Network
    - **AUC (0.984)** and **F1 (1.0)** show that the neural network is almost perfect in distinguishing between classes and balancing precision and recall.
    - **Precision (0.965)** and **Recall (1.0)** indicate excellent performance, with the model almost perfectly identifying all true positives.
    - **CA (0.92)** suggests slightly lower accuracy than some other models, possibly due to overfitting or difficulty in generalizing on certain instances.
6. AdaBoost
    - **AUC (1.0)** suggests that AdaBoost perfectly distinguishes between classes, making it the highest performing in terms of discriminatory ability.
    - **CA (0.88)** and **F1 (0.963)** show that while the model distinguishes well between classes, its overall accuracy and balance between precision and recall are somewhat lower than expected, given its AUC score.

- **Precision (0.892)** and **Recall (1.0)** indicate that AdaBoost excels at identifying true positives but has more false positives than other models.

Summary

- **AdaBoost** and **Neural Network** stand out with the highest **AUC** values (1.0 and 0.984, respectively), suggesting they are the best at distinguishing between classes.
- **SVM** and **Random Forest** also show strong performance, with **high accuracy** and balanced **precision-recall** scores.
- **Logistic Regression**, while having a high **F1 score**, has lower classification accuracy, which may suggest limitations in specific use cases.
- Each model has strengths depending on the desired metric (AUC for discriminatory power, F1 for balanced classification, or CA for overall accuracy), making the choice of model context-dependent based on the goals of the predictive analytics task.

6) SUMMARY OF KEY FINDINGS

As this chapter concludes, AI growth has been tremendous during the last few years, and its marketing applications across front and back-end activities are highly significant. Consumer behavior is evolving continuously in ways that are becoming increasingly difficult for organizations to predict and navigate in an actionable way. However, that challenge is being met with newer tools like predictive analytics. The focus of this text is on business analytics, and particularly on predictive analytics, how businesses are deploying these solutions, and which new applications are likely to emerge in view of these advancements. Predictive analytics-based marketing is today being applied across a wide spectrum of sectors, and integrating it with AI takes its undertakings even farther. Organizations can exploit predictive analytics and AI through a wider range of technologies than ever before, and customer data coupled with various analytical tools and capabilities, which are constantly evolving, offer novel opportunities for companies. AI is an integral part of modern business analytics and helps businesses not only to predict future outcomes but also to automate or semi-automate corrective action suggestions through various strategies.

7) PRACTICAL IMPLICATIONS FOR MARKETERS

This study has multiple implications for both strategic marketing management and industry. Indeed, it has increasingly been argued that big data analytics capabilities create a basis for innovation and competitive advantage for companies. Given rapidly increasing data volumes and the growing complexity and variety of data, associations apply data analytics to improve marketing operations. By clustering consumers based on past consumption behavior, predictive analytics identify group-specific patterns, test group differences regarding online and mobile device use, and estimate the influence of the results for strategic decision-making. Classification techniques allow the selection rules to discern two different clusters from the dataset, season-based attributes, and varying purchasing behavior within these clusters over time

8) BEST PRACTICES FOR LEVERAGING PREDICTIVE ANALYTICS IN MARKETING

The following best practices are distilled from successful case studies and are recommended for marketers wanting to leverage predictive analytics as a tool to enhance their marketing effectiveness: Prioritize data quality to create a deep understanding of consumers. Quality data leads to actionable insights that improve marketing effectiveness. Transparency fosters trust. A proprietary platform is only essential to control data. For substantial benefits, data confidentiality and exclusivity may be used. Materialize findings quickly and at low cost by means of test and learn deployments. Have a start and stop date on research projects to build in the understanding that this will not be the final answer and that there will be a continuing process of enhancement as the results commence and more findings are required. Expand competencies or initiate a partnership with universities or consultancies to enhance capacities for doing predictive analytics. Develop a company-wide culture of evidence-based decision-making in the long term. Open, clear, and focused structures encourage free exchanges of thoughts, exposing the company to fresh strategies and encouraging innovation. In conclusion, marketers and organizations must use this information to integrate predictive analytics into their marketing strategies. By doing so, they can enhance the capabilities of their marketing staff. This will make marketers more capable of executing marketing strategies and delivering customer value, which in turn will support the marketing concept objectives of enhancing consumer standings and, consequently, industry profitability. This study contributes to the still limited amount of academic research on the practice of predictive analytics in marketing. The study is unique because it aggregates best practices from influential business people, senior managers, and marketing personnel.

9) CONCLUSION:

Predictive models are a powerful tool to help companies achieve their goals, improve their operations, and create new opportunities, innovating and guiding managers to make better strategic and tactical decisions, reduce waste, and increase the return on profit. Companies that benefit from data to make data-driven decisions and operate on the basis of predictive models are responsive, work on the basis of data, and solve real-world problems by addressing the challenges of today. Data science is a powerful tool that adds value to the marketing strategies of any organization. It is not intended to replace managers, but rather to provide data with which managers can answer their questions. The research question " How do predictive analytics tools influence the analysis of consumer behavior in Society 5.0?" has been answered in the context of marketing innovation trends in Society 5.0. In this research, the invention of the term "marketing innovations" for holistic consideration of modern directions in the sphere of marketing communications, with respect to their focus, has been completed on consumers and society as a whole. The proposed classification of these innovations has been successfully applied to a case analysis. It was proved that it is possible to mix various marketing innovations experimentally due to the theoretically same focus on consumers and society, although practically there is much overlap. The proposed research model was used to describe the existing world marketing trends. The limitations and shortcomings of the research are that the concepts and categories developed in this scientific work are mainly of a theoretical nature and are more illustrative in the context of the practical aspects of the analysis of the experience of companies exploiting modern directions of marketing communications. The specific structure of the application of big data and predictive analytics tools is very much determined by the objectives and the set marketing effect of the marketing systems used. The theoretical and practical value of the work can possibly be increased by creating a case collection of the effective utilization of certain directions of big data in solving practical marketing problems. The developed ideas deserve further development that can be a prospective contribution to the field of innovative marketing.

10) FUTURE TRENDS AND IMPLICATIONS

It is estimated that by 2030, when purchasing power will no longer be the only way to gain social status and buy time, the economy will develop in the direction of welfare production, alternative work, digital currency circulation, and other modes, and enter into Society 5.0. Presently, the rapid development and comprehensive application of science and technology have fundamentally changed the production and life of human society, transforming it from unmanned meetings, online education, and e-books to paperless offices and network negotiations. With the advent of the era of big data, the analysis of consumer behavior, as a research category of big data, also shows corresponding trends. As an important means of modern enterprise marketing, consumer behavior analysis encompasses many research areas associated with artificial intelligence, including its application in the marketing field, and is also one of the research hotspots.Artificial intelligence and its applications, inspired by the operation and function of the human brain, have evolved from various aspects of science, engineering, and medicine, such as psychology, cognitive science, informatics, control science, computational neuroscience, and systems science, each with its effectiveness in different fields such as design, application, and simulation. We have indirect access to people's inner thoughts, mainly through information exchanged in various forms around them. The development trend of social networks supports structured data backed by big data technology, high-performance computing, and various machine learning and statistical methods, initially implementing the ambition of predictive analysis. Now, with a deeper internal emotion engine that can analyze and simulate various human emotions, the integration with social network information will make predictive analytics a powerful force in consumer research and business applications. This has also become a challenge and has attracted the attention and efforts of researchers and businesses around the world. Understanding consumer psychological needs and people's consumption behavior has garnered the attention of large companies and regulatory authorities. We will review the hot issues and current research progress in this area.

11) EMERGING TECHNOLOGIES IN PREDICTIVE ANALYTICS

Predictive analytics are being revolutionized by the latest technological breakthroughs that are further enhancing the capabilities of the analytics tools. Artificial intelligence (AI) and machine learning (ML) are revealing previously unseen patterns and consumer insights, thereby providing decision-makers with the guidance they need and drawing them to take customer insights to a much higher level. Marketing analytics is continually being shaped by its transformational technologies right from the age of the internet to date. It is important for marketers strategizing the future growth of their brands through analytics to know the technologies that are coming their way. AI and ML arc redefining the potential capabilities of predictive analytics that act as the best aids to decision-makers in the current scenario.

Moreover, a never-before-experienced ability of predictive analytics is to simultaneously track and analyze the internal customer behavior and countless influencing external factors and to guide the decision-making processes for the desired growth of the brand. The role of prediction with advanced capabilities has been reinforced through AI and ML due to the sophisticated designs of these innovations. The inherent nature of behavior and preferences of individuals or households in the market are more stagnantly framed as the features of interest. Hence, the application level of AI-based analytics, substituting demand estimation or demand forecasting, might be transformational. Marketers need to continuously recalibrate their learning scales and proactive marketing behavior to assimilate these ongoing technology-led changes into their best practices and growth-bearing techniques.

REFERENCES:

Andrejevic, M. (2011). THE WORK THAT AFFECTIVE ECONOMICS DOES. *Cultural Studies*, *25*(4–5), 604–620. DOI: 10.1080/09502386.2011.600551

Boone, T., Ganeshan, R., Jain, A., & Sanders, N. R. (2019). Forecasting sales in the supply chain: Consumer analytics in the big data era. *International Journal of Forecasting*, *35*(1), 170–180. DOI: 10.1016/j.ijforecast.2018.09.003

Chaudhary, K., Alam, M., Al-Rakhami, M. S., & Gumaei, A. (2021). Machine learning-based mathematical modelling for prediction of social media consumer behavior using big data analytics. *Journal of Big Data*, *8*(1), 73. DOI: 10.1186/s40537-021-00466-2

Chen, K., Lin, J., & Song, Y. (2019). Trading strategy optimization for a prosumer in continuous double auction-based peer-to-peer market: A prediction-integration model. *Applied Energy*, *242*, 1121–1133. DOI: 10.1016/j.apenergy.2019.03.094

Cheng, B. L., Cham, T. H., Dent, M. M., & Lee, T. H. (2019). Service innovation: Building a sustainable competitive advantage in higher education. *International Journal of Services. Economics and Management*, *10*(4), 289. DOI: 10.1504/IJSEM.2019.105014

Chiesa, V., & Frattini, F. (2011). Commercializing Technological Innovation: Learning from Failures in High-Tech Markets*: Commercializing Technological Innovation. *Journal of Product Innovation Management*, *28*(4), 437–454. DOI: 10.1111/j.1540-5885.2011.00818.x

Chong, A. Y.-L., Chan, F. T. S., & Ooi, K.-B. (2012). Predicting consumer decisions to adopt mobile commerce: Cross country empirical examination between China and Malaysia. *Decision Support Systems*, *53*(1), 34–43. DOI: 10.1016/j.dss.2011.12.001

Dahl, D. W., & Hoeffler, S. (2004). Visualizing the Self: Exploring the Potential Benefits and Drawbacks for New Product Evaluation. *Journal of Product Innovation Management*, *21*(4), 259–267. DOI: 10.1111/j.0737-6782.2004.00077.x

Diamantoulakis, P. D., Kapinas, V. M., & Karagiannidis, G. K. (2015). Big Data Analytics for Dynamic Energy Management in Smart Grids. *Big Data Research*, *2*(3), 94–101. DOI: 10.1016/j.bdr.2015.03.003

Fearne, A., & Hughes, D. (1999). Success factors in the fresh produce supply chain: Insights from the UK. *Supply Chain Management*, *4*(3), 120–131. DOI: 10.1108/13598549910279567

Garcia Martinez, M., & Walton, B. (2014). The wisdom of crowds: The potential of online communities as a tool for data analysis. *Technovation*, *34*(4), 203–214. DOI: 10.1016/j.technovation.2014.01.011

Gauba, H., Kumar, P., Roy, P. P., Singh, P., Dogra, D. P., & Raman, B. (2017). Prediction of advertisement preference by fusing EEG response and sentiment analysis. *Neural Networks*, *92*, 77–88. DOI: 10.1016/j.neunet.2017.01.013 PMID: 28254237

Goldsmith, R. E., Moore, M. A., & Beaudoin, P. (1999). Fashion innovativeness and self-concept: A replication. *Journal of Product and Brand Management*, *8*(1), 7–18. DOI: 10.1108/10610429910257904

Guha, S., & Kumar, S. (2018). Emergence of Big Data Research in Operations Management, Information Systems, and Healthcare: Past Contributions and Future Roadmap. *Production and Operations Management*, *27*(9), 1724–1735. DOI: 10.1111/poms.12833

Guiné, R. P. F., Florença, S. G., Barroca, M. J., & Anjos, O. (2020). The Link between the Consumer and the Innovations in Food Product Development. *Foods*, *9*(9), 1317. DOI: 10.3390/foods9091317 PMID: 32962007

Hinsch, C., Felix, R., & Rauschnabel, P. A. (2020). Nostalgia beats the wow-effect: Inspiration, awe and meaningful associations in augmented reality marketing. *Journal of Retailing and Consumer Services*, *53*, 101987. DOI: 10.1016/j.jretconser.2019.101987

Kang, J., & Park-Poaps, H. (2010). Hedonic and utilitarian shopping motivations of fashion leadership. *Journal of Fashion Marketing and Management*, *14*(2), 312–328. DOI: 10.1108/13612021011046138

Kim, H., Fiore, A. M., Niehm, L. S., & Jeong, M. (2010). Psychographic characteristics affecting behavioral intentions towards pop-up retail. *International Journal of Retail & Distribution Management*, *38*(2), 133–154. DOI: 10.1108/09590551011020138

Kim, W. B., Kim, C., & Kurata, K. (2025). Relationships among consumer innovativeness, learning, and global product purchases: Store manager perspectives in retailing. *Journal of Retailing and Consumer Services*, *82*, 104084. DOI: 10.1016/j.jretconser.2024.104084

Korfiatis, N., García-Bariocanal, E., & Sánchez-Alonso, S. (2012). Evaluating content quality and helpfulness of online product reviews: The interplay of review helpfulness vs. review content. *Electronic Commerce Research and Applications*, *11*(3), 205–217. DOI: 10.1016/j.elerap.2011.10.003

Lamey, L., Deleersnyder, B., Steenkamp, J.-B. E. M., & Dekimpe, M. G. (2012). The Effect of Business-Cycle Fluctuations on Private-Label Share: What has Marketing Conduct Got to do with it? *Journal of Marketing*, *76*(1), 1–19. DOI: 10.1509/jm.09.0320

Li, Y., Jin, X., & Xue, H. (2025). Avoid following individual shopping time habit to promote new products: Evidence from a randomized field experiment. *Journal of Retailing and Consumer Services*, *82*, 104058. DOI: 10.1016/j.jretconser.2024.104058

Magazzino, C., Mele, M., & Schneider, N. (2021). A machine learning approach on the relationship among solar and wind energy production, coal consumption, GDP, and CO2 emissions. *Renewable Energy*, *167*, 99–115. DOI: 10.1016/j.renene.2020.11.050

Mamun, A. A., Sohel, M., Mohammad, N., Haque Sunny, M., Dipta, D. R., & Hossain, E. (2020). A Comprehensive Review of the Load Forecasting Techniques Using Single and Hybrid Predictive Models. *IEEE Access : Practical Innovations, Open Solutions*, *8*, 134911–134939. DOI: 10.1109/ACCESS.2020.3010702

Martens, D., Provost, F., Clark, J., & Junqué de Fortuny, E.University of Antwerp. (2016). Mining Massive Fine-Grained Behavior Data to Improve Predictive Analytics. *Management Information Systems Quarterly*, *40*(4), 869–888. DOI: 10.25300/MISQ/2016/40.4.04

Olshavsky, R. W., & Spreng, R. A. (1996). An Exploratory Study of the Innovation Evaluation Process. *Journal of Product Innovation Management*, *13*(6), 512–529. DOI: 10.1111/1540-5885.1360512

Park, C., & Jun, J. (2003). A cross-cultural comparison of Internet buying behavior: Effects of Internet usage, perceived risks, and innovativeness. *International Marketing Review*, *20*(5), 534–553. DOI: 10.1108/02651330310498771

Pathare, P. B., Opara, U. L., & Al-Said, F. A.-J. (2013). Colour Measurement and Analysis in Fresh and Processed Foods: A Review. *Food and Bioprocess Technology*, *6*(1), 36–60. DOI: 10.1007/s11947-012-0867-9

Pham, T.-T. T., & Ho, J. C. (2015). The effects of product-related, personal-related factors and attractiveness of alternatives on consumer adoption of NFC-based mobile payments. *Technology in Society*, *43*, 159–172. DOI: 10.1016/j.techsoc.2015.05.004

Roach, G. (2009). Consumer perceptions of mobile phone marketing: A direct marketing innovation. *Direct Marketing (Bradford)*, *3*(2), 124–138. DOI: 10.1108/17505930910964786

Scuotto, V., Del Giudice, M., & Carayannis, E. G. (2017). The effect of social networking sites and absorptive capacity on SMES' innovation performance. *The Journal of Technology Transfer*, *42*(2), 409–424. DOI: 10.1007/s10961-016-9517-0

Sorescu, A. B., & Spanjol, J. (2008). Innovation's Effect on Firm Value and Risk: Insights from Consumer Packaged Goods. *Journal of Marketing*, *72*(2), 114–132. DOI: 10.1509/jmkg.72.2.114

Szmigin, I., & Foxall, G. (1998). Three forms of innovation resistance: The case of retail payment methods. *Technovation*, *18*(6–7), 459–468. DOI: 10.1016/S0166-4972(98)00030-3

Toubia, O., & Florès, L. (2007). Adaptive Idea Screening Using Consumers. *Marketing Science*, *26*(3), 342–360. DOI: 10.1287/mksc.1070.0273

Wang, Y., Chen, Q., Hong, T., & Kang, C. (2019). Review of Smart Meter Data Analytics: Applications, Methodologies, and Challenges. *IEEE Transactions on Smart Grid*, *10*(3), 3125–3148. DOI: 10.1109/TSG.2018.2818167

Williams, A. (2006). Tourism and hospitality marketing: Fantasy, feeling and fun. *International Journal of Contemporary Hospitality Management*, *18*(6), 482–495. DOI: 10.1108/09596110610681520

Zekić-Sušac, M., Has, A., & Knežević, M. (2021). Predicting energy cost of public buildings by artificial neural networks, CART, and random forest. *Neurocomputing*, *439*, 223–233. DOI: 10.1016/j.neucom.2020.01.124

Chapter 12
Future-Proofing Financial Institutions:
Strategies for Digital Transformation and Operational Excellence

Manjit Kour
https://orcid.org/0000-0003-1043-3187
Chandigarh University, India & North West University, South Africa

ABSTRACT

The financial sector is undergoing a transformation driven by fast-evolving digital technologies and changing customer expectations. AI, blockchain, cloud computing, and advanced analytics are changing the traditional models. This chapter talks about strategic levers for financial institutions to stay ahead in competition, focusing on operational excellence, customer-centric strategies, and data-driven decision-making to future-proof operations in a digitally disrupted landscape. Through qualitative research, the author used the following scholarly articles, industry reports, and case studies to discuss the actions along with the illustrative examples of best practices and challenges of digital transformation. Topics include optimizing workflow, changing management, and integrating advanced analytics for risk, fraud, and personalization. It extends to new technologies such as blockchain, AI, and IoT, and aligns digital transformation with sustainability and ethical goals

DOI: 10.4018/979-8-3373-1320-7.ch012

INTRODUCTION

The financial sector is undergoing a seismic transformation, driven by the rapid evolution of digital technologies and shifting customer expectations (Khatib et al. 2025). Financial institutions play an important role in the economic environment as they act as facilitators of financial flows, capital, and economic stability. Although the most recognized kind of FIs are banks, the sector consists of insurance companies, investment firms, non-banking financial companies, credit unions, and pension funds. The primary function of insurance companies is to help both individuals and business organizations reduce various kinds of risks that may emerge financially. Investment firms offer wealth management services, investment opportunities, and access to capital markets. NBFCs act as a crucial financial intermediary through providing credit, loans, and financial services to people and businesses who are not fully served by banks. Credit unions are member-owned financial cooperatives that provide savings and loan facilities with customer-friendly approaches. Pension funds manage long-term savings and investments, securing the retirement phase of people's lives. All these diverse entities are included under digital transformation within financial institutions beyond banking, reaching a wide cross-section of stakeholders in financial services.

Once dominated by legacy systems and traditional business models, financial institutions now find themselves at the crossroads of unprecedented change, requiring them to adapt or risk obsolescence. The forces driving this evolution-shaping forces of artificial intelligence (AI), blockchain, cloud computing, and advanced analytics-are not just efficiency-enabling but redefine the very design, delivery, and consumption of financial services (Rehan, 2024). In terms of digital transformation in finance, it is beyond the adoption of new technology; rather, it manifests a full-range shift to agility, customer-centricity, and data-driven strategies. With the incorporation of technology more and more deeply into the financial operation, competition would demand adjustments from institutions; the process required for such transformations calls for reframing operational designs that are going to focus more on flexibility over legacy systems supported by technology-based architecture. Also, it points to building a culture within an organization as more innovative and responsive to change.

There are multiple reasons for this change. Firstly, consumer behavior is changing fast, and the customers are looking for seamless, personalized, and secure financial services available anytime and anywhere. On the other hand, technology is helping financial institutions meet these expectations more efficiently. However, there are challenges with this progress. Financial institutions have the technical intricacies of advanced technologies, managing cybersecurity risks, ensuring regulatory compliance, and finally training the workforce to utilize the complete capability of the digital toolkit (Rehan, 2024). It is here that leadership assumes its importance. Visionary decision-makers guide companies through the passage of complexity in digital transformation (Soomro & Khan, 2024). They are responsible not only for the identification and implementation of the correct technologies but also for creating a culture that thrives on change. Strong leadership makes sure that the strategic objectives of the organization align with the requirements of a digital-first economy, thus giving it a robust basis for sustainable growth (Soomro & Khan, 2024). Another important dimension of digital transformation in finance is assessing and developing digital maturity (Teichert, 2019). The readiness of an institution to implement state-of-the-art technology and the objectives of such institutions must align with the changing demands of the dynamic marketplace. It is not only the technological capacity but also the degree of digital maturity that mirrors how well an organization can harness the power of technology to reach its strategic goals (Teichert, 2019). This chapter explores strategies for achieving digital maturity, providing insights into how financial institutions can position themselves as leaders in a competitive landscape.

It brings operational excellence through digital transformation, according to Voskerichyan & Baiming in 2024. To prosper amid a digital disruption trend, financial establishments must optimize work processes, utilize productive business models, and apply changes. These can ensure that any organization is more reactive to current market changes as well as shape the direction to which the entire industry is leaning. By building agile and efficient operational frameworks, financial institutions improve the ability to respond to customer needs, mitigate associated risks, and ultimately capitalize on emerging opportunities (Ogundipe et al. 2024). Data and analytics will be strategic assets in the digital era. Advanced analytics will allow the integration to deliver more holistic risk management as well as the capabilities of detecting fraud and offering differential customer experiences. By leveraging the power of data, organizations can drive growth and efficiency in operations with informed decisions. The chapter discusses how data and analytics will transform and shape the financial services industry going forward. It further talks about other emerging technologies including blockchain, AI, and the Internet of Things (IoT), which promise to revolutionize the financial sector. Blockchain is decentralized finance, but with transparency in the transactions to make them easier. AI empowers predictive analytics so that these institutions can even predict the demands of their clients and market flows. IoT supports real-time data processing, paving the way to innovate financial services in new arenas. These technologies will create significant challenges of implementation, scaling, and other ethical considerations that come with implementing such technologies.

Issues of sustainability and ethical considerations are now coming center stage in any discourse on financial digital transformation. As financials embrace digital transformation, they, too, cannot ignore societal or environmental responsibilities, and aligning digital strategies toward sustainability goals further enhances the institutional reputation while giving way to sustainable long-term value creation. What follows is the examination of ways in which such financial institutions integrate sustainability into the digital transformation that is responsible yet inclusive. Based on a comprehensive review of scholarly articles, industry reports, and case studies, the chapter is presented with a qualitative methodology. From theory to practice, this chapter looks at digital transformation through the theoretical and practical analysis of successful programs in global financial institutions. Using real-life scenarios, these cases provide the reader with some best practices; highlight challenges involved in implementing this kind of innovation; and develop solutions to mitigate the intricacies of digital transformation. The primary goal of this chapter is to find the strategic levers that financial institutions need to adapt to for survival in the digitally disrupted landscape. It looks to provide an all-rounded understanding of the drivers of digital transformation, the problems that financial institutions face, and the strategies leading to operational excellence and innovation. This chapter will address these aspects in an effort to arm financial professionals and policymakers with the knowledge and tools necessary to navigate the changing financial ecosystem.

The rest of the chapter is structured as follows: The next section puts light on existing literature on future proofing strategies in digitalization era. Next, in the subsequent section, is discussed the methodology of research that was adopted in the course of writing this research work. It is then followed by section looking into the effect of the digital revolution on financial institutions, looking specifically at legacy systems to agile frameworks. After this, discussion would turn to leadership roles in transformation and building an innovative culture. The chapter then goes on to address strategies of achieving digital maturity and optimizing operational excellence. The rest of the sections then explore the transformative potential of data and analytics and examine the opportunities and challenges that emerging technologies present. Finally, the chapter provides a strategic roadmap for financial institutions on how to achieve resilience and innovation in the digital age with practical recommendations.

LITERATURE REVIEW

Literature on technological progress, the regulatory framework, and organizational strategies presents diverse facets of coping with change and building resilience in the context of innovation and disruption. According to Divissenko (2023), it is timely and ambitious for the European Union's Markets in Crypto-Assets (MiCA) framework to address the issues in the fast-evolving crypto-asset market. The study delves into mechanisms that guarantee adaptability and relevance in regulatory frameworks, with an emphasis on the importance of regulatory capacity for risk anticipation and an open regulatory perimeter. The study is further developed by Ranchordás and Van't Schip (2019), who call for a future-proof approach to legislation that is adaptive and sustainable. Their interdisciplinary analysis does highlight the requirements for experimental legislation and regulatory impact assessments in terms of keeping law malleable in the face of fast-paced change in the digital world.

According to Setyawati and Muditomo (2023), in the context of fintech and big tech, human capital stands out as a crucial enabler. Their research demonstrates that developing innovation-related competencies, as well as adopting technology, mainly boosts banks' innovative abilities. Investing in digital infrastructure and customer-centric strategies is instrumental for digital transformation, while intrinsic motivation and user-friendly tools are both indispensable for developing digital talent. The human capital is core to this emphasis. De Waal and Linthorst (2020) reiterate that there are strategic approaches that characterize HPOs in dealing with megatrends and disruptors accordingly. Their study argues that continuous improvement, long-term orientation, and quality of employees can lead to sustainable performance, although they highlighted an absence of actionable strategies in the field of academic literature.

Such potential for artificial intelligence and machine learning for optimal liquidity, the enhancement of risk management, and to combat systemic risks among them money laundering, has been pointed out by Soramäki and Sissens (2019) in looking into data and advanced analytics related to FMIs. Similarly, Alami and Sergio (2024) study organizational resilience with machine learning models; they find R&D investments, employee satisfaction, and capital as important contributors of resilience in an emerging economy. The study gave huge importance to the interconnected nature of resilience factors and the applicability of ensemble models for predictive analysis.

The digitization of accounting and finance is also being transformed in some other directions. Fülöp et al. (2022) reviewed the present and developing tendencies within accounting digitization by employing TAM to test professional perceptions. Besides trust and perceived usefulness as being influential on digital service adoption, the study has reported that perceived risk does not notably prevent adoption. This clearly presents an increase in confidence towards digital tools provided by Industry 4.0 as relevant to reshaping the accounting profession.

From building regulatory frameworks prepared for the future to human capital benefits and building organizational resilience in the face of digitization, all these works hint at how organizations need forward-looking strategies that align innovation with sustainability. But the plethora of gaps seen in action towards actionable frameworks and more comprehensive types of guidance signals a thrust for more academic exercises in this space.

METHODS

This paper uses a qualitative research approach to utilize a summary of secondary data sources such as scholarly articles, industry reports, and case studies to explore strategic levers for digital transformation in the financial sector. Using case studies of successful global financial institutions explains best practices, implementation challenges, and actionable strategies. The thematic analysis extracts repeated trends in digital transformation regarding operation excellence, a customer-centric orientation, and leading-edge analytics-a solid understanding of what drives changes. The aggregate insights drawn from it suggest practical solutions to ensure proper alignment of an institution's strategies with technological advance and societal demand to promote sustainability in growth and innovation in financial sectors.

DIGITAL TRANSFORMATION IN FINANCIAL INSTITUTIONS: OPPORTUNITIES AND CHALLENGES

Digital transformation provides financial institutions with countless opportunities to innovate, streamline processes, and better their customers' experience. Of course, this transition is not without its challenges. Understanding these opportunities (Figure.1) and overcoming the challenges (Figure. 2) they bring are important for organizations to be competitive in an evolving market.

Figure 1. Opportunities of digital transformation in financial institutions

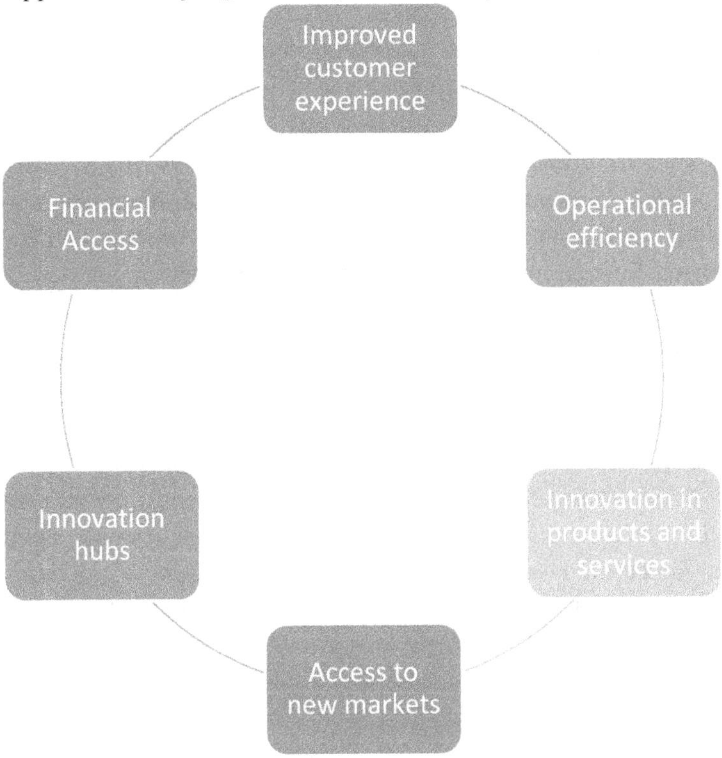

Opportunities

Improved Customer Experience- The use of digital transformation gives financial institutions the chance to provide personalized, seamless, and real-time services. Using advanced analytics and AI, the banks can know the needs of the customers better, predict preferences, and even offer tailored solutions (Kaluarachchi & Sedera, 2024). Mobile banking apps, AI-driven chatbots, and omnichannel support are prime examples of how institutions improve user engagement and satisfaction.

Operational Efficiency- Using the technological front of RPA, cloud computing, and blockchain, the financial institution would significantly cut its operational cost. This helps improve workflow and reduces the tedium of doing things through the automated nature of such technologies and back-office processes are accelerated to precision (Madakam et al. 2019).

Innovation in Products and Services- Digital transformation enables developing innovative financial products and services. For example, blockchain enables DeFi services, while AI enables robo-advisors and algorithmic trading (An et al. 2021).

These developments are opening wider horizons in terms of scope of services and accessibility to otherwise excluded markets. Advanced analytics and machine learning are revolutionizing risk management, allowing deeper insights into possible threats. Predictive models enable institutions to identify and mitigate risks proactively, whether related to credit, fraud, or market volatility (Addy et al. 2024). Real-time monitoring systems further enhance the ability to respond swiftly to emerging risks.

Access to New Markets- Digital channels enable traditional banks to leverage access to poor and distant segments. Mobile money and other e-payment services have facilitated financial access for the lower income groups across developing economies, such as Africa (Madan & Kour, 2024).

Innovation Hubs- Fintech companies and early-stage ventures help create a culture of innovation by allowing mainstream organizations to adopt new state-of-the-art solutions without research and development themselves (Mohan, 2020). Moreover, open banking policies promote novel products and offerings.

Financial Access- Digital technologies have made financial service access more widespread for low-income populations. Mobile payment systems, micro-lending applications, and digital wallets, among other innovations, now take banking closer to the remote and rural world, enhancing (Madan & Kour, 2024).

Figure 2. Challenges of digital transformation in financial institutions

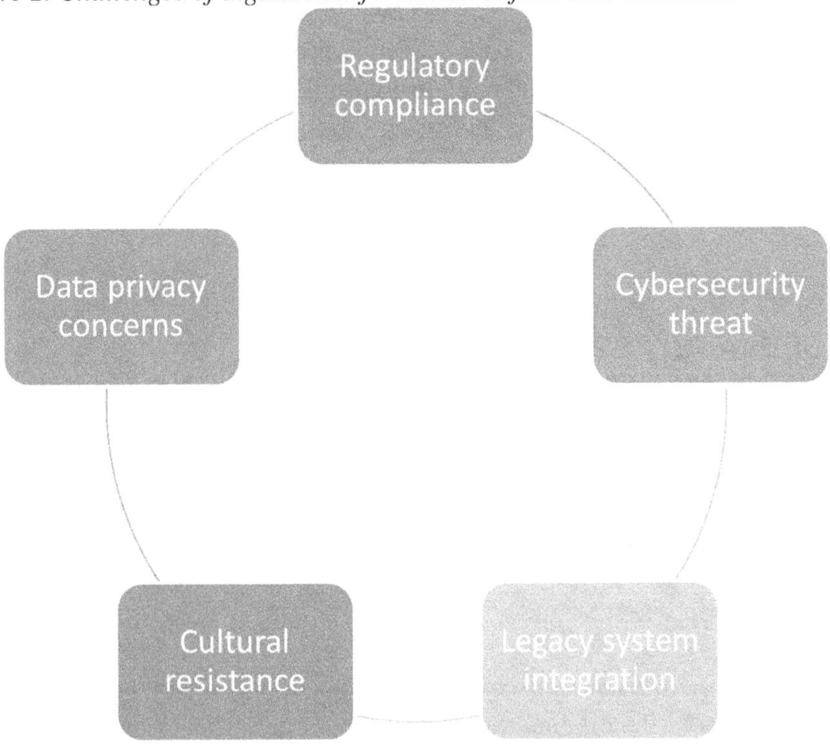

Challenges

Regulatory Compliance- The ever-changing digital world makes it challenging for financial institutions to operate in a complex regulatory environment (Kour, 2025). Compliance with data protection laws, anti-money laundering regulations, and consumer protection standards is resource-intensive and challenging.

Cybersecurity Threats- Increased dependence on digital technologies increases the risk of cyberattacks, such as data breaches, phishing schemes, and ransomware (Kour, 2024a). Institutions must continually update their cybersecurity frameworks to address these risks.

Legacy System Integration- Many financial organizations face challenges when trying to adapt modern technologies with legacy systems. This may bring about inefficiency, high cost, and a delay in digital transformation projects (Alinge et al. 2024).

Cultural Resistance- Organizational resistance to change is one of the biggest hindrances to digital transformation. Staff members may not be willing to embrace new technology or change well-established workflows and thus require intensive change management policies and training courses.

Data Privacy Concerns- The increasing dependence on data analytics raises concerns about customer data privacy and security (Kour, 2024). Institutions must ensure compliance with data protection regulations and build robust systems to safeguard.

FUTURE-READY STRATEGIES FOR DIGITAL TRANSFORMATION IN FINANCIAL INSTITUTIONS

Given the gradual escalation of complexity in financial institutions' digital transformation, building future-proof strategies is essential for the long-term sustainability, agility, and competitiveness of financial institutions. The section explains the key elements and actionable recommendations for designing and implementing digital-era appropriate strategies.

*Enhancing customer-centricity-*Financial institutions need to put the customer at the heart of their digital transformation journey. This means taking a proactive approach to understanding the needs, preferences, and pain points of customers. Advanced analytics, AI, and mechanisms for customer feedback can help institutions tailor their services to deliver personalized and seamless experiences (Kour, 2024a).

Omnichannel Engagement: Providing consistent and integrated customer experiences across digital and physical channels, such as mobile apps, web platforms, and in-branch interactions.

Hyper-Personalization: Using AI-driven insights to customize financial products and services based on individual customer profiles and behaviors.

Drive Agility through Cloud and DevOps- Financial institutions must adopt agility in the face of rapidly changing markets through the adoption of cloud computing and DevOps practices (Nagarajan, 2018). These technologies allow for quicker application deployment, scalability, and better collaboration among teams.

Cloud Adoption: Moving legacy systems to cloud platforms for flexibility, infrastructure cost savings, and easy access to data and applications.

DevOps Practice: Integration between development and operations teams aims to speed up the product's development cycle for digital services along with reducing the time-to-market.

Cybersecurity Resilience- Along with the progressive complexity of threats, financial companies need to use advanced cybersecurity steps to protect sensitive client information and stay trustworthy (Kour, 2024)
Proactive Threat detection: AI-based cyber tools for threat detection in the real-time monitor and identify them.
Employee Training: Improving staff's awareness about cybersecurity, thereby reducing human errors, for instance, phishing.
Compliance: Implementing international standards and best practices in cybersecurity, such as GDPR, PCI DSS, and ISO 27001.
Leverage Emerging Technologies- Blockchain, AI, and IoT are expected to transform the way financial institutions function and the experience of their customers (Kour, 2024b). The institutes should embrace these new technologies and find ways to apply them in the organization.
Blockchain for Transparency: Leverage blockchain to simplify processes such as cross-border payments, digital identity verification, and fraud prevention.
AI for Predictive Insights: Deploy AI models to manage risk, detect fraud, and predict customer behavior.
IoT for Real-Time Data: Use IoT devices to collect real-time data to be used in applications such as usage-based insurance and asset tracking.
An Innovation-Driven Culture- Culture that encourages experimentation and innovation will keep the digital transformation efforts going.
Leadership Commitment: Executives must champion digital initiatives and provide resources to innovation projects (Soomro & Khan, 2024).
Cross-Functional Collaboration: Encouragement of collaboration across different departments to fuse in diverse viewpoints and skills.
Innovation Labs: Innovation hubs dedicated to pilots and testing new technologies before launching on large scales
Prioritize Sustainability and Inclusion- Digital transformation strategies need to be aligned with larger societal objectives; its objective is to increase sustainability and participation.
Green IT Practices: Reduced environmental footprint from the digital operations with efficient energy use and sustainable technology practices.
Financial Inclusion: Enhanced access to finance through mobile banking, microfinance, and digital wallets.
Measure Success with KPIs- Defining explicit Key Performance Indicators will ensure digital transformation initiatives generate measurable results
Customer Satisfaction Measures: Net Promoter Scores (NPS), customer retention percentages, and ratings.
Operational Efficiency Measures: Tracking decreased processing times, error rates, and costs

Revenue Growth Measurement: Measuring the effects on revenue and profits from digital-based projects

Partner for Progress- Collaborations and Partnerships are imperative in speeding the digital transformation in the financial service industry. Therefore, banks will have to engage in a collaborative relationship with fintech, technology vendors and academia.

Fintech Partnerships: Engagements with fintechs to enhance innovation and consumer-centric solutions by the fintechs.

Industry Consortia: Engaging in industry forums to discuss best practices, set standards, and collectively solve common problems.

Continuous Learning and Upskilling- The speed of technological change requires continuous learning and upskilling of employees at all levels.

Digital Literacy Training: Organizational training initiatives to enhance employee's digital acumen and ability to adapt to change.

Leadership Development: Equipping leaders with the competencies and means to lead successful digital strategies.

Navigating Regulatory Complexities- An adaptation to changing regulatory environments is important for the success of digital endeavors.

Compliance Automation: Application of RegTech to automate compliance processes and thus reduce manual efforts.

Stakeholder Engagement: Collaboration with regulators to develop policies that encourage innovation but also protect the consumer.

Future-ready digital transformation strategies in financial institutions would need to have a balance between technological innovation, human-centric design, robust cybersecurity, and sustainability. A holistic and forward-looking strategy will guide financial institutions through the complexities of the digital age and position them to be leaders in an increasingly competitive market. Commitment, agility, and a willingness to change at all levels of the organization are the demands of the journey.

CASE STUDIES AND BEST PRACTICES

Digital transformation is best explained with practical examples. Case studies on successful transformations include the successful strategy and lessons that were learned on how to conquer challenges. Some financial institutions, having reached the critical milestones of digital transformation, are discussed as follows:

Case Study 1: DBS Bank (Singapore)

One of the digital transformations' leaders of the banking industries, DBS Bank has developed the core values based on client orientation that made it leverage its strength to digitize with customer-centricity- delivering technology that results in smooth personalized experiences. (Sia et al. 2021) Key initiatives under the banner: adoption of AI and machine learning by adopting AI-driven chatbots where customers were aided through query and answered within efficient and around-clock times while trying to minimize lengthy waits (Sia et al. 2021).

Cloud Migration: The bank scaled its operations, enhanced security, and improved cost efficiency by migrating to cloud infrastructure.
Agile Development Practices: DBS embraced agile methodologies across teams to speed up the development and deployment of new digital products (Sia et al. 2021).The bank has been named the "World's Best Digital Bank" because of its digital-first culture and innovative mindset.

Case Study 2: BBVA (Spain)

Banco Bilbao Vizcaya Argentaria (BBVA) is one of the most prominent global leaders that are embracing digital transformation. BBVA invested in fintech acquisitions and developed its in-house digital capabilities for maintaining its competitiveness at the forefront (Santomier at al. 2020). Some of the notable strategies include:

Open Banking Initiatives: BBVA was among the first bank to introduce an open banking platform so that third-party developers can integrate with its APIs for rich customers and services (Santomier at al. 2020).
Data-Driven Decision-Making: BBVA utilized big data analytics to enhance its credit scoring models, improving both the accuracy and the customer experience (Santomier et al. 2020).
Mobile-First Approach: The mobile app, launched by BBVA, is characterized by ease of use in features such as budget tracking and personalized financial advice, thereby ensuring high engagement and satisfaction.

Case Study 3: Goldman Sachs (United States)

Goldman Sachs has been able to redefine itself from being a traditional investment bank to becoming a technology-based financial institution. The key takeaways of this transformation are as follows:

Introduction of Marcus: Goldman Sachs introduced Marcus, which is an online savings account and personal loan, expanding its reach into the market for retail customers (Khatib et al. 2025).

Automation of Trading Operations: Goldman Sachs heavily invested in AI and algorithmic trading, which helped to optimize its operations, reduce costs, and improve efficiency (Khatib et al. 2025).

Partnership with Startups: Goldman Sachs partnered with fintech startups, which helped them explore new technologies and integrate innovative solutions into their offerings.

Best Practices for Digital Transformation

These case studies illustrate the following best practices that financial institutions can take on board:

Be customer-centric: Design products and services in such a way that convenience, personalization, and accessibility prevail.
Invest in talent and culture: Focus on building a digital-skilled workforce within an innovative and changing culture.
Leverage partnerships: Leverage fintech companies, technology providers, and other startups for accelerating innovation and getting the cutting-edge solutions.
Cybersecurity Focus: Implement stringent security measures to protect customer data and maintain trust in a digital ecosystem.
Continuous Innovation: Use agile methodologies and iterative processes to remain responsive to changes in the market and evolving needs of customers.

Case Study 4: HDFC Bank (India)

HDFC Bank has been one of the front-runners in digital banking through innovative technology adoption and enhancement of customer experience (Ray & Anirudhan, 2023). The key accomplishments include:

Digital Lending Platforms: HDFC Bank launched end-to-end digital loan applications, ensuring faster approvals and disbursements (Ray & Anirudhan, 2023).

AI-Powered Fraud Detection: The bank leveraged AI and machine learning to detect fraudulent transactions in real-time, thus increasing security.

Digital Payment Systems: HDFC Bank encouraged digital payment solutions, including UPI and contactless payments, which contributed to financial inclusion and convenience.

Case Study 5: Prudential Plc – Digital Transformation in Insurance

Prudential Plc is one of the world's leading insurance companies. The firm was facing significant challenges due to the changing nature of customer expectations, legacy systems, and the growing regulatory requirements. The company required a digital-first approach in order to improve operational efficiency, enhance customer experience, and maintain competitiveness in the rapidly changing financial landscape. Key Strategies Implemented:

- AI-powered claims processing to automate and expedite claim approvals.
- Cloud-based data management for enhanced security and scalability.
- AI-powered underwriting for customized insurance and better risk analysis.
- Mobile-first for engaging customers and greater self-service options.
- Blockchain for the integrity, security, and compliance requirements.

Case Study 6: Ping An Bank (China)

Ping An Bank's transformation into a digital powerhouse demonstrates the importance of integrating advanced technologies (Yao & Song, 2023). Major milestones include:

Ecosystem Development: Ping An developed a comprehensive ecosystem that integrates banking, insurance, and healthcare services, offering a one-stop solution for customers (Yao & Song, 2023).

AI and Blockchain Applications: The bank leveraged AI for customer service automation and blockchain for secure transaction processing.

Smart Branches: Ping An introduced "smart branches" powered by facial recognition and voice interaction, elevating the customer service experience.

CONCLUSION AND FUTURE DIRECTIONS

Digital technologies and what customers expect when it comes to services are fueling wide-scale changes for the financial services industry. Analysis of this change using the role of digital transformation towards restructuring financial institutions with the need for operational excellence and customer-centric strategies as well as data-driven decision-making can be well exhibited through studies of successful case studies and literature review to demonstrate how institutions can transition from legacy systems to agile, technology-enabled frameworks that allow for efficiency, promote a better customer experience, and stimulate innovation. Advanced technologies such as AI, blockchain, and analytics have been found to be crucial in achieving these goals, which help in better risk management, fraud detection, and personalized services.

As financial institutions move through this digital transformation, the chapter emphasizes visionary leadership, strategic alignment, and a culture of adaptability. Decentralized finance, real-time analytics, and IoT are only a few emerging technologies that carry immense promise with challenges that are to be rolled out thoughtfully and with ethical eyes. Concluding the chapter, the roadmap for future-proofing operations goes into strategic terms with an insistence on aligning digital transformation efforts with sustainability and societal needs. This discussion is meant to guide professionals and policymakers in fostering resilience, innovation, and sustainable growth in a digitally disrupted financial landscape.

It requires continued discovery and innovation. Future studies could be more focused on implications for financial systems in quantum computing, decentralized finance, and advanced robotics. In addition, it also calls for attention to how public policy can facilitate the digital journey so that innovations are not driven ahead of responsibility and consumer protection.

REFERENCES

Addy, W. A., Ugochukwu, C. E., Oyewole, A. T., Ofodile, O. C., Adeoye, O. B., & Okoye, C. C., (2024). Predictive analytics in credit risk management for banks: A comprehensive review. *GSC Advanced Research and Reviews*, *18*(2), 434–449. DOI: 10.30574/gscarr.2024.18.2.0077

Alami, R., & Sergio, R. (2024, April). Future-Proofing Strategies: Machine Learning Models for Organizational Resilience in Emerging Markets. In International Conference on Business and Technology (pp. 487-501). Cham: Springer Nature Switzerland. DOI: 10.1007/978-3-031-67431-0_47

Alonge, E. O., Dudu, O. F., & Alao, O. B. (2024). The impact of digital transformation on financial reporting and accountability in emerging markets. *International Journal of Science and Technology Research Archive, 7*(2), 025-049.

An, Y. J., Choi, P. M. S., & Huang, S. H. (2021). Blockchain, cryptocurrency, and artificial intelligence in finance. In *Fintech with artificial intelligence, big data, and blockchain* (pp. 1–34). Springer Singapore. DOI: 10.1007/978-981-33-6137-9_1

Colli, M., Cavalieri, S., Cimini, C., Madsen, O., & Vejrum Wæhrens, B. (2020). Digital transformation strategies for achieving operational excellence: a cross-country evaluation.

de Waal, A., & Linthorst, J. (2020). Future-proofing the high-performance organization. *Sustainability (Basel)*, *12*(20), 8507. DOI: 10.3390/su12208507

Divissenko, N. (2023). Regulation of Crypto-assets in the EU: Future-proofing the Regulation of Innovation in Digital Finance. *European Papers-A Journal on Law and Integration*, *2023*(2), 665–687.

Fülöp, M. T., Topor, D. I., Ionescu, C. A., Căpu neanu, S., Breaz, T. O., & Stanescu, S. G. (2022). Fintech accounting and Industry 4.0: Future-proofing or threats to the accounting profession? *Journal of Business Economics and Management*, *23*(5), 997–1015. DOI: 10.3846/jbem.2022.17695

Kaluarachchi, B. N., & Sedera, D. (2024). Improving Efficiency Through AI-Powered Customer Engagement by Providing Personalized Solutions in the Banking Industry. In Integrating AI-Driven Technologies Into Service Marketing (pp. 299-342). IGI Global Scientific Publishing.

Khatib, S. F., Mustafa, Z., & Abbas, A. F. (2025). Digital Transformation and Financial Sustainability. In Algorithmic Training, Future Markets, and Big Data for Finance Digitalization (pp. 33-74). IGI Global Scientific Publishing. DOI: 10.4018/979-8-3693-6386-7.ch002

Khatib, S. F., Mustafa, Z., & Abbas, A. F. (2025). Digital Transformation and Financial Sustainability. In Algorithmic Training, Future Markets, and Big Data for Finance Digitalization (pp. 33-74). IGI Global Scientific Publishing. DOI: 10.4018/979-8-3693-6386-7.ch002

Kour, M. (2024a). “Challenges and Opportunities of Machine Learning in the Financial Sector”, Algorithmic Approaches to Financial Technology: Forecasting, Trading, and Optimization (pp. 61-73). IGI Global Scientific Publishing. DOI: DOI: 10.4018/979-8-3693-1746-4.ch004

Kour, M. (2024b). Revolutionizing Finance: Unleashing Machine Learning’s Potential for a New Financial Era. In *2024 International Conference on Advances in Computing Research on Science Engineering and Technology (ACROSET)* (pp. 1-5). IEEE. DOI: 10.1109/ACROSET62108.2024.10743732

Kour, M. (2025). Navigating the Fintech Wave and Its Background and Challenges. In Algorithmic Training, Future Markets, and Big Data for Finance Digitalization (pp. 121-132). IGI Global Scientific Publishing DOI: 10.4018/979-8-3693-6386-7.ch005

Madakam, S., Holmukhe, R. M., & Jaiswal, D. K. (2019). The future digital work force: Robotic process automation (RPA). *Journal of Information Systems and Technology Management*, *16*, e201916001. DOI: 10.4301/S1807-1775201916001

Madan, V., & Kour, M. (2024), “Fintech and Mobile Payments in Transition: A Study of Adoption Challenges”, *2024 IEEE International Conference on Interdisciplinary Approaches in Technology and Management for Social Innovation (IATMSI)*, (2), pp. 1-6. IEEE. DOI: DOI: 10.1109/IATMSI60426.2024.10502648

Mohan, D. (2020). *The financial services guide to Fintech: Driving banking innovation through effective partnerships*. Kogan Page Publishers.

Nagarajan, A. D. (2018). DevOps implementation framework for Agile-based large financial organizations (Master's thesis).

Ogundipe, D. O., Odejide, O. A., & Edunjobi, T. E. (2024). Agile methodologies in digital banking: Theoretical underpinnings and implications for customer satisfaction. *Journal of Science and Technology, 10*(2), 021-030.

Ranchordás, S., & van't Schip, M. (2019). Future-proofing legislation for the digital age. Time, Law, and Change. University of Groningen Faculty of Law Research Paper, (36).

Ray, S., & Anirudhan, A. (2023). HDFC Bank's Digital Transformation Journey. *IUP Journal of Business Strategy, 20*(4).

Rehan, H. (2024). Modernizing Financial Markets With AI and Cloud Computing: Enhancing Efficiency, Precision, and Security Across Stocks, Crypto, Bonds, and Government Securities. *Distributed Learning and Broad Applications in Scientific Research*, *10*, 302–318.

Santomier, J., Söderman, S., & Kunz, R. (2020). Digital Transformation and Strategic Sponsorship: The Case of BBVA.

Setyawati, N., & Muditomo, A. (2023). FUTURE-PROOFING BANKS: LEVERAGING KEY HUMAN CAPITAL COMPETENCIES FOR SUCCESS IN THE AGE OF FINTECH AND BIG TECH. In Perbanas International Seminar on Economics, Business, Management, Accounting and IT *(1)* pp. 160-165).

Sia, S. K., Weill, P., & Zhang, N. (2021). Designing a future-ready enterprise: The digital transformation of DBS bank. *California Management Review*, *63*(3), 35–57. DOI: 10.1177/0008125621992583

Soomro, M. A., & Khan, A. N. (2024). Reimagining resilience: Visionary leadership, digital transformation, and strategic flexibility in small and medium enterprises in construction sector. *IEEE Transactions on Engineering Management*, *71*, 15070–15083. DOI: 10.1109/TEM.2024.3477629

Soramäki, K., & Sissens, D. (2019). Future-proofing the next generation of financial market infrastructures: The devil is in the data. *Journal of Digital Banking*, *4*(2), 125–130. DOI: 10.69554/NDHN6650

Teichert, R. (2019). Digital transformation maturity: A systematic review of literature. *Acta Universitatis Agriculturae et Silviculturae Mendelianae Brunensis*, *67*(6), 1673–1687. DOI: 10.11118/actaun201967061673

Voskerichyan, R. O., & Baiming, J. (2024). The Digital Transformation: Unlocking New Dimensions in Manufacturing Efficiency. Вестник Российского университета дружбы народов. Серия. *Ekonomika (Nis)*, *32*(2), 235–250.

Yao, T., & Song, L. (2023). Can digital transformation reduce bank systemic risk? Empirical evidence from listed banks in China. *Economic Change and Restructuring*, *56*(6), 4445–4463. DOI: 10.1007/s10644-023-09560-2

Chapter 13
Future Perspectives of Business in the Context of Digitalization

K. Poornima
https://orcid.org/0000-0003-1157-1901
Independent Researcher, India

ABSTRACT

Technology is the new mantra for Business nirvana. Too many businesses see digitization, mobile, cloud, social media, internet of things, artificial Intelligence, etc., as must-haves to keep up with the hip cloud. Digital technologies are about doing things in new ways. India is creating unique open digital networks, such as ONDC, designed to embed financial services from inception, fundamentally transforming the embedded finance landscape. Large consumer platforms, in categories such as travel, e-commerce, mobility & others, are embedding financial services to unlock significant value. The digitization of India's fragmented B2B supply chains is creating opportunities to deliver flow-based credit with profitable unit economics, especially for underserved last-mile MSMEs.

INTRODUCTION

Technology is the new mantra for Business nirvana. In the modern world, technological advancements have created new opportunities and demands for companies across all industries. Businesses must leverage the possibilities of digital tools and platforms like cloud computing, artificial intelligence, data analytics, Block chain, and cyber security to boost productivity, creativity, and customer happiness. Digital transformation employs modern digital technologies to modify an organization's

DOI: 10.4018/979-8-3373-1320-7.ch013

culture, business procedures, and customer experiences to satisfy shifting market and commercial expectations. Adopting new technologies or platforms is only one aspect; another is reassessing how an organization functions, provides value and competes in the digital age.

Digital transformation (DT) refers to the integration of digital technologies into all facets of business operations, fundamentally altering how organizations deliver value to customers and adapt to market dynamics. This process encompasses the adoption of technologies such as artificial intelligence (AI), the Internet of Things (IoT), block chain, and cloud computing, aiming to enhance efficiency, innovation, and customer engagement. Digital transformation can affect organizational culture, business models, processes, and customer interactions. While the concept has been evolving over the years, the urgency for transformation significantly increased after 2020 due to the global COVID-19 pandemic, which accelerated digital adoption worldwide.

RECENT DEVELOPMENTS IN DIGITAL TRANSFORMATION (2020–2025):

Acceleration Due to COVID-19: The COVID-19 pandemic significantly expedited digital transformation across various industries. A McKinsey survey from October 2020 revealed that the pandemic accelerated the adoption of digital technologies by several years, with many changes expected to persist long-term.

Integration of Advanced Technologies: Organizations are increasingly incorporating advanced technologies to drive innovation. A study published in 2023 examined the roles of IoT, social media, mobile applications, AI (including ChatGPT), augmented reality/virtual reality (AR/VR), and the metaverse in digital transformation, highlighting their unique challenges and opportunities.

Focus on Organizational Agility: Research indicates a bidirectional relationship between digitalization and organizational agility. A study from December 2021 identified three thematic clusters focusing on big-data analytics capabilities, supply chain agility, and IT capabilities, emphasizing the need for dynamic capabilities to enhance organizational agility.

Digital Transformation in Manufacturing: The manufacturing sector has recognized the value of digital transformation, especially during the pandemic. A Microsoft article from 2020 discussed how digital transformation became crucial for manufacturers, enabling them to adapt to new challenges and maintain operations.

Digital Economy Initiatives: Global organizations are supporting digital economy initiatives to promote digital inclusion and economic growth. The World Bank's Digital Economy for Africa (DE4A) initiative aims to digitally connect every individual, business, and government in Africa by 2030, propelling the continent forward in the digital age.

KEY ASPECTS OF DIGITAL TRANSFORMATION

Technology Adoption: Embracing cloud platforms, big data, and AI has been crucial for many businesses. The move to cloud infrastructures has provided greater flexibility, scalability, and collaboration across organizations. Similarly, AI and data analytics play a significant role in decision-making and enhancing customer experiences.

Business Model Innovation: Digital transformation enables the redesign of existing business models or the development of entirely new ones (Agasisti & Froumin, 2020). For example, companies in retail, banking, and healthcare have shifted from traditional brick-and-mortar approaches to e-commerce, digital banking, and telemedicine services.

Customer-Centricity: Digital technologies allow organizations to understand customers better through data analytics, enabling more personalized experiences. AI-powered customer service tools (e.g., chatbots) and omnichannel communication are critical to enhancing engagement.

Agile Organizational Structure: Adapting to digital trends requires organizations to become more agile. This involves restructuring teams, fostering innovation, and promoting a culture of continuous improvement.

Cybersecurity and Data Privacy: As organizations digitize, there is an increasing need for robust cybersecurity measures. Data protection is critical, and companies are investing more in securing digital assets and maintaining customer trust.

Global Trends in Digital Transformation (2020 and Beyond):

Impact of COVID-19: The pandemic acted as a significant catalyst for digital transformation, with businesses rushing to adopt remote working solutions, digital payment systems, and e-commerce platforms (Sankhe et al., 2020). This transition has been particularly pronounced in sectors like retail, healthcare, education, and finance.

E-Commerce and Retail Shift: The retail sector experienced a massive pivot to online shopping. Industry giants like Amazon and Alibaba (Asian Development Bank, 2020) have continued to grow, and smaller businesses have had to quickly adopt e-commerce platforms to survive. The rapid rise of digital payment solutions and logistics platforms is a direct outcome of this shift.

Telemedicine and Healthcare: Healthcare providers globally have been increasingly adopting digital solutions for consultations, diagnostics, and treatment delivery. Platforms like Teladoc (USA) and Doctolib (France) have gained widespread traction. The pandemic sped up the regulatory approval for telemedicine in many countries, making it a critical part of the healthcare system moving forward (Haleem et al., 2021).

Smart Manufacturing and Industry 4.0: The manufacturing industry has been incorporating advanced robotics, automation, and IoT to enhance production efficiency. The concept of smart factories, where digital systems autonomously monitor and optimize the production process, is becoming increasingly mainstream.

Artificial Intelligence and Automation: AI and automation have been significant drivers of digital transformation (Gartner, 2021). For instance, automation in financial services is being employed for everything from fraud detection to customer service. In the public sector, AI is being used for predictive analytics and improving citizen services.

Sustainability and Green Technologies: Digital technologies have enabled companies to innovate in areas of sustainability. In industries such as energy and manufacturing, digital tools are used to reduce waste, improve energy efficiency, and minimize carbon footprints (Deloitte, 2022).

Digitalization in Business: Business digitalization is the process of incorporating digital technologies into many facets of operations, which radically alters how companies function, provide value to clients, and maintain their competitiveness. In order to enhance corporate operations, decision-making, and customer experiences, digital tools, systems, and strategies must be implemented.

IMPORTANT DOMAINS IN WHICH DIGITALIZATION AFFECTS BUSINESSES

1. The Experience of Customers

   Personalization: Businesses employ AI and data analytics to provide clients with individualized marketing, services, and goods. Examples include customized customer service or e-commerce sites that offer individualized recommendations.

   Omnichannel Interaction: Companies can smoothly communicate with clients by using a variety of channels, including chatbots, email, social media, mobile apps, and websites.

2. Activities and Effectiveness

Automation: Software and robotics are used to automate many time-consuming and repetitive operations, including inventory management and data entry, which increases productivity and lowers human error.

Supply Chain Optimization: Digital tools provide real-time inventory tracking, demand forecasting, and logistics optimization for increased efficiency.

3. Data-Informed Decision Making

   Big data analytics are used by businesses to learn more about market trends, customer behavior, and operational efficiency. This makes it possible to make better decisions in real-time.

   Cloud computing: By keeping data in the cloud, businesses may access and evaluate information from a distance, facilitating improved decision-making flexibility and collaboration.

4. Promotion and Sales Digital marketing:

To connect with customers more successfully and reach a wider audience, businesses use digital platforms like social media, SEO, email campaigns, and online advertisements. Online storefronts and payment systems have revolutionized the way firms sell items by providing 24/7 access to goods and services.

5. **Innovation in Products and Services**: Businesses can create new goods and services or enhance their current offerings thanks to digital technologies. AI can be utilized for predictive maintenance in industrial equipment, while the Internet of Things (IoT) enables devices to communicate with people.
6. **Cooperation and Interaction:** Remote work is made possible and internal communication is improved by tools like video conferencing (e.g., Zoom, Teams), instant messaging, and project management software (e.g., Asana, Trcllo).

Cooperation with Partners: Businesses may work together with external partners, suppliers, and stakeholders more successfully thanks to cloud-based solutions.

7. **Risk management and cybersecurity**: Businesses must make significant investments in cybersecurity as they digitize to safeguard financial assets, data, and intellectual property from online attacks.
8. Talent management and human resources:

   HR Tech: Employers may better manage their workforce by utilizing digital technologies for hiring, managing employees, and monitoring performance.

Remote operation: By giving workers the means to operate remotely, digitalization promotes remote work and increases flexibility and work-life balance.

9. **Management of Finances:** Businesses can streamline their accounting, invoicing, and budgeting procedures with the use of digital tools and software (like Xero and QuickBooks), and they can access and analyze financial data instantly.

   Blockchain Technology: This is being incorporated into business models for safe and transparent financial transactions, particularly in the supply chain and finance sectors.

Figure 1. Digitalization Penetration in Business

This path's logical conclusion and forward-looking anticipation is to define "digital transformation," a phenomenon that encompasses not only technology but also professional, business, and cultural changes evident in every facet of social life. Digital transformation, for instance, has resulted in the use of smartphones to interact with the outside world.

Digital Growth:

Digital foundation (cost, speed, and dependability of internet service); digital reach (number of mobile devices, app downloads, and data consumption); and digital value (the extent to which consumers interact online through chat, tweeting, purchasing, or streaming) are the three components that make up the Digital Adoption Index.

Advantages of digitalization in business:

* Enhanced effectiveness of every operation
* Reduced human error
* Lower operating costs and safer cloud data storage
* Makes data analysis possible

Digital Growth and Indian Context:

There has been an increase in domestic IT firms in India, many of which have achieved global acclaim. With the help of government initiatives, venture funding, and a populace that is becoming more tech-savvy, the startup environment has expanded. Numerous “unicorns” (startups worth over $1 billion) have emerged from India., particularly in industries like fintech, e-commerce, edutech, and health tech.

ANTICIPATING THE NEXT BIG THING IN IT INNOVATION

By 2025, the entire world will be halfway through the present decade. And if the past ten years are any guide, some significant changes are in store.

Of course, in ten years, anything might happen. The predictions could all be completely inaccurate due to unforeseen disruptive or world-changing occurrences. However, because these theories are predicated on extrapolating current events in politics, society, and technology, they can still provide us with important insights into what the future might bring.

We are spending an increasing amount of time on the internet, utilizing digital services and virtual worlds. This tendency is unlikely to change over the next ten years as technology grows more affordable, pervasive, and immersive.

The following are some crucial areas where ***IT innovation*** may soon have a big influence:

1. Artificial Intelligence
2. Quantum Computing
3. The expansion of 5G
4. VR, or virtual reality 2.0
5. AR, or Augmented Reality
6. The Internet of Things
7. Agricultural Biotechnology
8. Self-Driving Automobiles
9. The blockchain
10. Cutting-Edge Computing
11. Customized Healthcare
12. Neuromorphic Computing
13. Technologies for Green Energy
14. Health Monitors That Are Wearable
15. Training with Extended Reality (XR)
16. Voice-Activated Technology
17. Space Travel
18. Synthetic Media

19. Advanced Robotics
20. Cyber Security and AI
21. Digital Twins
22. AI TRiSM
23. Telemedicine
24. Nano-Technology
25. Sustainable Technology

Figure 2. Digitalization Penetration in different Industry

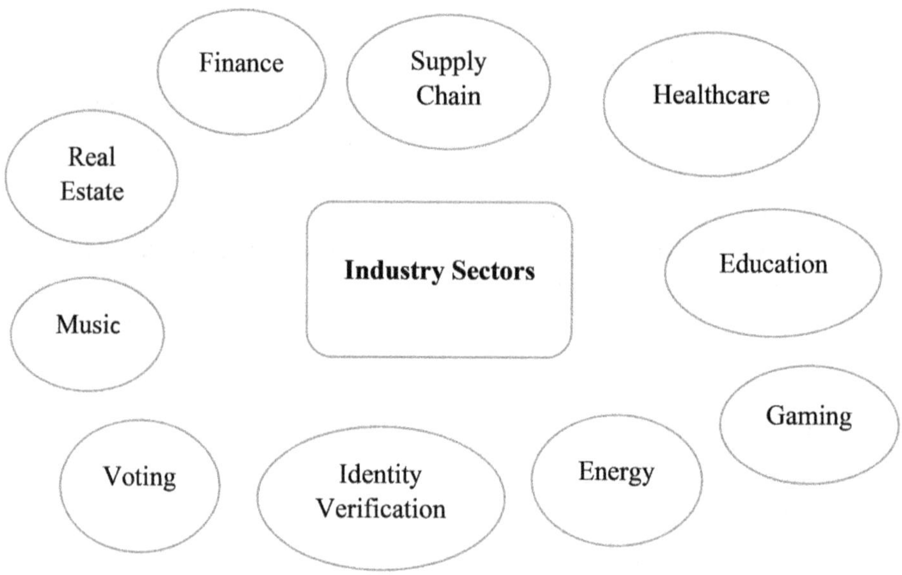

2024 has arrived following a year that has been full of setbacks, triumphs, and everything in between. Examining how the major Indian IT sectors will appear in the upcoming year, however, there are still some more significant problems that need to be addressed:

(i) Are technology-based and IPO-bound businesses and startups prepared for geopolitical challenges and setbacks?

(ii)Will laws and regulations on new technology encourage or impede innovation?

(iii) Which industries will be most challenged by the emergence of new technology and the laws governing them?

(iv) In 2025, how will generative AI alter the landscape for both tech giants and startups?

Established startups that have grown effectively over the last ten to twelve years will now have to contend with competition from new businesses that are implementing creative business plans and tactics. The middle-class consumer category is anticipated to be the main contributor, creating a larger base. The emergence of innovative concepts and prosperous business models in modern industries will propel growth in early-stage fundraising rounds across several industries.

Manufacturing, agritech, climate sustainability, healthcare, uncharted fintech territory, and anything on accelerating enterprise digital transformation are among the industries that are likely to become or stay in the national limelight.

Biotech has shown hints of a resurgence in the healthcare industry, suggesting a possible upturn, particularly in a worldwide setting. It is expected that the deep tech investment theme will continue to grow, with a particular emphasis on GenAI.

In 2034, continuous advancements and investments in areas such as stem cell research, genetic engineering, and personalized medicine will probably have a significant impact on how we treat and cure patients.

It is feasible that many genetic abnormalities may have been corrected before birth because of developments in genome editing tools like CRISPR-Cas9. This may lessen the occurrence of numerous inherited illnesses, including cystic fibrosis and muscular dystrophy. Even the general hereditary susceptibility to the harmful effects of high blood pressure or cholesterol may be lessened. This might have enormous societal repercussions, including longer lifespans for people.

Nowadays, whether we search on Google, choose which movies to watch on Netflix, or conduct online banking, we seldom ever consider how AI is working in the background. We won't think about it tomorrow because it keeps us healthy, powers our automobiles, and makes our jobs more efficient. Given the developments in speech production and natural language processing, voice control may replace human interaction with machines as the primary means of communication in ten years. Although the experience can be restricted and unreliable, we are already accustomed to conversing with machines such as Alexa or Siri. By 2034, however, it will seem perfectly banal to have fluent, fluid discussions with technology as natural language processing takes over. Our gadgets will also be far better at figuring out what we want and anticipating what will make us happy as a result of advanced behavioral analytics.

Over the last two to three years, the Private Equity/Venture Capital (PE/VC) ecosystem has seen a significant increase in domestic funds and a growing inflow of domestic capital. VCs now expect that in the years to come, international investors will create even more facilities in India. The multi-stage international venture capital fund MSA Novo most recently revealed a portfolio of six Indian startups, including Flint, Questbook, Apps For Bharat, Slice, Jar, and MPL. The potential to build in India for the global market has never been more apparent, since every sector in the country offers a chance for technology-led innovation. There is good reason to rejoice as long as innovation, consumer value, and sustainable expansion are the major priorities. Experts believe that now is the best time to incorporate a business or fund in India.

With a strong emphasis on preserving sound asset quality and taking a calculated strategy that rewards established category leaders, the venture loan market is changing. Venture debt is becoming more popular as the startup ecosystem develops, especially for startups with significant financial requirements and in new industries like electric vehicles (EVs).In the meantime, it's important to remember that investors will require more time than usual to assess firms, which will challenge founders' patience.

Some industry illustrations of big things in Business based on IT Innovation:

(i)The intraday trading of cyber security businesses reached a new all-time high, continuing their bull run. Their market capitalization skyrocketed at the end of the day's trading as a result. With a "BUY" rating and a target price of INR 640 per share, the local brokerage firm has started covering the food tech giant.

(ii) The Open Network for Digital Commerce is still growing in popularity. Since December 2024, the state-backed network's transaction volume has increased by over three times. By December 2025, it is anticipated that the number will have increased 7X-8X.

(iii)The SaaS-based accounting startup has raised the capital as part of its ongoing $15 Mn Series B funding round from Orange Orbit.

(iv) The Rise of AI-Powered Cooking Appliances The market for smart kitchen appliances is predicted to grow at a 19% CAGR from $20.9 billion in 2024 to an astounding $84 billion by 2031. Learn how this IIT alumnus team is fostering innovation with AI-powered cooking aids from upliance.ai.

(v) For as little as INR 500, customers of Lets Get Happi can get 24/7 therapy from psychologists. Through yoga classes, guided meditation, and therapeutic writing, users may also monitor their development on the platform.

(vi) Interpret Bags: Under the leadership of Canaan Partners, the firm has raised $20.8 million in its Series A financing. Interpret offers CXOs of different teams unified data and AI models to analyze massive amounts of unstructured consumer feedback from all channels.

(vii) Few startups have introduced AI-powered quality and produce-handling systems to ensure farmers receive the best possible prices for their produce.

Progress-in-line:

(i) Allen will purchase Unacademy: The offline coaching chain is in advanced talks to buy the edtech business led by Gaurav Munjal for an estimated $800 million, 75% less than its 2021 top valuation of $3.44 billion.
(ii) Through its first $75 million fund, the venture capital firm Antler says it has sponsored 30 Indian entrepreneurs so far this year, covering industries like fintech, consumer tech, and artificial intelligence. As a result, Antler now has 80 businesses in its India portfolio.
(iii) The international rating agency has kept its outlook on OYO constant and raised its corporate family rating from B3 to "B2." Moody's said the upgrade was due to the company's increased profitability in the last few quarters.
(iv) To introduce "The Growth Protocol," a blockchain-focused protocol for digital marketing and growth applications in the Web3 ecosystem, the publicly traded gaming behemoth has teamed up with blockchain startup Lysto.

The drone industry is undergoing a major shift, evolving from a niche technology into a critical tool across multiple sectors, including defense, agriculture, healthcare, and logistics.

The convergence of these technologies to produce whole new opportunities is probably going to be the next big thing in IT innovation. With blockchain, edge computing, AI, and quantum computing opening the door for discoveries, we are at the cusp of revolutionary shifts in every sector.

ETHICS AND CORPORATE RESPONSIBILITY IN THE CONTEXT OF DIGITAL BUSINESS

There is no escaping the fact that the digital world is here to stay. It is changing how we engage with clients and conduct business. Thus, digital ethics and corporate digital responsibility will also be crucial to the future of business.

Businesses are increasingly using digital technology to interact with clients, build connections, and reach a wider audience. In the process, they expose themselves to hazards as well as opportunities. Because of the intricacy of the digital world, companies must consider digital ethics and accountability if they want to succeed. The internet is an effective tool, and I think we should use it to improve the world.

But we also need to be aware of how our technologies are destroying the environment and violating people's rights and privacy.

There are four reasons why digital ethics are necessary. First and foremost, governments and consumers are becoming increasingly concerned about the misuse of personal data. This is followed by the effects and difficulties of automation and robotics. Other issues for firms to adopt digital ethics include the possibility of unethical use of new technologies and the so-called digital divide. To remain competitive, businesses must plan their digital transformation to increase public confidence in the digital economy while maintaining openness about the technologies they employ and the data they handle.

Technology at religious events: For example, the ongoing (January 2025) at the Mahakumbh Mela unravels the penetration and usage of technology for societal as well as business benefits.

With the help of blockchain solutions company Chaincode Consulting and the Indian Railway Catering and Tourism Corporation, the biggest religious event in the world will enable participants and devotees to interact with the event via a non-fungible tokens (NFT) platform. These NFTs will provide users with customized digital quizzes about Indian mythology and the festival, as well as rare digital souvenirs.

. Indian fintech firms raised $1.9 billion in 2024, a 33% decrease from $2.8 billion the previous year. India continued to be the world's third-largest fintech funding hub, behind the US and the UK, despite the drop. With this, we can say, that in this year's Maha Kumbh Mela, technology and tradition coexist.

More than 40 crore devotees are anticipated to attend the sacred event, and India Inc. is helping to make things easy for them. This traditional gathering is poised to embrace modernity with the help of AI-based monitoring and GPS tracking, insurance products from PhonePe, and the "Kumbh Sah'AI'yak" chatbot, which is powered by Ola's Krutrim AI. The Economic Times claims that businesses will spend at least Rs 3,000 crore alone on marketing and branding during the 45-day Maha Kumbh Mela in Prayagraj.

Figure 3. Ethical Concerns in Digitalization

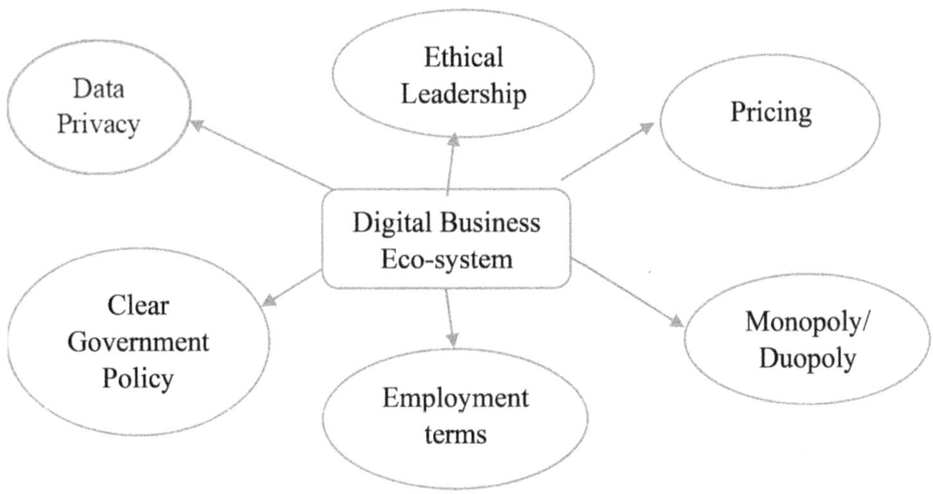

Significant development and innovation potential are presented by digital transformation, but it also presents moral dilemmas that businesses must resolve. The following are important factors to take into account when it comes to corporate responsibility and ethics in digital business:

DATA PRIVACY AND SECURITY

1. Security and Privacy of Data

Ethical Concerns: Digital companies frequently gather a great deal of personal information from their clients, including browsing and purchase patterns as well as private data. The following are some ethical issues with this data collection:

Making sure users are aware of the data being gathered and its intended use is known as informed consent. Protecting sensitive and personal information from abuse or breaches is known as data protection.

Transparency: Companies must provide users control over their data, have clear privacy policies, and be transparent about their data practices and policies.

Corporate Responsibility: Businesses have to put strong cyber security protections in place, adhere to data protection laws, and be open and honest about their data practices.

2. Automation and Artificial Intelligence

Ethical Concerns: AI and automation may give rise to ethical questions, particularly those about responsibility, bias, and justice.

Bias: AI systems have the potential to reinforce preexisting biases in training data, which could result in unfair hiring, credit scoring, or law enforcement outcomes.

Job displacement: Automation may result in the loss of jobs or their displacement, which raises concerns about how businesses should manage the well-being of their workforce.

Decision-making: Businesses need to guarantee accountability for the results of automated processes and be open and honest about how AI systems make decisions.

Corporate Responsibility: Companies should work to develop impartial and equitable AI systems, make investments in retraining or up-skilling workers affected by automation, and guarantee responsibility in decisions made using AI.

3. Environmental Impact and Sustainability

Moral Concerns: E-waste and data center energy consumption are two examples of how digital businesses may have a big influence on the environment. Businesses must balance environmental sustainability with corporate success, which presents ethical problems.

Energy consumption: Block chain technology, cloud computing, and data centers all need high energy usage. It is necessary to reduce the environmental impact of these technologies.

E-waste: Outdated electronic gadgets are disposed of due to the quick speed of technological advancement, which adds to the amount of e-waste.

Corporate Responsibility: Businesses are required to eliminate their carbon footprint, embrace sustainable practices, and be accountable for the lifecycle of their products, which includes recycling and cutting down on e-waste.

4. Fairness and Consumer Protection

Ethical Concerns: Consumers in the digital age may be at risk for exploitation, dishonest advertising, or fraudulent activities.

False advertising: Consumers may suffer from the ease with which false information or deceptive advertisements can be disseminated online.

Vulnerable group exploitation: Digital companies may use predatory marketing or exploitative commercial tactics to target vulnerable communities.

Corporate Responsibility: Businesses should safeguard vulnerable customers, maintain ethical marketing standards, and offer channels for resolving disputes. They must also be open and honest about the features, costs, and hazards of their products.

5. Digital Access and Inclusion

Ethical Concerns: Depending on how they are used, digital technologies have the potential to either improve or worsen inequality. There are moral questions about:

Technology access: Not all people or groups have equal access to digital tools or the internet, which could cause them to fall behind in the digital economy.

Affordability: Digital disparities may be exacerbated when certain digital services or goods are too expensive for lower-income populations.

Corporate Responsibility: Businesses should encourage digital inclusion by lowering costs, increasing accessibility, and improving the user experience of technology. To alleviate gaps in digital access, this may entail collaborations with NGOs or governments.

6. Transparency in the Supply Chain

Moral Concerns: Many digital enterprises depend on intricate international supply chains, which can give rise to questions about human rights, fair compensation, and working conditions in the procurement of raw materials or production processes.

Labor exploitation: In supply chains, there may be worries about subpar working conditions or unethical labor practices.

Environmental harm: Degradation of the environment may result from the extraction of resources for technology.

Corporate Responsibility: Digital companies need to make sure that their supply chains are open and compliant with human rights, environmental, and ethical labor norms. Sustainability programs and ethical sourcing are essential.

7. Transparency and Corporate Governance

Ethical Issues: In rapidly evolving digital settings, it can occasionally be difficult for digital enterprises to uphold the high norms of corporate governance.

Accountability: To prevent unethical behavior, it is essential to make sure that the company's leadership is held responsible for making moral decisions.

Regulatory compliance: Companies operating digitally are subject to both national and international restrictions.

Corporate Responsibility: Businesses must uphold sound governance procedures, encourage moral leadership, and guarantee operational openness. This entails following relevant laws and rules, disclosing information about corporate social responsibility (CSR) initiatives, and communicating openly and morally.

8. Ethical Advertising and Marketing

Ethical Concerns: Digital platforms provide a multitude of data that can be utilized for advertising that is highly targeted. Among the ethical issues are:

Manipulation: The possibility that businesses could utilize data to influence the behavior of customers, especially those who are more susceptible.

Privacy invasion: Advertising that uses excessive or intrusive data may violate the privacy of customers.

Corporate Responsibility: Businesses must make sure their methods are considerate and do not take advantage of users, and they must be open and honest about how they use customer data for marketing and advertising. Honesty, integrity, and respect for customers are the main focuses of ethical marketing tactics.

INDUSTRY ILLUSTRATIONS AND DISCUSSIONS

Personalized medicine may become the standard in ten years when patients anticipate that medical professionals will have access to nearly full genetic data so that they can develop treatments and cures that are unique to them. However, we will inevitably have to deal with a lot of ethical issues as a result of these technologies. It is important to carefully evaluate issues of consent and equal access to this technology when granting parents the ability to choose or alter genetic features that may be present in their children.

Furthermore, there may be unanticipated repercussions, such as the potential for these treatments to result in future health issues that we may not have anticipated. Anyone ready to give their genetic blueprint—even to their physician—would want to be quite certain that the right data protection measures are established. Given how this technology may alter the world in ten years, it is obvious that these are issues that must be resolved immediately if we are to attain a positive attitude.

A lot of financial services startups and businesses are taking tiny moves in 2024 in hopes of becoming the next big thing in Indian fintech. These financial services have been observed to pose a threat to fintech businesses throughout the past year. Every fintech startup, including those in banking, payments, digital lending, insurance, investments, and even AMCs, is being targeted by these firms. And it has a significant edge in entering the market because it is the highest capitalized fintech business in India, with almost $25 billion. On one hand, it is good growth on the other hand it may look like guerilla marketing. In addition, there were rumors that the well-established Indian conglomerate was negotiating with an Indian payments bank to buy a share in the business. The conglomerate's fintech aspirations would be strengthened by this.

Some Indian fintech companies that have been registered overseas since their founding are attempting to return to India. Much of this has to do with making it easy to interact with financial services authorities and to pay taxes. The opposite trend is anticipated to continue for lending platforms and payment businesses, where having a local presence is essential for regulatory compliance.

As witnessed by the changes seen by major SaaS giants, the distinction between traditional SaaS and AI is rapidly becoming less clear. To remain competitive in 2025, other Indian SaaS companies are also working hard to include AI capabilities in their product lines. The integration of AI with SaaS is especially noticeable in data analytics, marketing workflows, sales automation, HR, and customer support. To unlock the SaaS value that has thus far remained underutilized in India, companies will seek to utilize linguistic models unique to India in 2025. This may lead to unhealthy competition. Even if there were no major efforts taken to govern AI, there is undoubtedly a surge of startups attempting to self-regulate the industry by embracing ethical AI. Global AI legislation discussions began to take shape in 2024 and are probably going to get more specific in 2025. Among the seven initiatives the Indian government plans to concentrate on over the next five years, ethical AI has been identified as one of the main goals of the INR 10,372 Cr IndiaAI Mission.

The idea of "responsible AI" has also changed, and in 2025, it will continue to explore issues like compliance, transparency, data security, and real advantages to humanity.

However, one facet of the GenAI revolution will gain more attention in 2024: the drive for international or universal legislation to address the emergence of deepfakes, ethical AI, inauthentic or AI-generated material, and other undesirable elements in this context. As generative AI becomes more prevalent, startups and large corporations will also need to adapt to new realities. To address organizational readiness for emerging technologies, we anticipate that an increasing number of businesses will designate Chief AI Officers. Beyond boosting productivity and cutting down on turnaround times for customer service, content production, and user experience elements, businesses are still figuring out how to employ AI.

For a variety of reasons, numerous founders, including the CEO and co-founder of India's most valuable business (until recently), ended up entangled in legal issues. The Economic Offenses Wing and the Enforcement Directorate are looking into fraud claims in some of these more serious founder-related situations. These investigations and inquiries will conclude in 2025, yet issues related to fraud and corporate governance frequently go unnoticed for months before coming to light.

On the subject of e-commerce, the marketplace behemoths may face yet another challenge. The Indian government has already announced that the National E-commerce Policy will be implemented in 2024. In addition, the social security code for gig workers may finally be implemented after years of postponement. The holding structures of marketplace giants and the purported preferential treatment for certain large sellers may become even more complicated as a result of the e-commerce policy, which is generally anticipated to favor small sellers. In addition to this, the regulation may also disturb the incumbents by prohibiting predatory pricing and discounting, which would create space for new companies to emerge.

The social security code may provide much-needed relief for part-time delivery workers and mobility partners in 2023, while problems with gig economy platforms continue to exist. Startups and platforms would specifically need to set aside a portion of their earnings for gig worker insurance, pensions, gratuities, and provident fund payments. The increased reliance on platform fees makes sense in this regard as well. The reasoning for this is that businesses would charge customers to cover employee costs. It is yet unknown how this may affect platforms' profitability.

The much-discussed National AI Policy and the Digital Personal Data Protection Act are two more important pieces of the policy puzzle. The former is anticipated to cause fintech, social media, business tech, and services firms some issues, whilst the AI policy is probably going to impose more stringent guidelines to address AI-generated material in the media.

The increasing dependence on platform fees also makes sense in this regard. The idea is that businesses would bill customers to cover employee costs. It is still to be ascertained how this affects platform profitability.

Although the 15-minute delivery services appear to be brand-new and trendy, they have been developing for the past 12 months. In 2023, Swiggy and Zepto introduced "cafe" services in Bengaluru and Mumbai, respectively, but neither of these initiatives ever took off. Only until rapid commerce models became successful after this did these models get more traction. In any case, when Zomato and Swiggy embark on a new chapter in their long-running war, this is their new battlefield. Both of these businesses introduced identical 15-minute food delivery services without consulting the playbooks created by other eateries that have collaborated with them over the years if we are to accept their assertions. Considering that the warning signs have been there for a few months that seems a bit too. This has given restaurants a new headache and caused a massive upsurge in demand for quick food delivery. They must not just collaborate with companies like Zomato and Swiggy, but also somewhat compete with them.

The National Restaurant Association of India (NRAI) has a history of opposing the two titans, and we have seen the group rally against deep discounting and loyalty programs. The food delivery giants were accused by the association of anticompetitive activities in the past, including service bundling, excessive commissions, delayed payment cycles, and the imposition of one-sided terms.

Following user reports of stopped withdrawals, several crypto-currency companies have come under fire, leading to calls for stricter regulation of the industry and an investigation into the exchange's business practices. The petition was filed by users who claimed that the platform was mismanaging and hiding a February 2022 hack. The alleged illegal transfer of user assets overseas as a result of this hack heightened calls for responsible governance in the Bitcoin space. Additionally, the users have demanded that the Central Bureau of Inquiry (CBI) or a Special Investigation Team (SIT) conduct a comprehensive inquiry.

The NASDAQ-listed Cognizant and its CEO, Ravi Kumar S, are accused of participating in suspected anti-competitive actions in a counter-lawsuit filed in a US court by Infosys, which was accused of stealing trade secrets linked to its healthcare insurance software by a Cognizant subsidiary months ago. Cognizant allegedly purposefully hired its senior executives, according to Infosys. Meanwhile, as more business owners share their experiences of putting in long hours, the argument over the 90-hour norm continues to rage.

The flip side of India's digitalization boom: In India, many aspects of daily life are rapidly becoming digital. In addition, an excessive number of people are becoming victims of online fraud, with the elderly being a particularly vulnerable demographic. The nation's more than 138 million older folks could suffer even more from its significant advancements in digitalization if the government doesn't take action quickly.

THE NATURE OF GLOBALIZATION AND DIGITALIZATION ACROSS BORDERS

Globalization: The term "globalization" describes the growing interconnected ness and connectivity of nations, which is fueled by the flow of information, capital, people, products, and services across international borders. It is a complex process with social, cultural, political, and economic facets.

The Cross-Border Relationship Between Globalization and Digitalization

Globalization is both facilitated and driven by digitalization. It speeds up cross-border information, capital, and service flows, which facilitates cross-cultural idea sharing, company operations worldwide, and consumer access to goods and services.

Global Supply Chains: By allowing businesses to track shipments, improve logistics, and interact in real-time with suppliers and customers worldwide, digital technologies have simplified global supply chains.

Globalization is both facilitated and driven by digitalization. It speeds up cross-border information, capital, and service flows, which facilitates cross-cultural idea sharing, company operations worldwide, and consumer access to goods and services.

E-commerce & Digital Markets: Retail and commerce have been revolutionized by global digital platforms like Amazon, Alibaba, and eBay. These days, small enterprises don't need to be physically present in other nations to sell their goods.

Digital Divides: Although digitalization promotes globalization, it also draws attention to the disparities between areas with high and low levels of technological access. While developing nations may encounter difficulties with infrastructure, education, and internet access, developed nations typically gain more from digitization.

Cross-Border Collaboration: Individuals and organizations can work together across boundaries thanks to digital tools. This is seen in fields where teams from several nations can collaborate electronically, such as software development, scientific research, and the creative industries.

Cultural Exchange and International Content: As a result of digitalization, international streaming services like Netflix have grown in popularity, giving users access to content from around the world. Cross-border exchanges of news, information, and cultural items are more seamless than ever.

The Advantages of Globalization Through Digitalization

Spreads technology and knowledge: Technology and expertise are shared across nations thanks to digital globalization, which can stimulate economic progress.

Allows companies to enter new markets: Businesses may now access a larger range of markets and share ideas and information internationally thanks to digital globalization.

Empowers tiny companies: By participating in e-commerce marketplaces, small firms may now compete with larger multinational corporations thanks to digital globalization.

Drawbacks of Digital Globalization

Policies that protect: Protectionist measures, including online filtering and data localization mandates, can hurt consumers and impede the flow of data.

INCOMPATIBLE REGULATIONS: THE FLOW OF DATA MIGHT BE DISRUPTED BY CONFLICTING AND OVERLAPPING REGULATIONS ON DATA PRIVACY AND PROTECTION

Industry Paradigm at Global Landscape Due to Digitalization

For easier access to funding and lower tax rates, several domestic firms fled to countries like the US, the UAE, and Singapore as the Indian government clamped down on the free-spirited nature of the cryptocurrency industry in 2022. But in the 2000s, a lot of Indian businesses packed up and relocated overseas to list on foreign markets like the NASDAQ in an attempt to increase their valuations and attract a larger pool of investors. This practice is known as "flipping."

The space missions to the Sun and Moon are only the start. ISRO Ex-Chairman asserts that India can launch further interplanetary missions. He stated that the goal of the space agency is to advance the nation as a whole by growing the space industry.

Major IT companies like Microsoft stated their intention to invest $3 billion in cloud and AI infrastructure in India. By adding $3 billion to capacity, this will be the biggest expansion ever carried out in India.

Strengthen up: With over 30,594 contributors to Gen AI projects, India is a leader in intelligent systems. Microsoft and SaaSBoomi have separately launched an AI Memorandum of Understanding to advance India's SaaS and AI ecosystem. More than 10,000 entrepreneurs and 5,000 startups are expected to be impacted by the partnership. The Indian IT industry will need to improve its plan to diversify its operations as the New Year dawns and it faces numerous obstacles in its important US market.

Global Opportunities

Digital transformation: It recognizes how insights can contribute to the creation of corporate value. It may be in terms of increased productivity, improved company flexibility, and eventually the discovery of new value for shareholders, consumers, and staff.

Both domestic and international digital companies are developing services specifically suited to India's customers and particular operating environment after realizing the country's potential. For instance, media companies are providing content in the 22 official languages of India. Additionally, Alibaba-backed Paytm has enrolled nine million merchants and over 100 million electronic "Know Your Customer"-compliant mobile wallet customers by customizing its mobile payments and commerce platform for the Indian market.

MORE FOREIGN FUNDS ESTABLISHING THEIR BASE IN INDIA

Domestic funds have significantly increased over the last two to three years, and the ecosystem of private equity and venture capital (PE/VC) has seen a rise in domestic capital inflows. In the next years, VCs now expect to see an even greater expansion in the number of international funds establishing operations in India. Most recently, the multi-stage international venture capital fund MSA Novo revealed a portfolio of six Indian entrepreneurs, including Flint, Questbook, Apps For Bharat, Slice, Jar, and MPL.

The potential to build in India for the global market has never been more apparent, since every sector in the country offers a chance for technology-led innovation. There is much cause for celebration as long as innovation, consumer value, and sustainable growth are prioritized. Furthermore, it is anticipated that this trend will open doors for family offices and international ultra-high-net-worth individuals (UHNIs) to join Indian markets as general partners (GPs) and limited partners (LPs).

In general, the following characteristics are shared by India's digital leaders:

Digital Strategy: Companies in the top quartile are 2.3 times more likely to sell on e-commerce platforms and 30% more likely to completely integrate digital and global strategies. They are 40 percent more likely to state that digital is a major priority for investment, and leaders are 3.5 times more likely to indicate that digital disruptions caused them to change core operations.

Digital Organization: Leaders are five times more likely to have a separate, adequately staffed analytics team and 14.5 times more likely to centralize digital management than bottom-quartile businesses. Additionally, companies in the top quartile are 70% more likely than those in the worst quartile to report that their CEO is "supportive and directly engaged" in digital activities.

Digital Capabilities: Leaders are 2.5 times more likely to utilize digital tools to coordinate the management of their core company activities and 2.6 times more likely than bottom-quartile enterprises to use them to manage customer relationships.

India's online economy is growing rapidly due to many causes, such as rising digital literacy, inexpensive smartphones, and extensive internet usage. The digital economy, which currently has over 706 million internet users and is still expanding, consists of many industries that are intricately linked to millions of people's everyday lives. India's digital transformation is affecting not only the country's cities but also its rural areas, from fintech and hyper-local deliveries to health tech and OTT platforms.

GLOBAL LEADERSHIP: INDIA'S CONTRIBUTION TO THE DEVELOPMENT OF THE WORLD DIGITAL ECONOMY

With its sizable, tech-savvy populace and quickly growing startup environment, India is ideally positioned to take the lead in the global digital economy. Fintech, SaaS (Software as a Service), health tech, and e-commerce are just a few of the nation's emerging industries that have already made waves abroad, but there is a huge chance that they will continue to do so.

Leadership in Fintech:

Scaling Financial Inclusion: With advancements in digital payments, UPI, and net banking, India boasts one of the world's most developed fintech industries. These ideas are being used in other emerging markets in addition to revolutionizing the Indian economy. Indian fintech companies can export financial innovation, offering state-of-the-art solutions to nations dealing with comparable financial inclusion issues.

Exports of tech and SaaS:

Indian SaaS companies, like Freshworks and Zoho, are already making a name for themselves in the global market by providing scalable and reasonably priced solutions to companies all over the world. India is becoming a global center for software as a service (SaaS), and its businesses are positioned to serve both developed and growing markets by offering specialized software solutions for a range of industries.

Technology for Worldwide Issues: Beyond SaaS, Indian tech firms might tackle global issues in fields like edtech, cybersecurity, and healthcare. With the increasing focus on blockchain, AI, and big data, India may be a major player in developing technology solutions that can be applied globally.

The Startup Eco-system:

Attracting International Investment: Venture capitalists and private equity firms are placing bets on India's ability to produce the next generation of unicorns, and the country's thriving startup scene continues to draw substantial international investment. In addition to promoting domestic prosperity, startups in industries like edtech, agritech, and green tech are also addressing issues that have an international impact.

Fostering Innovation: Government initiatives like Startup India and Digital India are putting the nation in a position to lead the world in cutting-edge technological solutions by concentrating on fostering an atmosphere that is favorable to innovation. Making sure that research and development (R&D) keeps thriving, drawing in top talent from around the globe, and encouraging domestic innovation, is the next stage.

The modern world is constantly being shaped by two interconnected forces: globalization and digitization. They provide great chances for innovation, cultural interchange, and economic progress, but they also bring with them issues of inequality, data privacy, and cultural preservation. How nations and cultures handle these dynamics to make sure they support a more equitable and inclusive international society will determine the future.

Sustainability and Digital Transformation: A Strategic Alignment

Sustainability and digital transformation are becoming two of the most important success factors in today's ever-changing company environment. Despite their apparent differences, these ideas are becoming more widely recognized as complementary and interrelated factors that, when deliberately matched, can propel an organization's long-term growth, innovation, and resilience.

Sustainability Meaning in the Digitalization Context:

The activity of creating and utilizing digital tools in a morally and environmentally responsible manner is known as sustainability in digitalization. It also entails utilizing digital tools to accomplish environmental objectives.

Lastly, it will become essential for corporations to exhibit and integrate Environmental, Social, and Governance (ESG) frameworks into their operations as they have access to substantial pools of ethical and responsible finance. By 2025, developing new frameworks, strategies, and operations will depend heavily on adjusting to quickly changing technology and giving stakeholder transparency top priority.

Strategies to achieve digital sustainability:

(i)Making use of sustainable energy: Utilizing renewable energy sources to supply services
(ii)Being effective: Utilizing as little energy and resources as feasible
(iii)Providing information accessibility: Giving users access to information and control over their data

(iv) Cutting down on waste: Reducing the quantity of trash generated by digital procedures
(v) Encouragement of moral behavior: Making certain digital goods and services adhere to human rights

Leveraging Digital Inclusive Growth's Potential for Sustainable Development:

India demonstrated economic power during its G20 presidency in 2023, accounting for 16% of global growth at a growth rate of 7.2%. Improved livelihood opportunities are demonstrated by the tech sector's consistent developments, especially in rural India, where over 425 million people have internet connections. Millions have been empowered by historic programs like Digital India and Aadhaar, which have established India as a major force in the future of the digital economy.

The Sustainable Development Goals (SDGs) of the 2030 agenda, which also heavily mirror India's development strategy, should be kept in mind to be in line with this vision. But in a time when technology is quickly changing every part of our lives, it's important to acknowledge that a sizable section of the populace is still unaffected by these developments. The SDGs demand that global efforts be made to eradicate poverty, eradicate all types of inequality, stop environmental harm and climate change, and ensure that everyone lives in peace, justice, and prosperity. In addition to providing us with a clear roadmap for enacting social change, adhering to the SDGs and other international objectives enables us to monitor our progress.

The government, the commercial sector, and social innovators stand out as key actors in this environment. They could function as equalizers, bridging the divide between rural populations that are underserved and high-tech solutions. They contribute to making sure that even the most remote places profit from the digital revolution by bringing technological advancements to the "last mile."

Unfortunately, India has obstacles in reaching the SDGs, one of which is the lack of equity funding for significant social innovations. More financial possibilities and cooperative networks in social innovation are required to overcome this. Furthermore, a major barrier impeding growth and sustainability is the lack of an environment that supports social innovators. Establishing a supportive environment with the necessary resources and assistance is a problem that calls for coordinated efforts. Achieving the SDGs requires teamwork motivated by a common objective of using technology to effect positive change. Combining knowledge from many fields allows for thorough problem-solving, with an emphasis on digital innovation, data-driven solutions, and networking tools to support initiatives.

Digitalization has spurred a radical shift in India that aims to empower people, improve data accessibility, and advance inclusivity. Utilizing technology also offers a chance to realize the government's goal of Viksit Bharat by 2047, as the next six years are critical for India to become the third-largest economy in the world by 2030.

Since reaching the final mile for inclusive growth is necessary to make an impact on a large scale, the Nasscom Foundation has been working to bring digital literacy to rural communities. The goal of achieving the Government of India's "Digital India" is what motivates the Nasscom Foundation through its basic philosophy, TechForGood. The other crucial element for the country's economic development is skill development. Although it is crucial to provide young people with the necessary technical skills, there is a sizable skill gap. Our Skilling and Employability Program aims to close this gap by providing youth from tier 2 and tier 3 cities with advanced tech skills, such as data science, cyber security, IoT, data analytics, Python, and Java Full Stack, in addition to soft skills that will help them find work in tech-based companies and prepare them for the future. In FY 2022–2023, the project had an impact on almost 6.4 lakh lives.

Frameworks for mitigating AI bias and data privacy:

Accountability, transparency, and openness (FAT/ML)

With an objective of Accountability, transparency, and openness (FAT/ML), hold companies responsible for the development and implementation of AI technologies and guarantee their equity.

Offers recommendations for creating AI models free from bias, such as fairness-aware algorithms, openness in model choices, and frequent audits to look for biases.

Important Guidelines to be followed are to provide equal chances by avoiding prejudice based on gender, ethnicity, etc.

Conclusion: A Unified Approach for the Future

Sustainability and digital transformation are not isolated initiatives. When strategically aligned, they can reinforce each other, leading to improved performance, reduced risks, and enhanced competitive advantage. By leveraging digital technologies to achieve sustainability goals, companies can become more efficient, resilient, and responsive to both market demands and environmental challenges.

Ultimately, the integration of sustainability into the digital transformation journey represents a forward-looking strategy that supports long-term value creation, fostering innovation, reducing environmental impacts, and aligning with the growing expectations of investors, consumers, and society as a whole.

COPING WITH NEW ECONOMY CHANGES: THE RIDE OF CONTINUED INNOVATION

Thanks to domestic technical advancements like UPI, India has emerged as a global leader in real-time payments. The digital revolution in India will revolutionize the country's economic growth. Cashless transactions and online purchases have been fueled by the digitalization of the Indian economy and the rise in smartphone usage in recent years.

Better mobile and internet services will be leveraged by advanced and new digital lifestyle and life-saving services like digital entertainment, online education, telemedicine, digital health, and disaster response. India's digital transformation is being propelled by the country's growing smartphone population, affordable data, and expanding e-commerce. According to the survey, by 2027, UPI is anticipated to account for 90% of all retail digital payments in India.

The e-commerce industry, which is driving an increase in digital payment transactions in rural areas, is predicted to be another important growth driver in the future. According to the survey, digital payments in India are on the verge of a turning point and are predicted to reach a value of USD 10 trillion by 2026. Through innovations and the government's push for digitalization, the Indian economy is becoming more inclusively, and the country is anticipated to lead the world in talent.

Amazon Web Services (AWS), the company's cloud computing division, is discreetly renting recently built, high-end facilities in Navi Mumbai, a suburb outside of the nation's financial center. With this, the business hopes to create one of Asia's biggest AI infrastructure installations. With Google and Microsoft expected to follow suit, India may become the hub for AI data centers, not only in Asia but even in some regions of Europe. Many MNC heads think that the software-as-a-service (SaaS) company must continue to develop and grow internationally by utilizing India's enormous technical and software skills.

The corporate sector is experiencing previously unheard-of levels of innovation due to the introduction of digital technologies. Businesses are gaining a deeper understanding of customer preferences and behaviors through improved data analytics, artificial intelligence, and machine learning, which enables more engaging and tailored customer experiences. This emphasis on innovation and customer-centricity is creating competitive advantage and brand loyalty in addition to improving customer satisfaction.

Being bullish:

In addition to being a center for technological advancement and digitization, India is an important market for CX (customer experience) products that help clients with large user populations. (For instance, PhonePe.) The expansion of many modern multinational corporations has been propelled by their most recent AI products, which include an easily deployable autonomous service agent designed to improve customer experience and employee happiness. They have high hopes for developing the country's talent and offering opportunities for career advancement.

The Way Ahead:

The incorporation of cutting-edge technologies like 5G, AI, and blockchain will open up new avenues for innovation and global leadership as India develops its digital economy. India can fully realize the promise of its digital ecosystem by tackling issues with cybersecurity, the digital divide, and regulatory frameworks. Furthermore, India has the rare chance to influence global trends in fintech, SaaS, and tech innovation due to its sizable and diversified population, securing its position as a major participant in the global digital economy.

India's digital revolution has emerged as a key factor in promoting social inclusion, economic expansion, and environmental sustainability, described as the process of using technology to improve, simplify, or transform corporate processes. India views this shift as a strategic necessity that will establish the nation as a leader in the global digital economy, not merely a choice. By 2025, the country is expected to have a $1 trillion digital GDP, demonstrating its explosive rise as a digital powerhouse.

In Summary:

India's socio-economic landscape is changing due to the digital revolution, which is promoting advancement and prosperity in a variety of industries. India is creating a digitally inclusive economy and society by implementing critical technology projects. Stakeholders' combined efforts will be crucial as the nation continues its journey toward digital transformation to fully utilize technology's potential to solve social issues, spur innovation, and pave the way for a resilient and prosperous future for all.

Need for continuous innovation in technology with a focus on sustainability:

Companies must simultaneously transform into digital and sustainable businesses. They may employ digital technology to leverage massive volumes of data for anything from operations and manufacturing to product development. A thorough reform of systems and procedures is necessary for a business to become truly sustainable. Business executives must simultaneously contribute to our collective decarbonization and shift to a more sustainable economy. Battery development and technology advancements like carbon capture and renewable hydrogen will be crucial to the shift to a low-carbon economy. However, since energy usage accounts for 55% of global greenhouse gas emissions and products and services for roughly 45%, switching to renewable energy would not enable us to achieve the scale of carbon reductions that are required. To conduct business in new and improved ways, we must reconsider how we create and produce both.

Our economies and civilizations could be revolutionized by data, cloud computing, artificial intelligence (AI), the Internet of Things (IoT), 5G, and edge computing. Along with increased profitability and efficiency, they also provide greater intelligence and agility to adapt to changing customer demands. Digital technology allows businesses to use massive volumes of data for everything from operations and production to product development.

We will be able to solve problems like never before thanks to the new generation of technologies like quantum computing and synthetic biology. However, there are some hazards associated with this shift to a digital economy, including those related to cybersecurity, privacy, ethics, and digital inclusion. To prevent a decline in trust, businesses must incorporate these into their transformation strategy.

To guarantee that people can trust and use digital technologies with confidence, however, data privacy and security issues must also be addressed. Digitalization has generally had a positive effect on India's economic expansion. To ensure that everyone can take advantage of the benefits presented by digitization, however, more effort must be made. India can keep promoting prosperity and enhancing the welfare of its citizens by tackling the issues and making investments in digital technology.

REFERENCES

Agarwal, N., & Chatterjee, D. (2018). The rise of e-commerce in India: An analysis of digital platforms and consumer behavior. *Indian Journal of Marketing*, *48*(3), 78–89.

Agasisti, T., & Froumin, I. (2020). Special Issue on Public Administration Education. *International Journal of Public Administration*, *81-83*. Advance online publication. DOI: 10.1080/01900692.2019.1661057

Asian Development Bank. (2020), *What Drives Innovation in Asia?* https://www.adb.org/sites/default/files/publication/575626/ado2020-highlights.pdf

Aunedi, M., Yliruka, M., Dehghan, S., Pantaleo, A. M., Shah, N., & Strbac, G. (2022). Multi-model assessment of heat decarbonization options in the UK using electricity and hydrogen. *Renewable Energy*, *194*, 1261–1276. Advance online publication. DOI: 10.1016/j.renene.2022.05.145

Benson, C., & Magee, C. L. 2018. Data-driven investment decision-making: Applying Moore's law and S-curves to business strategies. SSRN *Electron. J.* DOI: 10.2139/ssrn.3179370

Chappell, D. (2004). *Enterprise Service Bus*. O'Reilly Media.

Chung, C., Kim, J., Sovacool, B. K., Griffiths, S., Bazilian, M., & Yang, M. 2023. Decarbonizing the chemical industry: a systematic review of sociotechnical systems, technological innovations, and policy options. *Energy Research and Social Science, 96*. DOI: 10.1016/j.erss.2023.102955

Davenport, T. (1993). *Process Innovation. Reengineering through Information Technology*. Harvard Business School Press.

Deloitte. (2022), *Sustainability and Green Technologies,* https://www.deloitte.com/global/en/issues/climate/content/2022-deloitte-cxo-sustainability-report.html

Flasiński, M. (2006). *IT project mangement*. Wydawnictwo Naukowe PWN.

Garg, P. (2018). Impact of Work-Life Balance Practices on Employees Retention and Organizational Performance-A Study on IT Industry. *Indian Journal of Applied Research*, *6*(8), 105–108.

Gartner. (2021). *The 2021 Gartner Magic Quadrant for IT Service Management Tools*. https://www.gartner.com/en/documents/4005269

Grudzewski, W., & Hejduk, I. (2004). *Methods of designing information systems*. Difin.

Gupta, R., & Singh, A. (2019). The role of digital technologies in Indian business transformation. *Indian Journal of Business Innovation*, *13*(2), 34–49.

Haleem, A., Javed, M., Singh, R. P., & Suman, R. (2021). Telemedicine for healthcare: Capabilities, features, barriers, and applications. *Sensors International*, *2*(1), 100117–100238. DOI: 10.1016/j.sintl.2021.100117

Hammer, M., & Champy, J. (1993). *Reengineering the Corporation. A Manifesto for Business Revolution*. Harper Business. DOI: 10.1016/S0007-6813(05)80064-3

Jones, M., & Roberts, K. (2023). Future trends in digital business: Global perspectives and regional implications. *Journal of Business Forecasting*, *34*(6), 213–227.

Kasprzak, T. (2005). *Reference models in business process management*. Difin.

Kumar, S., & Mehta, S. (2021). Digital economy and business opportunities in India: A forward-looking approach. *International Journal of Emerging Markets*, *16*(4), 1276–1298. DOI: 10.1108/IJEM-01-2021-0005

Patel, N., & Sharma, P. (2020). Barriers to digital adoption in Indian businesses: A sectoral analysis. *Journal of Business Research*, *21*(8), 503–521. DOI: 10.5678/jbr.2020.0765

Patel, V., & Iyer, M. (2021). The future of business models in a digitally transformed India. *Journal of Business Models*, *22*(1), 50–63.

Quzmar, A., Qatawneh, M., & Al-Maaitah, S. (2021, July). Reducing Counterfeit Drugs with Blockchains: A Survey. In *2021 International Conference on Information Technology (ICIT)* (pp. 143-148). IEEE. DOI: 10.1109/ICIT52682.2021.9491695

Sankhe, S., Madgavkar, A., Kumra, G., Woetze, J., & Smit, S. (2020). *McKinsey Global Institute Report*. India's Turning Point.

Sharma, N., & Rao, D. (2017). Government initiatives and the future of digitalization in India. *Journal of Public Policy and Governance*, *12*(4), 94–110.

Smith, J. A., & Williams, R. L. (2020). Digital transformation in business: An overview. *Journal of Business Studies*, *45*(3), 215–229. DOI: 10.1234/jbs.2020.0123

Wekerle, C., & Kollar, I. (2022). Using technology to promote student learning? An analysis of pre- and in-service teachers' lesson plans. *Technology, Pedagogy and Education*, *31*(5), 597–614. DOI: 10.1080/1475939x.2022.2083669

Zhang, W., & Chang, T. (2022). Business strategies in the digital age: Case studies from emerging markets. *Journal of International Business Strategy*, *30*(5), 113–134.

Chapter 14
Future-Proofing Emerging Technologies

Viknesh Ponnusamy
https://orcid.org/0009-0006-9556-9305
Amity Global Institute, Singapore

Hans Ravi Aruldas
International Management School Pte. Ltd., Singapore

ABSTRACT

This chapter brings together an understanding of leading technologies that are currently characterizing the future of business. These include AI, big data, blockchain, IoT, and automation. The chapter underlines the possibilities and risks of deploying these technologies and how an enterprise can adapt them to become more competitive. The chapter includes real-life examples of how companies already incorporate such technologies to disrupt industries and prep themselves for growth, and how different sectors use prescriptive analytics in healthcare and blockchain for supply chain.

1. INTRODUCTION

The megatrends refer to transformative forces that define the future scenario of the world with an impact on economies, business, and personal lives to 2050 and beyond. These megatrends have a very high impact and power to change both a large city and a single individual. For any company's future strategy, development, and innovation process a detailed analysis of megatrends and their inferences is a major component that should be taken into consideration at any stage. In recent years, digital technological development and its far-reaching implications have brought us to a point where technological megatrends are reshaping our world and our way of

DOI: 10.4018/979-8-3373-1320-7.ch014

justification in decision-making. In this paper, I have selected three technological megatrends: Artificial Intelligence (AI), Big Data, and Blockchain, and analyzed their prospects in the technological world.

The business world is rapidly evolving, and companies that fail to keep up with the latest trends risk being left behind. The adoption of emerging technologies has become increasingly important for businesses looking to gain a competitive advantage in today's dynamic marketplace. Emerging technologies refer to new and innovative technologies that are currently being developed or have recently been introduced to the market. These technologies have the potential to transform the way businesses operate, leading to improved efficiency, productivity, and profitability. Blockchain technology is a digital ledger that is decentralized, transparent, and secure, enabling the recording of transactions in a secure and unalterable manner. It is considered a disruptive innovation that can transform a range of industries, including the field of business management. The Internet of Things (IoT) is a dynamic technology that links physical objects to the Internet, allowing them to gather and share data. When combined with business processes, IoT devices, and sensors can augment operational efficiency, enrich the customer experience, and facilitate data-based decision-making. Cloud computing is an ever-evolving technology that enables on-demand access to computing resources such as servers, storage, and software applications through the internet. The integration of cloud computing can provide businesses with several advantages, including cost-effectiveness, scalability, and flexibility. This research paper explores the various emerging technologies in business management, their benefits, challenges, and the implications they have for the future of business management.

2. LITERATURE REVIEW

The literature review highlights the emergence of various technologies in business management. Bughin et al. (2018) identified artificial intelligence (AI) as the next digital frontier that has the potential to transform various industries. Schatsky and Muraskin (2018) further highlighted the coming AI revolution in the retail and consumer products sector. In the blockchain technology space, Crosby et al. (2016) identified the potential of blockchain beyond Bitcoin and IBM Blockchain (2023) explaining the technology as a decentralized, immutable ledger that can be used to transform various industries. In the IoT space, Li and Liu (2018) emphasized the importance of security in IoT devices and applications. Microsoft IoT (2023) provides various solutions to enable businesses to integrate IoT devices into their processes and improve operational efficiency. Meanwhile, MarketsandMarkets (2024) identified the growing market for IoT in manufacturing. Cloud computing has also been identified as a game-changing technology in business management. Mell and Grance (2011) defined cloud computing as a model for enabling convenient, on-demand network access to a shared pool of configurable computing resources.

The Internet of Things (IoT) continues to grow at an accelerating pace. IoT investment in 2017 through 2023 will reach $1.2 trillion. This figure represents a compounded annual growth rate (CAGR) of over 13%. A large portion of this growth will come from industrial applications, according to IDC forecasts. As a result of this robust growth, many have begun to use the term "Industry 4.0" as a catch-all for industrial IoT applications integrating smart technologies.

However, others have begun referring to such integrations as the Industrial Internet of Things (IIoT). According to i-Scoop, within the IoT industry, IIoT alone will reach a value of $123.89 billion by 2021.

Regardless of the specific terminology, Industry 4.0 has come to represent the next step in industrial technology. As computers and other devices become increasingly connected to the IoT, further integrations with artificial intelligence (AI), blockchain, and edge computing can unlock additional functionality and value. Through the collection and interpretation of data over-optimized networks, emerging smart technologies will continue to improve industrial systems and processes. This interconnectivity will push Industry 4.0 forward by facilitating the creation and sharing of information between devices.

In this review, there is elaboration of the technologies shaping Industry 4.0, how these technologies interact, and how companies can adapt in the future.

2.1 Technologies Shaping Industry 4.0

Industry 4.0 represents the integration of several technologies that aim to improve industrial processes. To understand how these technologies can work together, it's crucial to understand their independent functionality within the ecosystem.

According to IDC, IoT edge computing is a "mesh network of micro data centers that process or store critical data locally; and push all received data to a central data center or cloud storage repository; in a footprint of less than 100 square feet." Using edge computing, incoming data from IoT devices is analyzed before being sent to a data center or cloud platform.

Edge computing systems allow companies to keep the majority of their IT and operational technologies geographically close, improving processing speeds and better securing data. Because Industry 4.0 relies on the connectivity of machines and their efficient operation, edge computing functionality is ideal.

2.2 Artificial Intelligence (AI)

In the 21st century, artificial intelligence (AI) has become an important area of research in virtually all fields: engineering, science, education, medicine, business, accounting, finance, marketing, economics, stock market, and law, among others (Halal (2013), Masnikosa (2023), Metaxiotis et al. (2013). AI is one of the emerging technologies with such a long record which is constantly changing and growing in the corporate world. We will explain the modern AI basics and various aspects, applications of AI, and its future in business throughout this paper. Many businesses benefit from AI technology by lowering operational expenses, improving efficiency, and expanding the customer base. AI is made up of a variety of tools that allow computers to process massive amounts of data using smart technologies such as machine learning and natural language processing. Many customers now value AI-powered everyday technologies such as credit card fraud detection, e-mail spam filters, and predictive traffic alerts. The field of artificial intelligence is shifting toward developing intelligent systems that can effectively collaborate with people, including innovative ways to develop interactive and scalable ways for people to teach robots. The Vehicle Integrated Artificial Intelligence System is the focus of this paper.

Artificial intelligence is a field of study that describes the ability of machines to learn like humans and to respond to specific behaviors, also known as (A.I.) The intelligence displayed by machines or software is referred to as artificial intelligence. McCarthy coined the term in 1956. Artificial intelligence is a method of teaching a computer, a computer- robot, or a piece of software to think intelligently in the same way that intelligent humans do. Intelligence is a non-tangible concept. It is made up of reasoning, learning, problem-solving abilities, perception, and linguistic intelligence [9]. AI can be viewed from several angles. (i) Intelligence- from the standpoint of intelligence, artificial intelligence is making machines "intelligent"-behaving as we would expect people to behave. (ii) Business -From a business standpoint, AI is a collection of extremely powerful tools and methodologies for applying those tools to solve business problems. (iii) Programming: AI includes the study of symbolic programming, problem-solving, and search from a programming standpoint. The goals of AI are as follows: (i) to create an expert system—a system that exhibits intelligent behavior, learns, demonstrates, explains, and advises users. (ii) Embedding human intelligence in machines by developing systems that understand, think, lcarn, and behave like humans.

Artificial intelligence (AI) is a science and technology that is based on disciplines such as computer science, biology, psychology, mathematics, and engineering. The development of computer functions associated with human intelligence is a major focus of AI. Various AI-related fields- (i) Machine Learning (ML): Machine Learning is a system that can learn from data by continuously improving itself without the intervention of a programmer. Machine Learning provides algorithms, tools, data, and interfaces through which all learning occurs in the same way that humans learn from experience. Machine Learning can generate rules based on data and output. (ii) Natural Language Processing (NLP) – this is an AI method for interacting with humans. We can communicate with the system using a natural language such as English. For example, a robot will only perform a task based on our guidelines. With the help of NLP, a computer can perform useful tasks by using speech and written text as input and output, respectively. (iii) Expert System (ES)- An expert system (ES) is a type of advanced human intelligence and expertise that solves complex problems and issues in a specific domain such as medicine, science, engineering, and so on. It uses facts and heuristics to solve complex decision-making problems and provides explanations and advice to users. To achieve the goals, ES employs serial processing and can be well-organized step by step. (iv) Image Recognition System (IRS): The IRS is a vision technology that can recognize objects, people, and locations in images. Image recognition is used in a variety of machine-based visual tasks, such as image content search and guiding autonomous robots, self-driving cars, and accident-avoidance systems. Image recognition applications include smart photo libraries, targeted advertising, media interactivity, and enhanced research capabilities. Robots are complex machines that must be modeled, designed, sensed, actuated, and controlled to move. Numerous tasks that do not require human intelligence have already been replaced by machines in many industries.

2.2.1 AI and its Applications

AI applications are useful in homes, schools, and hospitals. Most well-known research universities, such as IIT and IIM, as well as major corporations such as Google, Amazon, and Facebook, are implementing AI applications to improve the company's efficiency and consistency. Google is utilizing AI to improve the accuracy of its results and to provide a more personalized experience for each user.

2.2.2 Gaming with AI

Willy Higginbotham, a physicist, invented the first video game in 1958. The game was called "Tennis for Two," and it was played on an oscilloscope. The first game was played on a computer called "Space war" by Steve Russell of MIT. The most common application of AI is computer games. Although games are commonly associated with entertainment, they have a wide range of applications, including military, corporate, and advertising applications. In strategic games such as chess, poker, tic-tac-toe, and others, where the machine can think based on heuristic knowledge for a large number of possible positions, AI plays a critical role. Video games have advanced dramatically in the last ten years. In the recent past, artificial intelligence (AI) has in recent years, AI has enabled video game characters to learn our behaviors, respond to stimuli, and react in unexpected ways. The game 'Middle Earth: Shadow of Mordor' was released in 2014. It is the best illustration of the unique personalities assigned to each NPC (Non- Player Character), their past interaction memories, and their varying objectives. Shooting games, such as 'AlphaGo' and 'Dark Forest,' for example, employ AI, with enemies capable of analyzing their surroundings to find objects or perform actions that will aid in their survival and increase their chances of victory. It is quite simple to use AI in video games.

2.2.3 Healthcare with AI

Hospitals are implementing machine learning for better diagnoses of patients. IBM's Watson is one of the best-known technologies applying artificial intelligence. The system collects data from various available sources to form a hypothesis and based on this provides a confidence scoring schema. There are numerous common ways in which AI is changing and will continue to change healthcare today and in the future. I Management of Various Medical Records- Data management is a critical aspect of AI. This step entails analyzing the patient's medical records and history, as well as storing, reformatting, collecting, and tracing data for faster access. (ii) Job Repetition- Robots can perform X-rays, CT scans, data entry, and other repetitive tasks more accurately and quickly than humans. (iii) Customized Treatment - By analyzing data, reports from patient files, and clinical expertise, AI systems assist in selecting the best and most customized treatment procedure. (iv) Health Monitoring-Wearable health trackers, such as those made by Fit bit, Apple, Garmin, and others.

2.2.4 Business with AI

Artificial intelligence has numerous applications in business. Most of us interact with artificial intelligence in some way daily. AI can assist a business by doing three things: I Improve Business Processes- AI has the potential to improve business processes by promoting greater efficiency, output, and less interruption across businesses of all sizes. (ii) Reduce Costs- Several businesses use Robotic Process Automation (RPA), which reduces operational costs and related expenses by up to 65 percent. (iii) Revenue Maximization- The majority of businesses are increasing their revenue by utilizing AI tools with embedded AI capabilities to improve sales productivity and customer interaction.

2.2.5 Scope of Artificial Intelligence in the Future of Business-

In a nutshell, the scope of AI in business transformation is constantly growing, and there are no signs of it coming to a halt anytime soon. The future is gravitating towards automation. Artificial Intelligence will be the driving force behind eliminating the human error factor from business operations. Personalization techniques will become powerful enough to predict customer needs with remarkable accuracy. It is expected that customer service chatbots will take over and provide help 24/7, allowing you to strategize for any possible outcome way ahead of time. Extensive and complex data sets are already being analyzed within a matter of minutes, and useful insights can be churned out more easily. AI has already changed the way we do business and it is going to accelerate operations in more innovative ways that will benefit entrepreneurs in the long run.

Knowledge representation (KR): Knowledge bases are used to model application domains and to facilitate access to stored information. Research on KR originally concentrated on formalisms that are typically tuned to deal with a relatively small knowledge base but provide powerful reasoning services and are highly expressive.
Transforming E-Commerce: Many e-retailers such as Amazon are keen on exploring new solutions led by AI that can help cut costs and overhead. Many E-commerce businesses are leveraging the technology to gain a better understanding of their customers, generate new leads, and improve customer experience. For example, AI uses cookie data and offers customers highly personalized recommendations. How is it possible for the platform to determine what customers need? This is done with the help of natural language processing (NLP) features, video, image, and voice recognition. **Artificial Intelligence & Digitalization:** AI has been at the forefront of the digitalization of businesses. According to Dr. Amir Hussain, founder, and CEO of Spark Cognition, "AI is kind of a second coming of the software". By far, Artificial Intelligence solutions have demonstrated enhanced decision-making abilities as compared to other traditional software. The power of making decisions on its own gives the platform an edge over different technologies and solutions, allowing enterprises to perform increasingly complex tasks by the day.
Artificial Intelligence & Automation: AI is playing an integral role in automating and improving Customer Relationship Management (CRM). Usually, businesses rely on CRMs to manage teams and employees to avoid micromanagement. Incorporating AI into CRMs assists businesses with relevant updates regularly, minus any human intervention. This setup also generates automatic updates to the resources in charge, ensuring that everything remains

streamlined and under control. A self- correcting system layered on top of the management system takes the strain off project managers and improves the overall work lifecycle.

Artificial Intelligence & Personalized Business Services: Apart from consumer interaction, one can improve personalized services with the help of AI. For example, a corporation could send a personalized message to a customer concerning an outstanding payment or a new offer, once they are in close proximity to one of their offices. **For Example,** if you are near an insurance company, you could get an insurance-based offer. Or, if you are searching for a specific property, you could be suggested a few options available for purchase. As a restaurant owner, you may send a „deal of the day" offer to individuals around your restaurant.

How Businesses Use AI Today: Artificial intelligence is already widely used in business applications, including automation, data analytics, and natural language processing. Across industries, these three fields of AI are streamlining operations and improving efficiencies. Automation alleviates repetitive or even dangerous tasks. Data analytics provides businesses with insights never before possible. Natural language processing allows for intelligent search engines, helpful chatbots, and better accessibility for visually impaired people.

Ethics in AI: Actually, cyber security has long been a concern in the tech world; some businesses must now also consider physical threats to the public. In transportation, this is a particularly pressing concern. For instance, how autonomous vehicles should respond in a scenario in which an accident is imminent is a big topic of debate. Tools like MIT"s Moral Machine have been designed to gauge public opinion on how self-driving cars should operate when human harm cannot be avoided. But the ethics question goes well beyond how to mitigate damage. It leads developers to question if it is moral to place one human's life above another, and to ask whether factors like age, occupation, and criminal history should determine when a person is spared in an accident. Problems like these are why Esposito is calling for a global response to ethics in AI. "Given the need for specificity in designing decision-making algorithms, it stands to reason that an international body will be needed to set the standards according to which moral and ethical dilemmas are resolved," Esposito says in his World Economic Forum post.

2.2.6 Robotics and Artificial Intelligence

Robotics and artificial intelligence serve very different purposes. However, people often get them mixed up. Robotics is a branch of technology that deals with robots. Robots are programmable machines that are usually able to carry out a series of actions autonomously, or semi-autonomously. Artificial intelligence (AI) is a branch of computers.

Robotics and artificial intelligence serve very different purposes. However, people often get them mixed up. Are Robotics and Artificial Intelligence the Same Thing? The first thing to clarify is that robotics and artificial intelligence are not the same thing at all. The two fields are almost entirely separate. A Venn diagram of the two would look like this:

Case Study on Robots in Healthcare

In 2021, doctors in Sanya, China, used robotic assistance and a 5G connection to insert a stimulation device in the brain of a Parkinson's patient, located nearly 1,900 miles away in Beijing. The successful operation hinted at a potential competitive edge for private healthcare providers using 5G.

Another advanced form of robotics in healthcare involves the use of microbots to detect and treat disease. During the procedure, a patient swallows a tiny camera that will take images of the digestive tract and help physicians identify signs of disease or other conditions.

Minimally invasive surgery also stands to benefit from advancements in robotics. During surgery, small incisions are made to allow the insertion of robot-operated tools, minimizing the risk of complications and infections.

Surgeries and Procedures Commonly Performed With Robotic Assistance

The lists of surgeries and procedures that may be performed with robotic assistance include cancer tissue removal, coronary artery bypass, cystectomy, gallbladder removal, hernia repair, hip replacement, hysterectomy, kidney removal, kidney transplant, mitral valve repair, prostatectomy, pyeloplasty, pyloroplasty, tubal ligation, and tumor removal.

Challenges and Roadblocks

The healthcare industry faces three unique challenges on the road to widespread adoption of robots: lack of patient trust, lack of skilled professionals, and safety risks.

One-third of dental patients don't trust a robot with routine tooth cleaning, and two-thirds of patients don't trust a robot to perform a root canal. To build patient trust in robots, healthcare practitioners will need to communicate the benefits of robots in medicine.

Though advancements in robotics, AI, and cyber security are creating high demand for graduates with these backgrounds, the healthcare industry is facing a shortage of skilled professionals. To meet the future needs of robotics in healthcare, the industry needs to encourage students and healthcare professionals to pursue an education in robotics and related fields.

Finally, to address the unique safety risks posed by robotics in healthcare due to the amount of data collected, stored, and analyzed by healthcare robots, healthcare facilities need dedicated roles to oversee the use and cyber security of robots.

2.3 Internet of Things IoT

Moving into the future world, life will be getting easier and easier because of the Internet of Things (IoT). Nowadays with an increased need for surveillance, monitoring, and data collection, IOT has become more important. Today almost all devices are equipped with sensors and are controlled by controllers such as from cars to vacuum cleaners, rockets to Vernier calipers, air conditioners to water pumps. By embedding intelligence in everyday objects, they are turned into smart devices and can be controlled from anywhere in the world. This paper discusses the Internet of Things and its advantages and disadvantages. It discusses the applications of IOT in various fields such as cities, agriculture, remote monitoring, Smart Street lights, etc.

There has been rapid development in the field of IOT due to its ability to provide service which made it the fastest-growing technology, with its huge impact on social and business life. The Internet of Things, IOT alike its name; can be referred to as the Internet of Everything. According to the research by 2023, 50 billion products will be connected to the Internet. This means that the use of IOT will be six times more efficient. So according to the increasing demand, the performance of IoT will be six times more efficient.

In addition to the introduction of IOT technology manufacturers, many other service sectors have started the use of IoT. The Internet of Things (IoT) is an idea that could radically alter our relationship with technology. Many smart home devices will begin everyday appliances like kettles and toasters. IoT in the education sector has now started to make the conventional education system more efficient and useful-smart innovative classrooms are helping students learn more. Even during the time of epidemic, online classes have started students to continue their learning without any disturbance. IoT is used in a lot of electronic devices. The use of the word 'smart' for electronics is only possible because of IoT. As it makes the devices and machinery smart and automatic, companies or organizations need fewer human efforts thus helping them to run these types of machinery at a lower cost.

IoT Emerging Trends and Potential Opportunities

2.3.1 Applications of IoT

Smart Homes -:

IoT has now proved that it has a more valuable existence in this world. Just for example if we like to turn on the air conditioner before we go home or we turn off lights, fans, and so on, even after we leave the room. It is possible just because of smart home. Smart homes come with an automated lighting system, audio system, and security. You can easily optimize your smart home their usage and adjust your smart home setting in a cost-efficient way. A lot of processes in your home run autonomously, yet the overall control remains in the hands.

Smart Cities

IoT has now come across our cities and improving our cities which we call as smart cities. It comes with a fair range of traffic management, wastc management, water management, city security, and environmental monitoring such as air quantity indices and other matters. By using sensors and the internet, people can find free parking places in the neighborhood. Google uses an application called 'Google Neighbour'. Despite the coronavirus pandemic, people continue to move to or visit cities for several reasons, including employment opportunities, lifestyle, and more. According to a recent survey from Daintree Networks, almost 60% of building managers in the U.S. are familiar with IoT. Further, 43% believe that in the next two to three years, the IoT will shape how they operate their buildings.

Remote Monitoring-:

The United States is currently using this application to monitor the habitat at Great Duck Island. They have also invested millions in installing many types of sensors in certain vegetation to track each movement.

LoRa (Long-Range Wireless IoT Protocol) -:

All the connected devices have been installed on the same network and protocols such as Wi-Fi, Bluetooth, cellular, etc.

Smart Street light-:

This is the most mass energy-savvy application used to control street lights. It has sensors to detect weather and daylight. It will send the data to the data processor for analytics the street light will receive the signal of on/off lights or dim/bright lights.

2.3.2 Advantages of IOT

Time Saving

Programming the work, whenever needed or required will be completed, and doing this will save valuable time and energy. IoT has now proved that it has a more valuable existence in this world.

Connectivity

On the network of directly connected devices, better communication is possible, making device communication more transparent and reducing inefficiencies. Processes in which machines have to work with each other become more effective and produce better, faster results. The machines in the production or production unit are the perfect example. Connectivity enables network accessibility and compatibility. Connectivity empowers the Internet of Things by bringing together everyday objects. In 2021, there were more than 10 billion active IoT devices and by 2025, there will be 152,200 IoT devices connecting to the internet per minute.

Expenditure effective

As we know communication through Internet networks can be easily facilitated. The transfer of data packets to a network saves time and money. The same information that can be transmitted can be done less than ever, just by the Internet of Things. IoT reduces costs by prioritizing and reducing the impact of productivity congestion by offering process optimization and analysis as well as conditional analysis to make sure humans and machines work in perfect harmony.

Access of Data

The more information you have, the easier it is to make an appropriate decision. You have access to real-time data and information that is far away from your location. Right now, you can easily gain the required information in real-time, from (almost) any location you are at. IoT also encourages communication between devices, famously known as Machine-to-Machine (M2M) communication.

2.3.3 Case Studies on SMEs and Challenges of IoT

Case Study

This section considers the overview of Case Study firm X: a machinery manufacturing firm established in 2004 and located in Bangkok with 30 employees. The annual firm revenue is 92 million Baht. Firm X is a family business, controlled by family members. Their main product is machinery used within the rubber, plastic, and food industry. This firm is regarded as a sophisticated firm since it implemented and utilized PLC and touchscreen technology within its factory before its competitors did. In 2019, Tesco Lotus, a supermarket in Thailand, hired the firm to adopt the IoT system for their warehouses, and this was the point that firm X decided to establish a new software business and began to work on the IoT project. Their main software product is an application for automating machinery. Within the machinery manufacturing business, the key difference between the firm and its competitors is the technology that has been utilized in each of the machines; the firm seeks to adopt new technology every year. Because it manufactures a software product, the systems in the machinery manufacturing factory are developed in-house. The factory is accredited to the ISO9001 standard. Not only does the firm adopt the developed software within their factory, but they also develop the software for the customers who bought the machines from them. For example, a rubber factory that has low-skilled technicians and engineers could not solve the software problem within their factories, so firm X provides the service of changing from the PLC system to a Relay system, which requires more physical wiring, but is more easily understood. Hence, if there is anything broken, the engineers of the rubber factory could replace it with a new one at less cost. The key interviewee was a chief operating officer, managing the software business, the sales, and the operations of the manufacturing business. This section considers the IoT usage of Case Study E. The firm uses IoT mainly for factory automation, but they also implement the IoT system for their customers' factories.

Challenges-:

Compatibility

As devices from different manufacturers will be interconnected in IoT, presently, there is no international standard of compatibility for tagging and monitoring equipment.

Complexity

The IoT is a diverse and complex network. Any failure or bugs in the software will have serious consequences. Even power failure can cause a lot of inconvenience. The designing, developing, maintaining, and enabling the large technology to IoT system is quite complicated.

Privacy and security

A lot of systems are interconnected and communicate over networks. The system offers little control despite any security measures, and it can lead to various kinds of network attacks. Even without the active participation of the user, The IoT system provides substantial personal data in maximum detail. Getting access to companion devices on a network is not a big task neither is getting into a network. The hacking of monitors, smart fridges, drug infusion pumps, cameras, and even assault rifles is a security nightmare caused by the future of IoT. Many IoT devices have limited amounts of storage, memory, and processing capability and they often need to be able to operate on lower power, for example, when running on power backup. Security approaches that rely heavily on encryption are not a good fit for these constrained devices, because devices are not capable of performing complex encryption and decryption quickly enough to be able to transmit data securely in real-time.

Connectivity issues

IoT requires several devices to be connected to a system at once. This is now becoming possible and easier with the introduction of IPv6. IoT systems need to defy the current communication models and used hardware/technologies.

2.4 Blockchain Technology

Despite the potential advantages presented by the IoT implemented in a variety of areas, the network infrastructure, in which all IoT objects are connected, handled, and controlled by a centralized server, faces a host of difficulties. These issues will stymie the development of IoT applications in the future. When the server breaks, this is a minor instance of failure whereby all IoT applications and services associated with it shall collapse, and thus affect the provision and quality of services offered by the IoT ecosystem (Atlam et al, 2020). Moreover, the centralized database which collects and stores all information generated by various IoT devices is a desirable target for any cyber-security threats. Furthermore, the security of personal data appears to be in jeopardy, as all IoT data entails delicate safety structures. Upgrading the IoT to one of the distributed ledger systems may be the best solution, given the centrally managed IoT architecture's severe challenges.

The most frequent and commonly used sort of such technology is blockchain. It is a decentralized, shareable, and incorruptible ledger that keeps track of all transactions in a peer-to-peer (P2P) network. A catalog of transactions gets consolidated and allotted to a block in the ledger. Thus, every block contains a log file with a date-time stamp and a hash function that is used to connect the prevailing block to the preceding block. As a consequence, a chain of blocks is formed, which is referred to as the blockchain. Sultan et al (2018) precisely defined it as "a decentralized database containing sequential, cryptographically linked blocks of digitally signed asset transactions, governed by a consensus model". To record a transaction in the public ledger, the majority of network participants in the blockchain platform must agree. Distributed ledger technology encourages the exchange of information by ensuring that all contributing users/nodes in the blockchain get a mockup of the authentic ledger, allowing all users to be kept up to date on newly added exchanges or blocks.

Incorporating IoT and blockchain offers significant advantages. For instance, because there is no need for a database controller to regulate IoT devices and their communication services with one another, utilizing the shared system features of blockchain can resolve vulnerabilities and a point of failure attributed to a centralized IoT ecosystem. Besides that, blockchain provides improved privacy and security because it employs intricate cryptography algorithms, hash functions, and timestamps, all of which provide such a secure cloud infrastructure. Furthermore, blockchain technology offers a tamper-proof and incorruptible ledger to protect data from malicious threats, so that any records change can only be cached in the ledger if the large percentages of attributing users substantiate it (Karafiloski et al, 2017). As a result, a reliable system is created in which the participating IoT applications seem to be the only entities that approve or disapprove a transaction based on their assent (Reyna, et al, 2018).

Advantages of Integrating Blockchain With IoT Functionalities

Because IoT devices create huge volumes of data that are difficult to analyze using standard approaches, the big data analytics tool helps the blockchain to provide effective web data storage and processing. Furthermore, many transactions are processed in formalized ledgers, necessitating additional data analysis. Smart contracts are another essential element of distributed ledger technology that enables artificial intelligence technology based on predetermined conditions. A smart contract is traditionally a component of source code remote execution on blockchain to carry out a series of actions when certain terms are fulfilled or authenticated.

As often narrowly associated with only Bitcoin by the masses, Blockchain is so much more powerful than it is utilized to secure banking records as well as in creditworthiness, mortgage lending, mass transit, online broadcasting, healthcare, and other fields. Information access control, Fintech, Internet of Things (IoT), Cloud Services, Multimedia, Education, and Tourism are all examples of the present status of the blockchain (Maiti and Ghosh, 2021; Yeasmin, 2019).

Supply chain and logistics are critical examples of integrating IoT with Blockchain. A worldwide value channel with various parties renders end-to-end monitoring more difficult. When Blockchain and IoT are coupled, they can substantially enhance dependability and provenance (Liu et al, 2021). IoT may be deployed to capture mobility, climate, Geolocation, and other data, which will then be recorded on the distributed ledger. Information obtained can be developed to identify and highlight major issues. Integrating unmanned cars produced with the use of IoT automobiles to the decentralized ecosystem of blockchain allows crucial data to be communicated securely and rapidly. It moreover helps in a variety of fields, such as automatic gasoline purchase, self-driving automobiles, intelligent parking management, and so forth. Smart buildings are another IoT solution. A smart home system allows customers to operate their household wirelessly via various electrical instruments such as cell phones. The ledger technology utilized in home automation, on the other hand, is centralized, making it insecure. The sharing economy is a type of market structure in which resources are exchanged among individual entities over the web. Blockchain enables the transfer of wealth in a safe and decentralized fashion, eliminating the requirement for a middleman. Any object may be leased, purchased, or exchanged through a cloud-based marketplace of linked objects (Reyna et al, 2018).

2.4.1 Blockchain in Finance

The financial services sector is projected to be valued at USD 2.6 trillion by 2023. The international economy handles trillions of dollars per day providing for billions across the world and along with it, carries with it a slew of problems that they've been pondering for a long time. From the hefty cost of stakeholders involved to disruption, exorbitant documentation, and privacy violations; these obstacles seem to be the underlying cause of the industry's significant losses each year. Per a PWC report, 45 percent of financial mediators such as stock exchanges, money transfer services, and payment networks face financial misconduct every year. The global financial system's issues may be addressed via blockchain. Blockchain technology has the potential to provide a solution to the global financial system's problems. Blockchain in the financial sector has led to the implementation of decentralized finance or DeFi which is a type of finance operated by a blockchain process that combines smart contracts to reduce intermediaries. Research on the impact of blockchain shows that blockchain can minimize costs and transform finances in the long run (Nguyen, 2023).

Proper implementation of blockchain can save billions of dollars that banks will spend in a variety of situations; such as with KYC norms, banks keep their customer databases and spend between US $ 60 million and US $ 500 million. Each bank can access a distributed database via a blockchain that avoids database duplication. (Nelito, 2019). The entire system is decentralized, and the repository is not controlled by a single institution, so there is no system failure. One can use smart contracts that eliminate the need for brokers. The microgrid is a good example of this. Microgrids with blockchain will change the ways to pay and use energy. Another use is in developing countries, which can be used to clean up the housing market and eliminate the so-called shadow economy (Harris, 2018).

Uses of Blockchain in the Finance Industry

As Blockchain technology becomes more prevalent, financial institutions aggressively research and deploy it to better the present centralized monetary industry. Banking institutions eliminate the intermediary by leveraging Cryptographic encryption, intractability, and transparency (Underwood, 2016). Classical banks' marketing strategies and technological qualities have been altered by distributed ledger innovation. For starters, it lowers expenses and revenue exchanges. Because endpoint upkeep and acquisition expenses are exorbitant, corporate bankers frequently have to invest a significant amount in a centralized registry. However, many reconciliations and settlement tasks increase labor expenses and personal procedural risk. Blockchain's mechanization and adoption of distributed ledger can start the development with minimal prices and visibility, with minimal expenditure (Nguyen, 2016). Secondly, it could precisely manage vulnerabilities. The banking sector prioritizes loan surveillance; however, the real functioning is less dependable and successful. Furthermore, worldwide control of capital flows might create challenges. The blockchain's multicentered characteristic sees every participant as a component in the Network, allowing effective peer-to-peer dealings amongst lenders and borrowers, removing the necessity of borrowing certificates by bankers as middlemen. The risk premium is significantly decreased as a result of knowledge asymmetries, and wealth administration capacity is increased. Ultimately, it looks for new methods to earn. Throughout the banking business, yet more corporate titans, particularly bankers' organizations, are engaging in or collaborating with Blockchain firms. Institutions must pursue creative business strategies to produce commercial offerings and expand opportunities in this intensely competitive climate (Chang et al, 2020).

As a result of the public autonomy, multi-centered, and non-tamper able attributes, the centralized financial sector business plan has radically altered, optimizing the treasury back-office and infrastructural facilities, enhancing customer experience, as well as providing a transition possible for the financial institution from conventional payment business to online banking space.

AI-powered blockchain technology plays a significant role in the context of Industry 4.0, which refers to the fourth industrial revolution characterized by the integration of smart technologies, automation, and data exchange in manufacturing and other industries (Javaid et al., 2021). The combination of AI and blockchain offers several advantages, enhancing the efficiency, security, and transparency of various processes (Shrivastava et al., 2023). It enables new business models, enhances automation, and fosters innovation in how industries operate and collaborate (Fernandez-Carames & Fraga-Lamas, 2019). AI algorithms can predict when machinery is likely to fail, enabling proactive maintenance. This information can be recorded on a blockchain, ensuring that the maintenance history is immutable (Zaman et al., 2023). This can lead to increased equipment reliability and reduced downtime (Rajmohan et al., 2020). Combining AI and blockchain in supply chain management can provide a transparent and traceable system. AI algorithms can analyze data from the supply chain, and this information can be stored securely on a blockchain. This ensures that every step in the supply chain is recorded, providing a tamper-proof and transparent history (Patil et al., 2023). Smart contracts, self-executing contracts with the terms of the agreement directly written into code, can be enhanced with AI. AI algorithms can help in the automatic execution of complex conditions, making transactions more efficient and reducing the need for intermediaries (Chauhan et al., 2023). AI and blockchain technologies can be integrated to improve interoperability between different systems and platforms (Narayan et al., 2022). This can facilitate seamless data exchange and collaboration across various components of Industry 4.0. A Main Industry 4.0 technology is shown in the Figure.

The processing of blockchain

Main Industry 4.0 technologies

Main benefits (green lines) and challenges (red lines) for blockchain-based Industry 4.0 applications

2.4.2 Smart contracts

Smart contracts, which are self-executing contracts with the terms of the agreement directly written into code, can automate various processes in manufacturing, supply chain, and other industries (Shojaeinasab et al., 2022). This automation reduces the need for intermediaries, streamlining operations and reducing costs (Nuttah et al., 2023). AI can enhance these contracts by making them more adaptive and intelligent, allowing for dynamic adjustments based on real-time data. This combination allows for automated execution of contracts based on predefined conditions, reducing the need for intermediaries and streamlining business processes (Timucin and BİROĞUL, 2021). AI can be used to optimize and automate these contracts based on predefined conditions. This can streamline processes, reduce errors, and enhance efficiency in various industrial applications (Grida and Mostafa, 2023). AI algorithms integrated with blockchain can provide real-time monitoring of smart contracts, allowing for better tracking of production processes, deliveries, and payments (ElMamy et al., 2020). Here's an overview of how AI-powered blockchain can enhance Industry 4.0 in domains of smart contracts and automation:

1. Automated and Trustworthy Execution: Smart contracts are self-executing contracts with the terms of the agreement directly written into code. AI algorithms can be integrated into smart contracts to automate decision-making processes. This ensures that the terms of the contract are executed without the need for intermediaries, reducing the risk of errors and fraud (Tadaka, 2020).
2. Dynamic and Adaptive Contracts: AI can enhance smart contracts by making them more dynamic and adaptive (Sultana et al., 2020). Machine learning algorithms can analyze data and adjust contract terms based on changing conditions, ensuring that agreements remain relevant and effective in dynamic business environments (Tan et al., 2022).
3. Enhanced Security: Blockchain's decentralized and cryptographic nature ensures the security and immutability of smart contracts. AI can further enhance security by providing advanced threat detection and prevention mechanisms, making it more challenging for malicious actors to exploit vulnerabilities (Lin et al., 2018).

2.4.3 Supply Chain Management

Combining AI and blockchain in supply chain management can lead to increased transparency efficiency, and security. AI can analyze the vast amount of data generated by the blockchain, providing valuable insights into the supply chain (Prause, 2019). This helps in identifying inefficiencies, reducing fraud, and optimizing the overall supply chain process (Mehta et al., 2021). Artificial intelligence, through image recognition and other techniques, can be employed to verify the authenticity of physical products by analyzing features such as packaging, labels, or serial numbers. Machine learning algorithms can optimize supply chain routes, scheduling, and inventory management by analyzing real-time data and making data-driven decisions (Chen et al., 2023). Machine learning models can be deployed at various points in the supply chain to make localized decisions based on real-time data (Kashem et al., 2023). Machine learning models can analyze historical data, including customer behavior, market trends, and external factors, to make accurate predictions about future demand (Fernández-Caramés et al., 2019). This helps in minimizing excess inventory or stockouts. By recording all transactions and events securely and transparently, blockchain supports the optimization efforts of AI by providing a reliable source of data. This reliable data is crucial for forecasting demand and optimizing inventory levels (Xu et al., 2022). The synergy between AI and blockchain enhances decision-making, automates processes, and improves overall supply chain performance (Faridi et al., 2020; Nuhiu and Aliu, 2023).

2.4.4 Predictive maintenance

Predictive maintenance involves using data, machine learning, and artificial intelligence (AI) to predict when equipment or machinery is likely to fail, allowing for maintenance to be performed just in time. AI algorithms can analyze data from sensors connected to machinery and equipment. This data can be stored securely on the blockchain (Sang et al., 2020). Predictive maintenance models can then use this information to predict when equipment is likely to fail, allowing for proactive maintenance, reducing downtime, and optimizing operations (Sang et al., 2021). By utilizing machine learning models, AI can identify patterns and anomalies in the data to predict when equipment is likely to fail (Rahman et al., 2021)

Applying predictive maintenance to AI-powered blockchain Technology in the context of Industry 4.0 can offer several advantages.

1. Increased Reliability and Uptime: Predictive maintenance helps in identifying potential issues before they result in system failure. This is crucial for AI-powered blockchain technology in Industry 4.0, where continuous operation and reliability are essential (Oladapo et al., 2023).
2. Cost Reduction: Performing maintenance only when needed reduces unnecessary downtime and maintenance costs. AI algorithms can analyze historical data, system logs, and real-time performance metrics to optimize maintenance schedules and minimize costs (ElMamy et al., 2020).

3. Data-Driven Insights: AI can analyze large volumes of data generated by AI-powered blockchain systems. By monitoring patterns and anomalies, it can provide valuable insights into the health of the system, helping operators make informed decisions (El Akrami et al., 2023).

2.4.5 Data security and privacy by immutable records

1. Immutable Records on Blockchain: Blockchain is a decentralized and distributed ledger that records transactions across a network of computers. Once data is recorded in a block and added to the chain, it becomes extremely difficult to alter. This immutability ensures the integrity of the data (Joshi et al., 2022).
2. Smart Contracts for Automation: Smart contracts are self-executing contracts with the terms of the agreement directly written into code. These contracts can automate various processes in Industry 4.0, such as supply chain management, quality control, and more. AI algorithms can be integrated into smart contracts to enable intelligent decision-making based on predefined rules (Timuçin and Biro ul, 2023).
3. Enhanced Security with Cryptography: Blockchain relies on cryptographic techniques to secure transactions. Each participant in the network has a unique private key, and the data is encrypted. This cryptographic layer adds an extra level of security, making it challenging for unauthorized entities to access or tamper with the data.
4. Decentralization for Resilience: Traditional centralized systems are vulnerable to single points of failure. In a decentralized blockchain network, there's no single point of control, making it more resistant to attacks. This decentralized nature enhances the resilience of the system, reducing the risk of data breaches (Silva et al., 2020).
5. Permissioned Blockchains: Industry 4.0 may require permissioned blockchains where access to the network is restricted to authorized participants. This helps in maintaining control over who can read or write to the blockchain, contributing to data privacy (Mendoza Arvizo et al., 2023).
6. Privacy-Preserving AI Techniques: Employ privacy-preserving AI techniques to ensure that sensitive data used in AI algorithms remains confidential. This includes methods like federated learning, homomorphic encryption, and secure multi-party computation, which allow AI models to be trained on decentralized data without exposing individual data points (Erdem et al., 2019).
7. Auditability and Transparency: Blockchain provides a transparent and auditable history of transactions. Every change made to the data is recorded and time-stamped, allowing for a complete audit trail. This transparency ensures

accountability and helps in compliance with regulations (Mushtaq and Haq, 2019).

8. Consensus Mechanisms: Blockchain networks use consensus mechanisms to validate transactions and agree on the state of the ledger. Consensus adds another layer of security by ensuring that all participants in the network agree on the validity of transactions (Tanwar et al., 2020).

2.4.6 Collaborative Manufacturing

Collaborative manufacturing in the context of AI-powered Blockchain Technology in Industry 4.0 represents a convergence of advanced technologies to enhance efficiency, transparency, and trust in the manufacturing process (Wan et al., 2022). Collaborative manufacturing involves the integration of various stakeholders, including manufacturers, suppliers, customers, and even machines, to work together seamlessly (Carey, 2022). This collaborative approach enhances flexibility, efficiency, and responsiveness in the production process. AI and blockchain can enable collaborative manufacturing where different entities within a supply chain can securely share information and collaborate in a trusted environment, improving efficiency and reducing the risk of fraud. Blockchain facilitates decentralized and secure collaboration between different parties in the manufacturing process. AI algorithms can optimize production processes, and the decentralized nature of blockchain ensures data integrity and security (Huang et al., 2021). Collaborative manufacturing for AI-powered Blockchain Technology in Industry 4.0 holds the promise of creating more efficient, transparent, and trustworthy manufacturing ecosystems (Bellavista et al., 2021). However, addressing the associated challenges and ensuring seamless integration is key to realizing these benefits.

2.4.7 Challenges of ai-powered blockchain technology in Industry 4.0

1. Scalability: Blockchain networks often face scalability issues, especially when handling a large number of transactions. The integration of AI could potentially increase the computational load, making it essential to address scalability concerns (Akram et al., 2020).
2. Interoperability: Ensuring seamless integration with existing systems and technologies within Industry 4.0 can be challenging. AI-powered blockchain solutions need to be compatible with various protocols and standards (Kumar et al., 2023a).

3. Data Privacy and Security: Industry 4.0 involves the exchange of sensitive data. Ensuring the privacy and security of this data is crucial (Ur et al., 2023). AI algorithms operating on the blockchain must comply with data protection regulations and maintain the confidentiality of business-critical information (Taherdoost, 2022; Das et al., 2023).
4. Smart Contracts Complexity: Smart contracts play a vital role in automating processes on the blockchain. However, creating complex smart contracts that involve AI algorithms introduces challenges in terms of reliability, auditing, and debugging (Demertzis et al., 2020).
5. Energy Consumption: Blockchain networks, particularly those using proof-of-work consensus mechanisms, can be energy-intensive. Incorporating AI algorithms may increase computational requirements, adding to the overall energy consumption. Developing energy-efficient solutions is essential (Sasikumar et al., 2023).

2.5 Cloud Computing

Nowadays cloud computing plays a major role in IT. It is not used for business activities but it is now also used for educational purposes. Cloud computing is just internet-based computing you can say it is another property of the internet. customers plug into the "cloud "and access applications, and services which is priced and on demand. In this paper, we will discuss some case studies of cloud computing and the reasons why cloud computing came into the IT market.

There are three features of cloud computing: first, cloud computing provides services on demand, and these resources are scalable over multiple data centers. Second, these services are easily accessible and location-independent, meaning that a user can access the suite of services worldwide through internet browsers. Third, service quality is guaranteed for bandwidth and memory capacity. These features give rise to three benefits: first, cloud computing provides a more palatable cost structure. It requires no upfront hardware cost compared with traditional data warehouses with high capital investments. The use of cloud computing is recorded as operating expenses and significantly reduced. Second, business risk is reduced as it is outsourced to providers who are better equipped to manage these risks. Third, organizations that are starting their digital journey can leverage the latest computing technology to leapfrog their capabilities compared to a traditional step-wise build-up.

2.5.1. Five Key Drivers of Singapore's Cloud Computing Adoption

2.5.1.1- Public Demand for and Satisfaction with E-Government Services

Singapore has done well in ICT infrastructure development as evidenced by the extraordinarily high mobile subscription penetration rate of 155.6% and, high internet penetration rate of 135.1% in 2023. In addition, 83% of the Singapore population are internet users, exposed to the information deluge on the world wide web. The statistic is comparable to the average internet users of above 70% in developed countries. On the contrary, fixed telephone line usage has plunged to 36.4% and 25.7% for wired broadband subscriptions. These numbers underscore the high levels of connectivity and mobility of residents in Singapore.

2.5.1.2. Focus on Whole-of-Government Policies and Practices-

The public service is moving towards a whole-of-government approach to policy making and operations, providing a key driver for whole-of-government cloud adoption. This approach reflects the reality that many contemporary issues cannot be siloed. **For example**, aging is not only a health issue; it also impacts the agencies in transport, environment, society, and family. To create opportunities for silos to "talk" to each other, the government organized forums for top leaders within different sectors to discuss and craft policies. For example, the Social Forum brings together senior leaders of all the social agencies to build consensus for social policies. This is also a great platform to commission important analytics projects that require data from multiple agencies— a typically arduous process. When these projects are debated and benefits delineated among senior leaders across agencies, they are more willing to share data that contributes to the common good. Such initiatives provide use cases for whole-of-government cloud adoption.

2.5.1.3. Restructuring Technology Agencies to Integrate Strategy and Implementation

To enable whole-of-government innovation and implementation, technology-related agencies are also being restructured. Telco and media regulation are now brought under one agency—the Info-communications Media Development Authority of Singapore (IMDA). Merging both the InfoComm Development Authority of Singapore (IDA), and the Media Development Authority of Singapore (MDA), IMDA will focus on leveraging new technologies for better regulation and application to improve lives.

2.5.1.4. Building the Smart Nation Platform

A key catalyst of cloud computing is the Smart Nation Platform (SNP) involving a nationwide communications and sensor infrastructure. This platform will enable centralized data aggregation and sharing, as well as provide new capabilities to derive refined insights from cross-referencing a vast array of Government datasets. Datasets can also be shared with the community and industry for co-creation and self-enablement. The SNP will be a key enabler for Singapore to remain on the cutting edge of Government operations and service delivery.

2.5.1.5. Purpose-Driven Cloud Applications in Healthcare

Healthcare Transformation Within the healthcare space, cloud computing is applied at scale to healthcare transformation. Signaling the government's commitment to transformation, the Ministry of Health (MOH) recently appointed outgoing National University of Singapore President and former MOH Director of Medical Services, Professor Tan Chorh Chuan, to direct the inaugural Healthcare Transformation Office as Singapore's first Chief Health Scientist.

Tele-Medicine Another application of cloud computing is tele-medicine. The aim of tele-health is to shift from institution-based care towards home and community care, augmenting healthcare resources (e.g., allied health professionals, mental health psychiatrists, etc.) in the system. Essentially such technologies shift from a doctor-centric model to a team-based model to manage the shortage of doctors and promote holistic care.

2.6 Quantum Computing and Robotics

Quantum computing harnesses the power of quantum mechanics to help us overcome the limitations that today's computing world has been encountering. However, because the quantum state of a particle is a natural phenomenon, it means that it can be reproduced in many different ways. It has the potential to transform many industries, as such computers solve problems exponentially faster and with less power consumption than traditional computers.

Quantum Computing Advantages

Quantum computing has significant applications that offer prospects of scalability. It can accelerate various fields such as:

- **Search and optimization**: **Grover's algorithm** for database search offers **quadratic speedup** - a database of 4 entries takes 1 lookup on a quantum computer versus 4 lookups classically.

- This could enable ultra-fast search of large real-time financial, healthcare or classified datasets.

- **Chemical simulation:** Quantum computers with **50-100 qubits** could precisely simulate **complex enzyme mechanisms** and **chemical processes** for the industrial production of fertilizers, polymers, and biomass.
  - This may enable the designing of better catalysts and materials.
- **Artificial intelligence:** Japanese researchers designed a quantum neural network that detects **credit card fraud with 63% accuracy versus 50%** for classical networks.
  - This demonstrates the potential for quantum machine learning.
    - **Financial analysis:** Massively enhanced portfolio optimization, price modeling, risk analysis, and economic forecasting can be achieved to alter financial services, investment strategies, and economic policy.
    - **Secure communication:** All widespread public-key **cryptography** is vulnerable to attacks using quantum algorithms negating the security of communication channels and protocols like online banking, messaging, and sensitive database access.
  - **China** recently demonstrated **(QUESS)** satellite-based quantum key distribution for **hack-proof military communications.**
    - **Healthcare advancement:** Precision modeling of protein folding mechanisms, gene sequencing, disease pathology simulations, personalized medicine, and drug discovery can transform healthcare and life sciences through quantum computing.
    - **Climate change science:** NASA plans to use quantum computing for very high-resolution regional and global scale climate modeling by assimilating large datasets using quantum neural networks.

Quantum Computing Limitations

Quantum computing emerged as one of the most promising technologies, however, there are the following limitations:

- **Hardware fragility:** Google's 72-qubit quantum computer requires temperatures near absolute zero inside vacuum chambers, with vibration dampening and magnetic shielding increasing complexity.
- **Decoherence:** IBM researchers reported the coherence time for a 27-qubit system was 47 microseconds - too short for practical error correction schemes.
- **Lack of standards:** Currently no standards exist for quantum programming languages and software tools.

- Recently IBM and Microsoft clashed over how fidelity benchmarks are defined across superconducting, ion trap, and photonics hardware.

• **Qubit scalability:** Intel's 49-qubit superconducting quantum chip Pushan is a high watermark.

- However, the company hasn't elaborated a roadmap for the millions of qubits likely needed for quantum advantage in domains like machine learning or chemical simulation.

3. METHODOLOGY

The methodology employed in this research paper involves a comprehensive review of existing literature on emerging technologies in business management. The study also analyses case studies of companies that have successfully implemented these technologies to improve their operations. Additionally, the research paper includes a survey of business managers' opinions on the impact of emerging technologies on their businesses. The survey was conducted through online questionnaires, and the data collected was analyzed using statistical methods.

4. RESULTS

The findings of the study reveal that emerging technologies offer numerous benefits, including improved customer experience, cost savings, and increased efficiency. Artificial intelligence is particularly useful in business management, with applications in customer service, marketing, and data analysis. Blockchain technology has been shown to enhance supply chain management, improve transparency, and reduce transaction costs. The Internet of Things has been used to improve asset tracking and supply chain management, while cloud computing has enabled businesses to store and access data remotely, reducing costs associated with physical data storage. Robotic process automation has been shown to improve process efficiency and reduce errors.

5. CONCLUSION

The adoption of emerging technologies has become increasingly important for businesses looking to gain a competitive advantage in today's dynamic marketplace. The research paper highlights the various emerging technologies in business management and their potential benefits. However, the implementation of these technologies requires careful planning, management, and investment. The paper concludes with suggestions for future research, including the need to examine the potential ethical implications of these technologies, the impact of emerging technologies on employment, and the role of government regulation in managing emerging technologies.

6. FUTURE SCOPE

The future of business management is heavily reliant on emerging technologies, and as such, there is a need for continued research on the subject. The potential implications of emerging technologies on the future of work, employment, and job security are areas that require further exploration. Additionally, the ethical implications of emerging technologies, such as the impact on privacy and data protection, require careful consideration. The role of government regulation in managing emerging technologies is also an area that requires further research. Future research should also focus on the development of frameworks to guide businesses.

REFERENCES

Abdulrahman, Y., Arnautović, E., Parezanović, V., & Svetinovic, D. (2023). AI and Blockchain Synergy in Aerospace Engineering: An Impact Survey on Operational Efficiency and Technological Challenges. *IEEE Access: Practical Innovations, Open Solutions*, *11*, 87790–87804. DOI: 10.1109/ACCESS.2023.3305325

Abidi, M. H., Mohammed, M. K., & Alkhalefah, H. (2022). Predictive maintenance planning for industry 4.0 using machine learning for sustainable manufacturing. *Sustainability (Basel)*, *14*(6), 3387. DOI: 10.3390/su14063387

Adhikari, A., & Winslett, M. (2019) A hybrid architecture for secure management of manufacturing data in industry 4.0, in: 2021 IEEE International Conference on Pervasive Computing and Communications Workshops (PerCom Workshops), IEEE, pp. 973–978

Adru, S. L., Johnson, S., & Hemalatha, M. (2023). AI and Blockchain Based Platform for Empowering Content Creators: Enabling NFT–Driven Content Sharing, Inspiration Generation, and Collaborative Interaction, in. 4th International Conference on Smart Electronics and Communication (ICOSEC) 2023. IEEE, pp. 666–673.

Ahmad, T., Zhu, H., Zhang, D., Tariq, R., Bassam, A., Ullah, F., AlGhamdi, A. S., & Alshamrani, S. S. (2022). Energetics Systems and artificial intelligence: Applications of industry 4.0. *Energy Reports*, *8*, 334–361. DOI: 10.1016/j.egyr.2021.11.256

Aiello, G., Benítez, J., Carpitella, S., Certa, A., Enea, M., Izquierdo, J., & La Cascia, M. (2021). A decision support system to assure high-performance maintenance service. *Journal of Quality in Maintenance Engineering*, *27*(4), 651–670. DOI: 10.1108/JQME-11-2019-0107

Akram, S. V., Malik, P. K., Singh, R., Anita, G., & Tanwar, S. (2020). Adoption of blockchain technology in various realms: Opportunities and challenges. *Security and Privacy*, *3*(5), e109. DOI: 10.1002/spy2.109

Al-Rakhami, M., & Al-Mashari, M. (2022). Interoperability approaches of blockchain technology for supply chain systems. *Business Process Management Journal*, *28*(5/6), 1251–1276. DOI: 10.1108/BPMJ-04-2022-0207

Alsamhi, S. H., Shvetsov, A. V., Shvetsova, S. V., Hawbani, A., Guizani, M., Alhartomi, M. A., & Ma, O. (2022). Blockchain-empowered security and energy efficiency of drone swarm consensus for environment exploration. *IEEE Transactions on Green Communications and Networking*, *7*(1), 328–338. DOI: 10.1109/TGCN.2022.3195479

Althabatah, A., Yaqot, M., Menezes, B., & Kerbache, L. (2023). Transformative procurement trends: Integrating industry 4.0 technologies for enhanced procurement processes. *Logistics (Basel)*, *7*(3), 63. DOI: 10.3390/logistics7030063

Aoun, A., Ilinca, A., Ghandour, M., & Ibrahim, H. (2021). A review of Industry 4.0 characteristics and challenges, with potential improvements using blockchain technology. *Computers & Industrial Engineering*, *162*, 107746. DOI: 10.1016/j.cie.2021.107746

Atlam, H. F., & Wills, G. B. (2021). IoT security, privacy, safety and ethics. In *Digital twin technologies and smart cities* (pp. 123–149). Springer.

Bandari, V., & Comprehensive, A. (2021). Review of AI applications in automated container orchestration. Predict. Maint., Secur. Compliance, Resour. Optim., Contin. Deploy. Test. *Int. J. Intell. Autom. Comput.*, *4*, 1–19.

Barton, M., Budjac, R., Tanuska, P., Gaspar, G., & Schreiber, P. (2022). Identification overview of Industry 4.0 essential attributes and resource-limited embedded artificial-intelligence-of-things devices for small and medium-sized enterprises. *Applied Sciences (Basel, Switzerland)*, *12*(11), 5672. DOI: 10.3390/app12115672

Bellagarda, J. S., & Abu-Mahfouz, A. M. (2022). An updated survey on the convergence of distributed ledger technology and artificial intelligence: Current state, major challenges and future direction. *IEEE Access : Practical Innovations, Open Solutions*, *10*, 50774–50793. DOI: 10.1109/ACCESS.2022.3173297

Bellavista, P., Esposito, C., Foschini, L., Giannelli, C., Mazzocca, N., & Montanari, R. (2021). Interoperable blockchains for highly-integrated supply chains in collaborative manufacturing. *Sensors (Basel)*, *21*(15), 4955. DOI: 10.3390/s21154955 PMID: 34372191

Benzidia, S., Makaoui, N., & Subramanian, N. (2021). Impact of ambidexterity of blockchain technology and social factors on new product development: A supply chain and Industry 4.0 perspective. *Technological Forecasting and Social Change*, *169*, 120819. DOI: 10.1016/j.techfore.2021.120819

Bodkhe, U., Tanwar, S., Parekh, K., Khanpara, P., Tyagi, S., Kumar, N., & Alazab, M. (2020). Blockchain for industry 4.0: A comprehensive review. *IEEE Access: Practical Innovations, Open Solutions*, *8*, 79764–79800. DOI: 10.1109/ACCESS.2020.2988579

Borowski, P. F. (2021). Digitization, digital twins, blockchain, and industry 4.0 as elements of management process in enterprises in the energy sector. *Energies*, *14*(7), 1885. DOI: 10.3390/en14071885

Brilliantova, V., & Thurner, T. W. (2019). Blockchain and the future of energy. *Technology in Society*, *57*, 38–45. DOI: 10.1016/j.techsoc.2018.11.001

Carey, B. (2022). Metaverse technologies, behavioral predictive analytics, and customer location tracking tools in blockchain-based virtual worlds. Rev. Contemp. Philos. 188–204.

Chae, B., & Olson, D. (2022). Technologies and applications of Industry 4.0: Insights from network analytics. *International Journal of Production Research*, *60*(12), 3682–3704. DOI: 10.1080/00207543.2021.1931524

Chauhan, S., Singh, R., Gehlot, A., Akram, S. V., Twala, B., & Priyadarshi, N. (2023). Digitalization of supply chain management with industry 4.0 enabling technologies: A sustainable perspective. *Processes (Basel, Switzerland)*, *11*(1), 96. DOI: 10.3390/pr11010096

Chen, K., Golhar, D.Y., & Banerjee, S. (2023). Blockchain applications and challenges for supply chain and Industry 4.0: a literature review. *Int. J. Appl. Decis. Sci. 16* (1), 41.

Darwish, D. (2023). Blockchain and Artificial Intelligence for Business Transformation Toward Sustainability. In *Blockchain and its Applications in Industry 4.0* (pp. 211–255). Springer. DOI: 10.1007/978-981-19-8730-4_8

Das, L., Bibhu, V., Logeswaran, R., Dadhich, K., & Sharma, B. (2023). *AI Model for Blockchain Based Industrial IoT and Big Data*. Concepts, Methodologies, Tools, and Applications. AI Models for Blockchain-Based Intelligent Networks in IoT Systems Springer. DOI: 10.1007/978-3-031-31952-5_3

Dastres, R., & Soori, M. (2021). Artificial Neural Network System. *Int. J. Imaging Robot*, *21*, 13–25.

Dastres, R., & Soori, M. (2021a). Advances in web-based decision support systems. *Int. J. Eng. Future Technol.*, *19*, 1–15.

Dastres, R., & Soori, M. (2021b). The role of information and communication technology (ICT) in environmental protection. *Int. J. Tomogr. Simul.*, *35*, 24–37.

de Almeida, P. G. R., dos Santos, C. D., & Farias, J. S. (2021). Artificial intelligence regulation: A framework for governance. *Ethics and Information Technology*, *23*(3), 505–525. DOI: 10.1007/s10676-021-09593-z

Demertzis, K., Iliadis, L., Tziritas, N., & Kikiras, P. (2020). Anomaly detection via blockchained deep learning smart contracts in industry 4.0. *Neural Computing & Applications*, *32*(23), 17361–17378. DOI: 10.1007/s00521-020-05189-8

El Akrami, N., Hanine, M., Flores, E. S., Aray, D. G., & Ashraf, I. (2023). Unleashing the potential of blockchain and machine learning: Insights and emerging trends from bibliometric analysis. *IEEE Access: Practical Innovations, Open Solutions*, *11*, 78879–78903. DOI: 10.1109/ACCESS.2023.3298371

ElMamy, S. B., Mrabet, H., Gharbi, H., Jemai, A., & Trentesaux, D. (2020). A survey on the usage of blockchain technology for cyber-threats in the context of industry 4.0. *Sustainability (Basel)*, *12*(21), 9179. DOI: 10.3390/su12219179

Esmaeilian, B., Sarkis, J., Lewis, K., & Behdad, S. (2020). Blockchain for the future of sustainable supply chain management in Industry 4.0. Resour. *Resources, Conservation and Recycling*, *163*, 105064. DOI: 10.1016/j.resconrec.2020.105064

Esmat, A., de Vos, M., Ghiassi-Farrokhfal, Y., Palensky, P., & Epema, D. (2021). A novel decentralized platform for peer-to-peer energy trading market with blockchain technology. *Applied Energy*, *282*, 116123. DOI: 10.1016/j.apenergy.2020.116123

Faridi, M. S., Ali, S., Duan, G., & Wang, G. (2020). Blockchain and IoT based textile manufacturing traceability system in industry 4.0. in: International conference on security, privacy and anonymity in computation, communication and storage. Springer, pp. 331–344.

Fernandez-Carames, T. M., & Fraga-Lamas, P. (2019). A review on the application of blockchain to the next generation of cybersecure industry 4.0 smart factories. *IEEE Access: Practical Innovations, Open Solutions*, *7*, 45201–45218. DOI: 10.1109/ACCESS.2019.2908780

Gad, A. G., Mosa, D. T., Abualigah, L., & Abohany, A. A. (2022). Emerging trends in blockchain technology and applications: A review and outlook. *Journal of King Saud University. Computer and Information Sciences*, *34*(9), 6719–6742. DOI: 10.1016/j.jksuci.2022.03.007

Gaikwad, L., & Khang, A. (2023). AI Powered Workforce Management in Industry 4.0 Era. In *Designing Workforce Management Systems for Industry 4.0* (pp. 35–46). CRC Press. DOI: 10.1201/9781003357070-3

Gami, B., Agrawal, M., Mishra, D. K., Quasim, D., & Mehra, P. S. (2023). Artificial intelligence-based blockchain solutions for intelligent healthcare: A comprehensive review on privacy preserving techniques. *Transactions on Emerging Telecommunications Technologies*, *34*(9), e4824. DOI: 10.1002/ett.4824

Garces, E., Li, S., & Daim, T. U. (2023). Cybersecurity and Technology Convergence: Analysis of AI, Blockchain, and IoT Using SNA. In: A Technology Landscape Analysis. Cybersecurity Springer, pp. 39–70.

Ghasemzadeh, B., Celik, T., & Karimi Ghaleh Jough, F.. (2022). Road map to BIM use for infrastructure domains: Identifying and contextualizing variables of infrastructure projects. *Scientia Iranica*, *29*(6), 2803–2824. DOI: 10.24200/sci.2022.56935.4998

Ghobakhloo, M., & Fathi, M. (2021). Industry 4.0 and opportunities for energy sustainability. *J. Clean. Prod. 295*, 126427.

Grida, , MMostafa, , N.A. (2023). Are smart contracts too smart for Supply Chain 4.0? A blockchain framework to mitigate challenges. *Journal of Manufacturing Technology Management*, *34*, 644–665.

Gupta, R., Tanwar, S., Kumar, N., & Tyagi, S. (2020). Blockchain-based security attack resilience schemes for autonomous vehicles in industry 4.0: A systematic review. *Computers & Electrical Engineering*, *86*, 106717. DOI: 10.1016/j.compeleceng.2020.106717

Habbal, A., Ali, M. K., & Abuzaraida, M. A. (2024). Artificial Intelligence Trust, Risk and Security Management (AI TRiSM): Frameworks, applications, challenges and future research directions. *Expert Systems with Applications*, *240*, 122442. DOI: 10.1016/j.eswa.2023.122442

Jena, A. K., & Dash, S. P. (2021). Blockchain technology: introduction, applications, challenges. In *Blockchain Technology* (pp. 1–11). Applications and Challenges Springer.

Jha, M. K., Yogeshwari, A., Rubini, P., & Singh, M. (2023). Converge of IoT and AI in metaverse: challenges and opportunities. In: 5th International Conference on Inventive Research in Computing Applications (ICIRCA) 2023. IEEE, pp. 1462–1467 DOI: 10.1109/ICIRCA57980.2023.10220628

Joshi, S., Pise, A. A., Shrivastava, M., Revathy, C., Kumar, H., Alsetoohy, O., & Akwafo, R. (2022). Adoption of blockchain technology for privacy and security in the context of industry 4.0. *Wireless Communications and Mobile Computing*. DOI: 10.1155/2022/4079781

Karimi Ghaleh Jough, F., & Ghasemzadeh, B. (2023). Uncertainty Interval Analysis of Steel Moment Frame by Development of 3D-Fragility Curves Towards Optimized Fuzzy Method. *Arabian Journal for Science and Engineering*, 1–18.

Kashem, M. A., Shamsuddoha, M., Nasir, T., & Chowdhury, A. A. (2023). Supply chain disruption versus optimization: A review on artificial intelligence and blockchain. *Knowledge (Beverly Hills, Calif.)*, *3*, 80–96.

Lal, B., Kumar, M. A., Chinthamu, N., & Pokhriyal, S. (2023). Development of Product Quality with Enhanced Productivity in Industry 4.0 with AI Driven Automation and Robotic Technology. In: Second International Conference on Augmented Intelligence and Sustainable Systems (ICAISS) 2023. IEEE, pp. 184–189

Latif, H. (2023). Advancing data integrity in banking: AI/ML solutions and best practice. Int. J. Comput. Sci. Technol. 7, 185–203

Pfeiffer, A., & Bugeja, M. (2021). Introd. Concept "Digital-Agent Signatures": How SSI Can. Be Expand. Needs Ind. 4. 0, Artif. Intell. Ind. 4. 0: A Collect. Innov. Res. CaseStud. that are Reworking Way We Look. Ind. 4. 0 Thanks. *Artificial Intelligence*, 213–233.

Picone, M., Cirani, S., & Veltri, L. (2021). Blockchain security and privacy for the Internet of Things. MDPI 892.

Popkova, , E.GBogoviz, , A.VLobova, , S.VVovchenko, , N.GSergi, , B.S. (2023). Blockchain, sustainability and clean energy transition. *Global Transitions*, *5*, 64–78.

Prause, G. (2019). Smart contracts for smart supply chains. *IFAC-Pap.*, *52*, 2501–2506.

Pustišek, M., Živić, N., & Kos, A. (2021). *Blockchain: Technology and Applications for Industry 4.0*. Smart Energy, and Smart Cities, Walter de Gruyter GmbH & Co KG.

Raja, G., Manaswini, Y., Vivekanandan, G. D., Sampath, H., Dev, K., & Bashir, A. K. (2020) AI-powered blockchain-a decentralized secure multiparty computation protocol for IoV, in: IEEE INFOCOM 2020-IEEE Conference on Computer Communications Workshops (INFOCOM WKSHPS), IEEE, pp. 865–870. DOI: 10.1109/INFOCOMWKSHPS50562.2020.9162866

Rajmohan, R., Kumar, T. A., Pavithra, M., Sandhya, S., Julie, E., Nayahi, J., & Jhanjhi, N. (2020). Blockchain: Next-generation technology for industry 4.0, Blockchain. *Technology*, 177–198.

Rathore, B. (2023). Digit. Transform. 4. 0: Integr. Artif. Intell. Metaverse Mark., Eduzone: Int. Peer Rev. *Referee Multidiscip. J.*, *12*, 42–48.

Rosa Righi, R., Alberti, A. M., & Singh, M. (2020). Blockchain technology for Industry 4.0. Springer.[REMOVED IF= FIELD]

Saheb, T., Jamthe, S., & Saheb, T. (2022). Developing a conceptual framework for identifying the ethical repercussions of artificial intelligence: A mixed method analysis. *J. AIRobot. Workplace Autom.*, *1*(4), 371–398. DOI: 10.69554/FFZO4203

Salah, K., Rehman, M. H. U., Nizamuddin, N., & Al-Fuqaha, A. (2019). Blockchain for AI: Review and open research challenges. *IEEE Access: Practical Innovations, Open Solutions*, *7*, 10127–10149. DOI: 10.1109/ACCESS.2018.2890507

Salama, R., Al-Turjman, F., Altrjman, C., Kumar, S., & Chaudhary, P. (2023). A Comprehensive Survey of Blockchain-Powered Cybersecurity-A survey. In: International Conference on Computational Intelligence, Communication Technology and Networking (CICTN) 2023. IEEE, pp. 774–777. DOI: 10.1109/CICTN57981.2023.10141282

Sang, G. M., Xu, L., & de Vrieze, P. (2021). A predictive maintenance model for flexible manufacturing in the context of industry 4.0. *Frontiers in Big Data*, *4*, 663466. DOI: 10.3389/fdata.2021.663466 PMID: 34514378

Sang, G. M., Xu, L., de Vrieze, P., Bai, Y., & Pan, F. (2020). Predictive maintenance in industry 4.0.: Proc. 10th Int. Conf. Inf. Syst. Technol. 1–11.

Sasikumar, R., Karthikeyan, P., & Murugan, T. (2023). Privacy and Security Through Blockchain in Industry 4.0. In: An Industrial Cybersecurity Perspective. Handbook of Research on Data Science and Cybersecurity Innovations in Industry 4.0 Technologies IGI Global, pp. 315–334.

Schiliro, D. (2022). Digital Economy and Digital Transformation. In: Digital Technologies for Entrepreneurship in Industry 4.0 IGI Global, pp. 26–42. DOI: 10.4018/978-1-6684-4265-4.ch002

Ur, R. K., Mian, Z., & Yasir, A. (2023). *Implementation of Reliability Centered Maintenance (RCM) in the background of industry 4.0–Issues*. Challenges, and Opportunities.

Usmani, U. A., Happonen, A., & Watada, J. (2023). Advancements in Industry 4.0 Asset Management: Interoperability and Cyber Security Challenges and Opportunities. in: *Proceedings of the Future Technologies Conference*. Springer, pp. 468–488. DOI: 10.1007/978-3-031-47448-4_35

Varadam, D., Shankar, S. P., Bharadwaj, A., Saxena, T., Agrawal, S., & Dayananda, S. (2024). Enhancing Industrial Robotics Performance and Security With AI and Blockchain Technologies. In: AI and Blockchain Applications in Industrial Robotics IGI Global, pp. 58–81.

Vashishth, T. K., Sharma, V., Sharma, K. K., Kumar, B., Chaudhary, S., & Panwar, R. (2024). Intelligent Resource Allocation and Optimization for Industrial Robotics Using AI and Blockchain. In: AI and Blockchain Applications in Industrial Robotics IGI Global, pp. 82–110.

Zarrin, J., Wen Phang, H., Babu Saheer, L., & Zarrin, B. (2023). 2021 Blockchain for decentralization of internet: Prospects, trends, and challenges. *Cluster Computing*, *24*, 2841–2866.

Zhang, L., Xie, Y., Zheng, Y., Xue, W., Zheng, X., & Xu, X. (2020). The challenges and countermeasures of blockchain in finance and economics. Syst. Res. Behav. Sci. 37, 691–698. M. Soori, R. Dastres and B. *Arezoo Journal of Economy and Technology*, *1*, 222–241.

Compilation of References

Abdulrahman, Y., Arnautović, E., Parezanović, V., & Svetinovic, D. (2023). AI and Blockchain Syncrgy in Aerospace Engineering: An Impact Survey on Operational Efficiency and Technological Challenges. *IEEE Access: Practical Innovations, Open Solutions, 11*, 87790–87804. DOI: 10.1109/ACCESS.2023.3305325

Abidi, M. H., Mohammed, M. K., & Alkhalefah, H. (2022). Predictive maintenance planning for industry 4.0 using machine learning for sustainable manufacturing. *Sustainability (Basel), 14*(6), 3387. DOI: 10.3390/su14063387

Accenture. (2018). *Making it personal: Shoppers demand a hyper-relevant experience*. https://www.accenture.com

Accenture. (2018). *Personalization Pulse Check*. Accenture.

Addy, W. A., Ugochukwu, C. E., Oyewole, A. T., Ofodile, O. C., Adeoye, O. B., & Okoye, C. C., (2024). Predictive analytics in credit risk management for banks: A comprehensive review. *GSC Advanced Research and Reviews, 18*(2), 434–449. DOI: 10.30574/gscarr.2024.18.2.0077

Adhikari, A., & Winslett, M. (2019) A hybrid architecture for secure management of manufacturing data in industry 4.0, in: 2021 IEEE International Conference on Pervasive Computing and Communications Workshops (PerCom Workshops), IEEE, pp. 973–978

Adobe. (2021). Transitioning to a subscription-based model: Lessons in digital transformation. Harvard Business Review. https://hbr.org

Adru, S. L., Johnson, S., & Hemalatha, M. (2023). AI and Blockchain Based Platform for Empowering Content Creators: Enabling NFT–Driven Content Sharing, Inspiration Generation, and Collaborative Interaction, in. 4th International Conference on Smart Electronics and Communication (ICOSEC) 2023. IEEE, pp. 666–673.

Agarwal, N., & Chatterjee, D. (2018). The rise of e-commerce in India: An analysis of digital platforms and consumer behavior. *Indian Journal of Marketing*, *48*(3), 78–89.

Agasisti, T., & Froumin, I. (2020). Special Issue on Public Administration Education. *International Journal of Public Administration, 81-83*. Advance online publication. DOI: 10.1080/01900692.2019.1661057

Ahmad, T., Zhu, H., Zhang, D., Tariq, R., Bassam, A., Ullah, F., AlGhamdi, A. S., & Alshamrani, S. S. (2022). Energetics Systems and artificial intelligence: Applications of industry 4.0. *Energy Reports*, *8*, 334–361. DOI: 10.1016/j.egyr.2021.11.256

Aiello, G., Benítez, J., Carpitella, S., Certa, A., Enea, M., Izquierdo, J., & La Cascia, M. (2021). A decision support system to assure high-performance maintenance service. *Journal of Quality in Maintenance Engineering*, *27*(4), 651–670. DOI: 10.1108/JQME-11-2019-0107

Ajitha, S., Geevarathna, , & Huxley, S. (2024). Exploring public perception, awareness, and satisfaction with AI applications in Karnataka, India: The role of individual characteristics and media influence. *International Journal of System Assurance Engineering and Management*, 1–16. DOI: 10.1007/s13198-024-02594-3

Akram, S. V., Malik, P. K., Singh, R., Anita, G., & Tanwar, S. (2020). Adoption of blockchain technology in various realms: Opportunities and challenges. *Security and Privacy*, *3*(5), e109. DOI: 10.1002/spy2.109

Akter, S., Michael, K., Uddin, M. R., McCarthy, G., & Rahman, M. (2022). Transforming business using digital innovations: The application of AI, blockchain, cloud and data analytics. *Annals of Operations Research*, *308*(1-2), 1–33. DOI: 10.1007/s10479-020-03620-w PMID: 35935743

Alami, R., & Sergio, R. (2024, April). Future-Proofing Strategies: Machine Learning Models for Organizational Resilience in Emerging Markets. In International Conference on Business and Technology (pp. 487-501). Cham: Springer Nature Switzerland. DOI: 10.1007/978-3-031-67431-0_47

Alibaba. (2021). *Annual report 2021.* Alibaba Group.

Alonge, E. O., Dudu, O. F., & Alao, O. B. (2024). The impact of digital transformation on financial reporting and accountability in emerging markets. *International Journal of Science and Technology Research Archive, 7*(2), 025-049.

Al-Rakhami, M., & Al-Mashari, M. (2022). Interoperability approaches of blockchain technology for supply chain systems. *Business Process Management Journal*, *28*(5/6), 1251–1276. DOI: 10.1108/BPMJ-04-2022-0207

Alsamhi, S. H., Shvetsov, A. V., Shvetsova, S. V., Hawbani, A., Guizani, M., Alhartomi, M. A., & Ma, O. (2022). Blockchain-empowered security and energy efficiency of drone swarm consensus for environment exploration. *IEEE Transactions on Green Communications and Networking*, *7*(1), 328–338. DOI: 10.1109/TGCN.2022.3195479

Althabatah, A., Yaqot, M., Menezes, B., & Kerbache, L. (2023). Transformative procurement trends: Integrating industry 4.0 technologies for enhanced procurement processes. *Logistics (Basel)*, *7*(3), 63. DOI: 10.3390/logistics7030063

Alt, R. (2020). Electronic markets on Blockchain applications. *Electronic Markets*, *30*(1), 1–5. DOI: 10.1007/s12525-020-00413-8

Amazon. (1999). Amazon annual report 1999. https://www.amazon.com/annual-reports

Amazon. (2021). *Sustainability at Amazon.* Amazon.com.

Amazon. (2021). Upskilling 2025: Preparing for a future-ready workforce. Harvard Business Review. https://hbr.org

Amazon. (2023). Amazon annual report 2023 https://www.amazon.com/annual-reports

Amazon. (2025). Amazon sustainability and technology report 2025. https://www.amazon.com/sustainability-2025

Anderson, D., & Ackerman Anderson, L. (2010). *Beyond Change Management: How to Achieve Breakthrough Results Through Conscious Change Leadership*. Jossey-Bass.

Anderson, H. J. (2021). Managing Gen Z: Work Expectations and the New Workforce. *Journal of Business Research*, *124*, 464–475.

Anderson, J., & Sun, P. (2021). Impacts of transformational leadership on organizational change capability: The role of trust and learning. *European Journal of Management and Business Economics*, *30*(2), 191–204. DOI: 10.1108/EJMBE-06-2021-0180

Andoni, M., Robu, V., Flynn, D., Abram, S., Geach, D., Jenkins, D., McCallum, P., & Peacock, A. (2019). Blockchain technology in the energy sector: A systematic review of challenges and opportunities. *Renewable & Sustainable Energy Reviews*, *100*, 143–174. DOI: 10.1016/j.rser.2018.10.014

Andrejevic, M. (2011). THE WORK THAT AFFECTIVE ECONOMICS DOES. *Cultural Studies*, *25*(4–5), 604–620. DOI: 10.1080/09502386.2011.600551

An, Y. J., Choi, P. M. S., & Huang, S. H. (2021). Blockchain, cryptocurrency, and artificial intelligence in finance. In *Fintech with artificial intelligence, big data, and blockchain* (pp. 1–34). Springer Singapore. DOI: 10.1007/978-981-33-6137-9_1

Aoun, A., Ilinca, A., Ghandour, M., & Ibrahim, H. (2021). A review of Industry 4.0 characteristics and challenges, with potential improvements using blockchain technology. *Computers & Industrial Engineering*, *162*, 107746. DOI: 10.1016/j.cie.2021.107746

Armbrust, M., Fox, A., Griffith, R., Joseph, A. D., Katz, R., Konwinski, A., Lee, G., Patterson, D., Rabkin, A., Stoica, I., & Zaharia, M. (2010). A view of cloud computing. *Communications of the ACM*, *53*(4), 50–58. DOI: 10.1145/1721654.1721672

Arthur, W. B. (2009). *The nature of technology: What it is and how it evolves*. Free Press.

Asian Development Bank. (2020), *What Drives Innovation in Asia?* https://www.adb.org/sites/default/files/publication/575626/ado2020-highlights.pdf

AT&T Archives. (2020). The history of the telephone: 1920s adoption statistics. https://www.att.com/archives

Atlam, H. F., & Wills, G. B. (2021). IoT security, privacy, safety and ethics. In *Digital twin technologies and smart cities* (pp. 123–149). Springer.

Atzori, L., Iera, A., & Morabito, G. (2010). The Internet of Things: A survey. *Computer Networks*, *54*(15), 2787–2805. DOI: 10.1016/j.comnet.2010.05.010

Aunedi, M., Yliruka, M., Dehghan, S., Pantaleo, A. M., Shah, N., & Strbac, G. (2022). Multi-model assessment of heat decarbonization options in the UK using electricity and hydrogen. *Renewable Energy*, *194*, 1261–1276. Advance online publication. DOI: 10.1016/j.renene.2022.05.145

Bain & Company. (2021) *Overcoming resistance to change: The role of leadership*. https://www.bain.com (Accessed: 20 December 2024).

Ball, M. (2022). *The metaverse: And how it will revolutionize everything*. W.W. Norton & Company. DOI: 10.15358/9783800669400

Bandari, V., & Comprehensive, A. (2021). Review of AI applications in automated container orchestration, Predict. Maint., Secur. Compliance, Resour. Optim., Contin. Deploy. Test. *Int. J. Intell. Autom. Comput.*, *4*, 1–19.

Bandura, A. (1977). *Social Learning Theory*. Prentice Hall.

Barbosa, B., Saura, J. R., Zekan, S. B., & Ribeiro-Soriano, D. (2024). RETRACTED ARTICLE: Defining content marketing and its influence on online user behavior: a data-driven prescriptive analytics method. *Annals of Operations Research, 337*(S1, Suppl 1), 17–17. DOI: 10.1007/s10479-023-05261-1

Barton, M., Budjac, R., Tanuska, P., Gaspar, G., & Schreiber, P. (2022). Identification overview of Industry 4.0 essential attributes and resource-limited embedded artificial-intelligence-of-things devices for small and medium-sized enterprises. *Applied Sciences (Basel, Switzerland), 12*(11), 5672. DOI: 10.3390/app12115672

Bass, B. M. (1999). *Transformational Leadership: Industry, Military, and Educational Impact*. Erlbaum.

Beck, R., Müller-Bloch, C., & King, J. L. (2018). Governance in the Blockchain economy: A framework and research agenda. *Journal of the Association for Information Systems, 19*(10), 1020–1034. DOI: 10.17705/1jais.00518

Bellagarda, J. S., & Abu-Mahfouz, A. M. (2022). An updated survey on the convergence of distributed ledger technology and artificial intelligence: Current state, major challenges and future direction. *IEEE Access : Practical Innovations, Open Solutions, 10*, 50774–50793. DOI: 10.1109/ACCESS.2022.3173297

Bellavista, P., Esposito, C., Foschini, L., Giannelli, C., Mazzocca, N., & Montanari, R. (2021). Interoperable blockchains for highly-integrated supply chains in collaborative manufacturing. *Sensors (Basel), 21*(15), 4955. DOI: 10.3390/s21154955 PMID: 34372191

Benson, C., & Magee, C. L. 2018. Data-driven investment decision-making: Applying Moore's law and S-curves to business strategies. SSRN *Electron. J.* DOI: 10.2139/ssrn.3179370

Benzidia, S., Makaoui, N., & Subramanian, N. (2021). Impact of ambidexterity of blockchain technology and social factors on new product development: A supply chain and Industry 4.0 perspective. *Technological Forecasting and Social Change, 169*, 120819. DOI: 10.1016/j.techfore.2021.120819

Bersin, J. (2022). "The Future of Work: Understanding Gen Z in the Workplace." *Harvard Business Review*. https://www.hbr.org

Bharat, K. (2003). *Google News: How and Why*. Google Blog. https://googleblog.blogspot.com

Bock, L. (2015). *Work Rules!* Insights from Inside Google That Will Transform How You Live and Lead. Twelve.

Bodkhe, U., Tanwar, S., Parekh, K., Khanpara, P., Tyagi, S., Kumar, N., & Alazab, M. (2020). Blockchain for industry 4.0: A comprehensive review. *IEEE Access: Practical Innovations, Open Solutions, 8*, 79764–79800. DOI: 10.1109/ACCESS.2020.2988579

Boone, T., Ganeshan, R., Jain, A., & Sanders, N. R. (2019). Forecasting sales in the supply chain: Consumer analytics in the big data era. *International Journal of Forecasting, 35*(1), 170–180. DOI: 10.1016/j.ijforecast.2018.09.003

Borowski, P. F. (2021). Digitization, digital twins, blockchain, and industry 4.0 as elements of management process in enterprises in the energy sector. *Energies, 14*(7), 1885. DOI: 10.3390/en14071885

Boston Consulting Group. (2021). The digital acceleration index: How to assess digital maturity and prioritize next steps. https://www.bcg.com

Boyd, D. M., & Ellison, N. B. (2007). Social network sites: Definition, history, and scholarship. *Journal of Computer-Mediated Communication, 13*(1), 210–230. DOI: 10.1111/j.1083-6101.2007.00393.x

Brian Arthur. (2009). *The Nature of Technology: What It Is and How It Evolves*. Free Press.

Brilliantova, V., & Thurner, T. W. (2019). Blockchain and the future of energy. *Technology in Society, 57*, 38–45. DOI: 10.1016/j.techsoc.2018.11.001

Brin, S., & Page, L. (1998). The anatomy of a large-scale hypertextual web search engine. *Computer Networks and ISDN Systems, 30*(1–7), 107–117. DOI: 10.1016/S0169-7552(98)00110-X

Brown, T., & Bessant, J. (2013). Case study findings on digital transformation in European banks illustrate the role of open communication in reducing resistance.

Brown, J. S., & Duguid, P. (2021). *The Social Life of Information* (2nd ed.). Harvard Business Review Press.

Brown, P., Tan, W., & Zhang, K. (2023). Digital Banking Transformation: Employee Adaptation and Resistance. *International Journal of Financial Innovation, 8*(2), 133–150.

Brown, T., & Jones, L. (2022). Navigating Change in Manufacturing: Overcoming Resistance to IoT. *Journal of Organizational Change Management, 35*(4), 245–260.

Brown, T., & Katz, B. (2015). *Change by Design: How Design Thinking Transforms Organizations and Inspires Innovation*. Harper Business.

Brynjolfsson, E., & McAfee, A. (2014). *The Second Machine Age: Work, Progress, and Prosperity in a Time of Brilliant Technologies*. W.W. Norton & Company.

Brynjolfsson, E., & McAfee, A. (2017). *Machine, platform, crowd: Harnessing our digital future*. W. W. Norton & Company.

Brynjolfsson, E., & McAfee, A. (2017). *The second machine age: Work, progress, and prosperity in a time of brilliant technologies*. W.W. Norton & Company.

Buchheit, P. (2009). *The Origin of Gmail*. [Blog post on Gmail's development] https://paulbuchheit.blogspot.com

Bughin, J., Hazan, E., Lund, S., Chui, M., Dahlström, P., & Wiesinger, A. (2018). Artificial intelligence: The next digital frontier. McKinsey Global Institute. https://www.mckinsey.com/business-functions/mckinsey-digital/our-insights/artificial-intelligence-the-next-digital-frontier

Burnes, B. (2004). Kurt Lewin and the planned approach to change: A re-appraisal. *Journal of Management Studies*, *41*(6), 977–1002. DOI: 10.1111/j.1467-6486.2004.00463.x

California Consumer Privacy Act (CCPA), Cal. Civ. Code §1798.100 et seq.

Cameron, K. S., & Quinn, R. E. (2022). *Diagnosing and Changing Organizational Culture: Based on the Competing Values Framework* (4th ed.). Wiley.

Cappelli, P. (2021). Reshaping Talent Acquisition Strategies for a Digital Generation. *HRMagazine*, *66*(4), 32–39.

Carey, B. (2022). Metaverse technologies, behavioral predictive analytics, and customer location tracking tools in blockchain-based virtual worlds. Rev. Contemp. Philos. 188–204.

Cerf, V. G. (1995). The internet: Past, present, and future. *Communications of the ACM*, *38*(11), 33–35. DOI: 10.1145/219717.219744

Chae, B., & Olson, D. (2022). Technologies and applications of Industry 4.0: Insights from network analytics. *International Journal of Production Research*, *60*(12), 3682–3704. DOI: 10.1080/00207543.2021.1931524

Chaffey, D. (2020). *Global social media statistics*. Smart Insights.

Chan, K., Wong, L., & Tan, S. (2021). Pulse checks in digital transformation: A case study of DBS Bank. *Journal of Change Management*, *18*(3), 210–224. DOI: 10.1080/14697017.2021.2023

Chappell, D. (2004). *Enterprise Service Bus*. O'Reilly Media.

Chaudhary, K., Alam, M., Al-Rakhami, M. S., & Gumaei, A. (2021). Machine learning-based mathematical modelling for prediction of social media consumer behavior using big data analytics. *Journal of Big Data*, *8*(1), 73. DOI: 10.1186/s40537-021-00466-2

Chauhan, S., Singh, R., Gehlot, A., Akram, S. V., Twala, B., & Priyadarshi, N. (2023). Digitalization of supply chain management with industry 4.0 enabling technologies: A sustainable perspective. *Processes (Basel, Switzerland)*, *11*(1), 96. DOI: 10.3390/pr11010096

Cheng, B. L., Cham, T. H., Dent, M. M., & Lee, T. H. (2019). Service innovation: Building a sustainable competitive advantage in higher education. *International Journal of Services. Economics and Management*, *10*(4), 289. DOI: 10.1504/IJSEM.2019.105014

Chen, K., Golhar, D.Y., & Banerjee, S. (2023). Blockchain applications and challenges for supply chain and Industry 4.0: a literature review. *Int. J. Appl. Decis. Sci. 16* (1), 41.

Chen, K., Lin, J., & Song, Y. (2019). Trading strategy optimization for a prosumer in continuous double auction-based peer-to-peer market: A prediction-integration model. *Applied Energy*, *242*, 1121–1133. DOI: 10.1016/j.apenergy.2019.03.094

Chen, W., Xu, Z., & Shi, S. (2021). Understanding cryptocurrency fraud: A systematic review. *IEEE Transactions on Computational Social Systems*, *8*(5), 1138–1150. DOI: 10.1109/TCSS.2021.3071234

Chia, A. (2018). SkillsFuture: Enhancing Singapore's Workforce Adaptability. *Journal of Asian Public Policy*, *11*(3), 292–305.

Chiesa, V., & Frattini, F. (2011). Commercializing Technological Innovation: Learning from Failures in High-Tech Markets*: Commercializing Technological Innovation. *Journal of Product Innovation Management*, *28*(4), 437–454. DOI: 10.1111/j.1540-5885.2011.00818.x

Choi, T. M., Dolgui, A., Ivanov, D., & Pesch, E. (2022). OR and analytics for digital, resilient, and sustainable manufacturing 4.0. *Annals of Operations Research*, *310*(1), 1–6. DOI: 10.1007/s10479-022-04536-3

Chong, A. Y.-L., Chan, F. T. S., & Ooi, K.-B. (2012). Predicting consumer decisions to adopt mobile commerce: Cross country empirical examination between China and Malaysia. *Decision Support Systems*, *53*(1), 34–43. DOI: 10.1016/j.dss.2011.12.001

Christensen, *The Innovator's Dilemma*, pp. 24-27.

Christensen, C. M. (1997). *The innovator's dilemma: When new technologies cause great firms to fail.* Harvard Business Review Press.

Chung, C., Kim, J., Sovacool, B. K., Griffiths, S., Bazilian, M., & Yang, M. 2023. Decarbonizing the chemical industry: a systematic review of sociotechnical systems, technological innovations, and policy options. *Energy Research and Social Science, 96*. DOI: 10.1016/j.erss.2023.102955

Cisco. (2020). *Cybersecurity series: Data privacy and security trends in 2020.* https://www.cisco.com

Clayton, M. (1997). *Christensen, The Innovator's Dilemma: When New Technologies Cause Great Firms to Fail.* Harvard Business Review Press.

Clohessy, T., & Acton, T. (2019). Investigating the influence of organizational factors on blockchain adoption: An innovation theory perspective. *Industrial Management & Data Systems, 119*(7), 1457–1491.

CoinMarketCap. (2025). Decentralized finance (DeFi) market overview: Q1 2025. https://coinmarketcap.com/defi-report-2025

Colli, M., Cavalieri, S., Cimini, C., Madsen, O., & Vejrum Wæhrens, B. (2020). Digital transformation strategies for achieving operational excellence: a cross-country evaluation.

Crosby, M., Pattanayak, P., Verma, S., & Kalyanaraman, V. (2016). Blockchain technology: Beyond Bitcoin. *Applied Innovation Review*, *2*, 6–19.

Czvetkó, T., Kummer, A., Ruppert, T., & Abonyi, J. (2022). Data-driven business process management-based development of Industry 4.0 solutions. *CIRP Journal of Manufacturing Science and Technology*, *36*, 117–132. DOI: 10.1016/j.cirpj.2021.12.002

Dahl, D. W., & Hoeffler, S. (2004). Visualizing the Self: Exploring the Potential Benefits and Drawbacks for New Product Evaluation. *Journal of Product Innovation Management*, *21*(4), 259–267. DOI: 10.1111/j.0737-6782.2004.00077.x

Dart Project Management. (2024) *AI-Powered Stakeholder Analysis: Improving Project Outcomes.* https://www.itsdart.com/blog/ai-powered-stakeholder-analysis.

Darwish, D. (2023). Blockchain and Artificial Intelligence for Business Transformation Toward Sustainability. In *Blockchain and its Applications in Industry 4.0* (pp. 211–255). Springer. DOI: 10.1007/978-981-19-8730-4_8

Das, L., Bibhu, V., Logeswaran, R., Dadhich, K., & Sharma, B. (2023). *AI Model for Blockchain Based Industrial IoT and Big Data*. Concepts, Methodologies, Tools, and Applications. AI Models for Blockchain-Based Intelligent Networks in IoT Systems Springer. DOI: 10.1007/978-3-031-31952-5_3

Dastres, R., & Soori, M. (2021). Artificial Neural Network System. *Int. J. Imaging Robot*, *21*, 13–25.

Dastres, R., & Soori, M. (2021a). Advances in web-based decision support systems. *Int. J. Eng. Future Technol.*, *19*, 1–15.

Dastres, R., & Soori, M. (2021b). The role of information and communication technology (ICT) in environmental protection. *Int. J. Tomogr. Simul.*, *35*, 24–37.

Davenport, T. (1993). *Process Innovation. Reengineering through Information Technology*. Harvard Business School Press.

Dawson, P. (2021). Organisational change as a nonlinear dynamic. In *Global Encyclopedia of Public Administration, Public Policy, and Governance*. Springer., https://link.springer.com/referenceworkentry/10.1007/978-3-319-31816-5_73-1,

de Almeida, P. G. R., dos Santos, C. D., & Farias, J. S. (2021). Artificial intelligence regulation: A framework for governance. *Ethics and Information Technology*, *23*(3), 505–525. DOI: 10.1007/s10676-021-09593-z

de Waal, A., & Linthorst, J. (2020). Future-proofing the high-performance organization. *Sustainability (Basel)*, *12*(20), 8507. DOI: 10.3390/su12208507

Deloitte. (2020). *Automation with intelligence: Reimagining the organization in the 'Age of With'*. https://www2.deloitte.com

Deloitte. (2020). *The ethics of data: A survey of consumer attitudes towards data privacy and protection.*https://www2.deloitte.com

Deloitte. (2022), *Sustainability and Green Technologies,* https://www.deloitte.com/global/en/issues/climate/content/2022-deloitte-cxo-sustainability-report.html

Deloitte. (2022). Digital maturity and the agile transformation journey. https://www2.deloitte.com

Deloitte. (2022). Digital maturity assessment framework. https://www2.deloitte.com

Deloitte. (2022). Digital transformation through a learning culture: The importance of upskilling and innovation. https://www2.deloitte.com

Deloitte. (2022). Diversity and innovation in the digital age. https://www2.deloitte.com

Deloitte. (2022). Leadership for the digital era: Fostering innovation and collaboration. Deloitte Insights. https://www2.deloitte.com

Deloitte. (2022). Leadership strategies for digital transformation. https://www2.deloitte.com

Deloitte. (2022). Tesla's innovation ecosystem: Driving sustainability and leadership in the automotive industry. https://www2.deloitte.com

Deloitte. (2023) *2023 Global Human Capital Trends: Leading in a Boundaryless World.* Available at: https://www2.deloitte.com

Deloitte. (2023). "Global Gen Z and Millennial Survey 2023." https://www.deloitte.com

Deloitte. (2023). Blockchain's role in ethical supply chains. https://www2.deloitte.com

Demertzis, K., Iliadis, L., Tziritas, N., & Kikiras, P. (2020). Anomaly detection via blockchained deep learning smart contracts in industry 4.0. *Neural Computing & Applications*, *32*(23), 17361–17378. DOI: 10.1007/s00521-020-05189-8

Deutsche Bank. (2021). *Innovation and technology strategy: Key developments and progress.* https://annualreport.deutsche-bank.com/files/documents/other-presentations-and-events/4_IDD_2020_CTO_final.pdf.

Diamantoulakis, P. D., Kapinas, V. M., & Karagiannidis, G. K. (2015). Big Data Analytics for Dynamic Energy Management in Smart Grids. *Big Data Research*, *2*(3), 94–101. DOI: 10.1016/j.bdr.2015.03.003

Digiconomist. (2025). Bitcoin energy consumption index 2025. https://digiconomist.net/bitcoin-energy-consumption

Divissenko, N. (2023). Regulation of Crypto-assets in the EU: Future-proofing the Regulation of Innovation in Digital Finance. *European Papers-A Journal on Law and Integration*, *2023*(2), 665–687.

E&Y. (2023). "Gen Z at Work: Expectations and Engagement in a Digital World." https://www.ey.com

Edmondson, A. C. (1999). Psychological Safety and Learning Behavior in Work Teams. *Administrative Science Quarterly*, *44*(2), 350–383. DOI: 10.2307/2666999

El Akrami, N., Hanine, M., Flores, E. S., Aray, D. G., & Ashraf, I. (2023). Unleashing the potential of blockchain and machine learning: Insights and emerging trends from bibliometric analysis. *IEEE Access: Practical Innovations, Open Solutions*, *11*, 78879–78903. DOI: 10.1109/ACCESS.2023.3298371

El Khatib, M. M., Alzoubi, H. M., Ahmed, G., Kazim, H. H., Al Falasi, S. A. A., Mohammed, F., & Al Mulla, M. (2022, February). Digital transformation and SMART-the analytics factor. In *2022 International Conference on Business Analytics for Technology and Security (ICBATS)* (pp. 1-11). IEEE.

Ellis, R. K. (2009). *A Field Guide to Learning Management Systems*. ASTD Learning Circuits.

ElMamy, S. B., Mrabet, H., Gharbi, H., Jemai, A., & Trentesaux, D. (2020). A survey on the usage of blockchain technology for cyber-threats in the context of industry 4.0. *Sustainability (Basel)*, *12*(21), 9179. DOI: 10.3390/su12219179

Ertz, M., & Boily, É. (2019). The rise of the digital economy: Thoughts on blockchain technology and cryptocurrencies for the collaborative economy. *International Journal of Innovation Studies*, *3*(4), 84–93. DOI: 10.1016/j.ijis.2019.12.002

Esmaeilian, B., Sarkis, J., Lewis, K., & Behdad, S. (2020). Blockchain for the future of sustainable supply chain management in Industry 4.0. Resour. *Resources, Conservation and Recycling*, *163*, 105064. DOI: 10.1016/j.resconrec.2020.105064

Esmat, A., de Vos, M., Ghiassi-Farrokhfal, Y., Palensky, P., & Epema, D. (2021). A novel decentralized platform for peer-to-peer energy trading market with blockchain technology. *Applied Energy*, *282*, 116123. DOI: 10.1016/j.apenergy.2020.116123

Ethereum. (2025). Ethereum proof-of-stake update 2025. https://ethereum.org/en/upgrades

European Commission. (2020). Digital Skills and Jobs Coalition: What is Digital Literacy? https://ec.europa.eu/digital-strategy/our-policies/digital-skills

European Parliament and Council of the European Union. (2016). *General Data Protection Regulation (GDPR), Regulation (EU) 2016/679 on the protection of natural persons with regard to the processing of personal data and on the free movement of such data.*

Everledger. (2025). Everledger diamond tracking report 2025. https://www.everledger.io/report-2025

Facebook. (2012). Facebook annual report 2012. https://www.facebook.com/annual-reports

Faridi, M. S., Ali, S., Duan, G., & Wang, G. (2020). Blockchain and IoT based textile manufacturing traceability system in industry 4.0. in: International conference on security, privacy and anonymity in computation, communication and storage. Springer, pp. 331–344.

Fearne, A., & Hughes, D. (1999). Success factors in the fresh produce supply chain: Insights from the UK. *Supply Chain Management*, *4*(3), 120–131. DOI: 10.1108/13598549910279567

Fernandez-Carames, T. M., & Fraga-Lamas, P. (2019). A review on the application of blockchain to the next generation of cybersecure industry 4.0 smart factories. *IEEE Access: Practical Innovations, Open Solutions*, *7*, 45201–45218. DOI: 10.1109/ACCESS.2019.2908780

Fischer, C. S. (1992). *America calling: A social history of the telephone to 1940*. University of California Press. DOI: 10.1525/9780520915008

Fitbit. (2025). Fitbit user statistics 2025 https://www.fitbit.com/stats-2025

Flasiński, M. (2006). *IT project mangement*. Wydawnictwo Naukowe PWN.

Floridi, L., Cowls, J., Beltrametti, M., Chatila, R., Chazerand, P., Dignum, V., & Vayre, P. (2018). AI4People—An ethical framework for a good AI society: Opportunities, risks, principles, and recommendations. *Nature Machine Intelligence*, *1*(1), 1–7. DOI: 10.1038/s42256-018-0002-2 PMID: 30930541

Floyd, S. W., & Wooldridge, B. (2021). *Building Strategy from the Middle: Reconceptualizing Middle Management's Role in Strategy Execution*. Routledge.

Fokkema, T., Kooiman, T. J. M., Krijnen, W. P., Van Der Schans, C. P., & De Groot, M. (2020). Reliability and validity of consumer-grade fitness trackers: A systematic review. *Sensors (Basel)*, *20*(15), 4212. DOI: 10.3390/s20154212 PMID: 32751189

Forbes Insights. (2023) *Transformational leadership and organizational success*. https://www.forbes.com

Forrester. (2023). Consumer trust in AI recommendations: 2023 survey. https://www.forrester.com/ai-trust-2023

Fuller, A., Fan, Z., Day, C., & Barlow, C. (2020). Digital twin: Enabling technologies, challenges, and open research. *IEEE Access : Practical Innovations, Open Solutions*, *8*, 108952–108971. DOI: 10.1109/ACCESS.2020.2998358

Fülöp, M. T., Topor, D. I., Ionescu, C. A., Căpu neanu, S., Breaz, T. O., & Stanescu, S. G. (2022). Fintech accounting and Industry 4.0: Future-proofing or threats to the accounting profession? *Journal of Business Economics and Management*, *23*(5), 997–1015. DOI: 10.3846/jbem.2022.17695

Gad, A. G., Mosa, D. T., Abualigah, L., & Abohany, A. A. (2022). Emerging trends in blockchain technology and applications: A review and outlook. *Journal of King Saud University. Computer and Information Sciences, 34*(9), 6719–6742. DOI: 10.1016/j.jksuci.2022.03.007

Gaikwad, L., & Khang, A. (2023). AI Powered Workforce Management in Industry 4.0 Era. In *Designing Workforce Management Systems for Industry 4.0* (pp. 35–46). CRC Press. DOI: 10.1201/9781003357070-3

Gallup. (2022) *State of the Global Workplace Report.* https://www.gallup.com

Gami, B., Agrawal, M., Mishra, D. K., Quasim, D., & Mehra, P. S. (2023). Artificial intelligence-based blockchain solutions for intelligent healthcare: A comprehensive review on privacy preserving techniques. *Transactions on Emerging Telecommunications Technologies, 34*(9), e4824. DOI: 10.1002/ett.4824

Gammon, K. (2018). Experimenting with blockchain: Can one technology boost both data integrity and patients' pocketbooks? *Nature Medicine, 24*(4), 378–382. PMID: 29634691

Garces, E., Li, S., & Daim, T. U. (2023). Cybersecurity and Technology Convergence: Analysis of AI, Blockchain, and IoT Using SNA. In: A Technology Landscape Analysis. Cybersecurity Springer, pp. 39–70.

Garcia Martinez, M., & Walton, B. (2014). The wisdom of crowds: The potential of online communities as a tool for data analysis. *Technovation, 34*(4), 203–214. DOI: 10.1016/j.technovation.2014.01.011

Garg, P. (2018). Impact of Work-Life Balance Practices on Employees Retention and Organizational Performance-A Study on IT Industry. *Indian Journal of Applied Research, 6*(8), 105–108.

Gartner. (2011). Mobile app ecosystem trends 2011. https://www.gartner.com/app-trends-2011

Gartner. (2021). *AI in compliance: How AI is transforming compliance management.* https://www.gartner.com

Gartner. (2021). *The 2021 Gartner Magic Quadrant for IT Service Management Tools.* https://www.gartner.com/en/documents/4005269

Gartner. (2022) *The critical role of managers in organizational transformation.* https://www.gartner.com

Gartner. (2022). Leadership challenges in digital transformation: Addressing skills gaps and resistance. https://www.gartner.com

Gartner. (2022). Predicting digital transformation success: Leadership and market readiness. https://www.gartner.com

Gartner. (2022). The future of healthcare: How Mayo Clinic is leading with digital transformation. https://www.gartner.com

Gartner. (2022). The strategic importance of digital literacy in leadership. https://www.gartner.com

Gartner. (2022). The strategic role of leadership in digital transformation. https://www.gartner.com

Gartner. (2022). Transformational leadership and digital transformation. https://www.gartner.com

Gartner. (2023). Overcoming silos in the hybrid workplace: Best practices for seamless collaboration. *Gartner Research Insights*.https://www.gartner.com/en/insights/overcoming-silos.

Gartner. (2023). The role of data-driven decision-making in digital transformation: Insights from Netflix and beyond. https://www.gartner.com

Gartner. (2023). The role of scenario planning in navigating digital disruptions. Gartner Research. https://www.gartner.com

Gartner. (2023). Top IT and business priorities for 2023–2024. https://www.gartner.com

Gartner. (2024). Top strategic technology trends for 2025. https://www.gartner.com/technology-trends-2025

Garvin, D. A., Edmondson, A. C., & Gino, F. (2022). Building a Learning Organisation. *Harvard Business Review*, *100*(2), 45–60.

Gauba, H., Kumar, P., Roy, P. P., Singh, P., Dogra, D. P., & Raman, B. (2017). Prediction of advertisement preference by fusing EEG response and sentiment analysis. *Neural Networks*, *92*, 77–88. DOI: 10.1016/j.neunet.2017.01.013 PMID: 28254237

General Electric. (2022). Digital leadership development program: Building future-ready leaders. McKinsey & Company. https://www.mckinsey.com

Geoffrey, G. (2016). *Parker, Marshall W. Van Alstyne, and Sangeet Paul Choudary, Platform Revolution: How Networked Markets Are Transforming the Economy—and How to Make Them Work for You*. W.W. Norton & Company.

Ghasemzadeh, B., Celik, T., & Karimi Ghaleh Jough, F.. (2022). Road map to BIM use for infrastructure domains: Identifying and contextualizing variables of infrastructure projects. *Scientia Iranica*, *29*(6), 2803–2824. DOI: 10.24200/sci.2022.56935.4998

Ghobakhloo, M., & Fathi, M. (2021). Industry 4.0 and opportunities for energy sustainability. *J. Clean. Prod. 295*, 126427.

Ghobakhloo, M., & Fathi, M. (2021). Industry 4.0 and opportunities for energy sustainability. *Journal of Cleaner Production*, *295*, 126427. DOI: 10.1016/j.jclepro.2021.126427

Goldsmith, R. E., Moore, M. A., & Beaudoin, P. (1999). Fashion innovativeness and self-concept: A replication. *Journal of Product and Brand Management*, *8*(1), 7–18. DOI: 10.1108/10610429910257904

Google, A. I. (2021). AI principles: Ethical guidelines for artificial intelligence development. https://ai.google/principles

Google. (2004). Google annual report 2004. https://www.google.com/annual-reports

Google. (2025). Google ad revenue report 2025. https://www.google.com/ad-revenue-2025

GovTech Singapore. (2023). *Tech Kaki: Enhancing citizen engagement in digital services*. https://www.tech.gov.sg/.

Grand View Research. (2025). AI in healthcare market size, share & trends analysis report. https://www.grandviewresearch.com/ai-healthcare-2025

Grida, , MMostafa, , N.A. (2023). Are smart contracts too smart for Supply Chain 4.0? A blockchain framework to mitigate challenges. *Journal of Manufacturing Technology Management*, *34*, 644–665.

Grigo, J., Hansen, P., & Pantelimon, S. (2020). Smart contracts: Legal implications and future perspectives. *Computer Law & Security Review*, *38*, 105431. DOI: 10.1016/j.clsr.2020.105431

Grudzewski, W., & Hejduk, I. (2004). *Methods of designing information systems*. Difin.

Gubbi, J., Buyya, R., Marusic, S., & Palaniswami, M. (2013). Internet of Things (IoT): A vision, architectural elements, and future directions. *Future Generation Computer Systems*, *29*(7), 1645–1660. DOI: 10.1016/j.future.2013.01.010

Guha, S., & Kumar, S. (2018). Emergence of Big Data Research in Operations Management, Information Systems, and Healthcare: Past Contributions and Future Roadmap. *Production and Operations Management*, *27*(9), 1724–1735. DOI: 10.1111/poms.12833

Guiné, R. P. F., Florença, S. G., Barroca, M. J., & Anjos, O. (2020). The Link between the Consumer and the Innovations in Food Product Development. *Foods*, *9*(9), 1317. DOI: 10.3390/foods9091317 PMID: 32962007

Gui, Y., Cai, Z., Yang, Y., Kong, L., Fan, X., & Tai, R. H. (2023). Effectiveness of digital educational game and game design in STEM learning: A meta-analytic review. *International Journal of STEM Education*, *10*(1), 36. DOI: 10.1186/s40594-023-00424-9

Gupta, R., & Singh, A. (2019). The role of digital technologies in Indian business transformation. *Indian Journal of Business Innovation*, *13*(2), 34–49.

Gupta, R., Tanwar, S., Kumar, N., & Tyagi, S. (2020). Blockchain-based security attack resilience schemes for autonomous vehicles in industry 4.0: A systematic review. *Computers & Electrical Engineering*, *86*, 106717. DOI: 10.1016/j.compeleceng.2020.106717

H&M. (2024). H&M sustainability and technology report 2024. https://www.hm.com/sustainability-2024

Habbal, A., Ali, M. K., & Abuzaraida, M. A. (2024). Artificial Intelligence Trust, Risk and Security Management (AI TRiSM): Frameworks, applications, challenges and future research directions. *Expert Systems with Applications*, *240*, 122442. DOI: 10.1016/j.eswa.2023.122442

Haddud, A., DeSouza, A., Khare, A., & Lee, H. (2017). Examining potential benefits and challenges associated with the Internet of Things integration in supply chains. *Journal of Manufacturing Technology Management*, *28*(8), 1055–1085. DOI: 10.1108/JMTM-05-2017-0094

Haleem, A., Javed, M., Singh, R. P., & Suman, R. (2021). Telemedicine for healthcare: Capabilities, features, barriers, and applications. *Sensors International*, *2*(1), 100117–100238. DOI: 10.1016/j.sintl.2021.100117

Hamel, G. (2007). *The Future of Management*. Harvard Business School Press.

Hammer, M., & Champy, J. (1993). *Reengineering the Corporation. A Manifesto for Business Revolution*. Harper Business. DOI: 10.1016/S0007-6813(05)80064-3

Hanggoro, D., Windiatmaja, J. H., & Sari, R. F. (2022, July). Blockchain-based Attendance Management and Payroll System using Hyperledger Composer Framework. In *2022 IEEE Region 10 Symposium (TENSYMP)* (pp. 1-6). IEEE.

Hardy, A. (2022). *Securing e-Estonia: challenges, insecurities, opportunities. Royal Holloway*. University of London.

Harvard Business Review. (2021). Agile and iterative leadership in digital transformation. https://hbr.org

Harvard Business Review. (2021). Amazon's supply chain mastery: Automation and customer focus. https://hbr.org

Harvard Business Review. (2021). Building a growth mindset for digital leadership. https://hbr.org

Harvard Business Review. (2021). Emotional intelligence and collaborative cultures. https://hbr.org

Harvard Business Review. (2021). How Satya Nadella transformed Microsoft into a tech powerhouse. https://hbr.org

Harvard Business Review. (2021). How Satya Nadella transformed Microsoft: Leadership lessons for the digital age. https://hbr.org

Harvard Business Review. (2021). Procter & Gamble's digital transformation journey: Insights from employee engagement strategies. https://hbr.org

Harvard Business Review. (2022) *Supportive managers and employee adaptability in transformation*. https://www.hbr.org

Harvard Business Review. (2022). Emotional intelligence and leadership in digital transformation. https://hbr.org

Harvard Business Review. (2022). Emotional intelligence and leadership in transformation. https://hbr.org

Harvard Business Review. (2022). Microsoft's digital transformation under Satya Nadella. https://hbr.org

Harvard Business Review. (2022). Microsoft's leadership under Satya Nadella: Embracing empathy and cultural reinvention. https://hbr.org

Harvard Business Review. (2022). Sustainability in the digital age: A roadmap for responsible innovation. https://hbr.org

Hayes, S. C., Luoma, J. B., Bond, F. W., Masuda, A., & Lillis, J. (2020). Acceptance and Commitment Therapy: Model, Processes and Outcomes. *Behaviour Research and Therapy*, *60*, 1–25. DOI: 10.1016/j.brat.2019.03.002 PMID: 16300724

Hess, E. D. (2014). *Learn or Die: Using Science to Build a Leading-Edge Learning Organization*. Columbia Business School Publishing.

Hinsch, C., Felix, R., & Rauschnabel, P. A. (2020). Nostalgia beats the wow-effect: Inspiration, awe and meaningful associations in augmented reality marketing. *Journal of Retailing and Consumer Services*, *53*, 101987. DOI: 10.1016/j.jretconser.2019.101987

Hofstede, G. (2021). *Culture's Consequences: Comparing Values, Behaviours, Institutions, and Organizations Across Nations*. Sage Publications.

Huy, Q. N. (2022). Emotional balancing of organizational continuity and radical change: The role of middle managers. *Administrative Science Quarterly*, *67*(1), 132–169. DOI: 10.1177/00018392211024671

Ibid., pp. 89-92.

IBM. (2021). *Cost of a data breach report 2021.* IBM Security. https://www.ibm.com/security/data-breach

IBM. (2021). *SkillsBuild programme: Empowering workforce transformation through learning and development*. https://www.ibm.com/skillsbuild.

IBM. (2024). IBM Watson Health annual report 2024. https://www.ibm.com/watson-health-2024

IBM. (2025). IBM Food Trust impact report 2025. https://www.ibm.com/food-trust-2025

IDC. (2025). Worldwide smart manufacturing spending guide. https://www.idc.com/smart-manufacturing-2025

Inform, I. T. (2024). *Stakeholder Analysis with AI*. https://www.informit.com/articles/article.aspx?p=3192418&seqNum=3.

Infosys (2023) *Building trust through co-creation: ESG sustainability report*. https://www.infosys.com/sustainability/documents/esg-2022-23/soc-story4.pdf.

International Energy Agency (IEA). (2022). Data centers and energy consumption: Trends and strategies for sustainability. https://www.iea.org

International Energy Agency. (2022). IoT in optimizing energy grids. https://www.iea.org

Internet World Stats. (2020). Internet growth statistics: 1995–2000. https://www.internetworldstats.com/stats.htm

IoT Analytics. (2025). State of IoT 2025: Global device forecast. https://iot-analytics.com/state-of-iot-2025

IRENA. (2019). *Renewable energy statistics 2019*. International Renewable Energy Agency.

Isaacson, W. (2011). *Steve Jobs*. Simon & Schuster.

Jabbour, C. J. C., Jabbour, A. B. L. S., Sarkis, J., & Godinho Filho, M. (2023). Unlocking the potential of Industry 4.0 technologies to address resistance to change in sustainable supply chains. *Journal of Cleaner Production*, *368*, 133017. DOI: 10.1016/j.jclepro.2022.133017

James, F. (1996). *Moore, The Death of Competition: Leadership and Strategy in the Age of Business Ecosystems*. HarperBusiness.

Javaid, M., Haleem, A., Singh, R. P., Suman, R., & Rab, S. (2021). Blockchain technology applications for Industry 4.0: A literature-based review. Blockchain. *Research and Applications*, *2*(4), 100027. DOI: 10.1016/j.bcra.2021.100027

Jena, A. K., & Dash, S. P. (2021). Blockchain technology: introduction, applications, challenges. In *Blockchain Technology* (pp. 1–11). Applications and Challenges Springer.

Jeong, W., & Choi, M. (2019). Design of recruitment management platform using digital certificate on blockchain. *Journal of Information Processing Systems*, *15*(3), 707–716. DOI: 10.3745/jips.03.0121

Jha, M. K., Yogeshwari, A., Rubini, P., & Singh, M. (2023). Converge of IoT and AI in metaverse: challenges and opportunities. In: 5th International Conference on Inventive Research in Computing Applications (ICIRCA) 2023. IEEE, pp. 1462–1467 DOI: 10.1109/ICIRCA57980.2023.10220628

Jones, A., & Smith, R. (2022). A systematic review of organizational culture literature: Lessons from the past and avenues for future research. *Journal of Organizational Studies*, *45*(1), 56–80. DOI: 10.1080/JoS.2022.467853

Jones, M., & Roberts, K. (2023). Future trends in digital business: Global perspectives and regional implications. *Journal of Business Forecasting*, *34*(6), 213–227.

Joseph, A. (1942). *Schumpeter, Capitalism, Socialism, and Democracy*. Harper & Row.

Joshi, S., Pise, A. A., Shrivastava, M., Revathy, C., Kumar, H., Alsetoohy, O., & Akwafo, R. (2022). Adoption of blockchain technology for privacy and security in the context of industry 4.0. *Wireless Communications and Mobile Computing*. DOI: 10.1155/2022/4079781

Kaluarachchi, B. N., & Sedera, D. (2024). Improving Efficiency Through AI-Powered Customer Engagement by Providing Personalized Solutions in the Banking Industry. In Integrating AI-Driven Technologies Into Service Marketing (pp. 299-342). IGI Global Scientific Publishing.

Kang, J., & Park-Poaps, H. (2010). Hedonic and utilitarian shopping motivations of fashion leadership. *Journal of Fashion Marketing and Management*, *14*(2), 312–328. DOI: 10.1108/13612021011046138

Karimi Ghaleh Jough, F., & Ghasemzadeh, B. (2023). Uncertainty Interval Analysis of Steel Moment Frame by Development of 3D-Fragility Curves Towards Optimized Fuzzy Method. *Arabian Journal for Science and Engineering*, 1–18.

Kashem, M. A., Shamsuddoha, M., Nasir, T., & Chowdhury, A. A. (2023). Supply chain disruption versus optimization: A review on artificial intelligence and blockchain. *Knowledge (Beverly Hills, Calif.)*, *3*, 80–96.

Kasprzak, T. (2005). *Reference models in business process management*. Difin.

Kelly, K. (2016). *The inevitable: Understanding the 12 technological forces that will shape our future*. Viking Press.

Khan, A. A., Laghari, A. A., Shaikh, A. A., Bourouis, S., Mamlouk, A. M., & Alshazly, H. (2021). Educational Blockchain: A secure Degree Attestation and Verification Traceability Architecture for Higher Education Commission. *Applied Sciences (Basel, Switzerland)*, *11*(22), 10917. DOI: 10.3390/app112210917

Khatib, S. F., Mustafa, Z., & Abbas, A. F. (2025). Digital Transformation and Financial Sustainability. In Algorithmic Training, Future Markets, and Big Data for Finance Digitalization (pp. 33-74). IGI Global Scientific Publishing. DOI: 10.4018/979-8-3693-6386-7.ch002

Kim, H., Fiore, A. M., Niehm, L. S., & Jeong, M. (2010). Psychographic characteristics affecting behavioral intentions towards pop-up retail. *International Journal of Retail & Distribution Management*, *38*(2), 133–154. DOI: 10.1108/09590551011020138

Kim, W. B., Kim, C., & Kurata, K. (2025). Relationships among consumer innovativeness, learning, and global product purchases: Store manager perspectives in retailing. *Journal of Retailing and Consumer Services*, *82*, 104084. DOI: 10.1016/j.jretconser.2024.104084

Kişi, N. (2022). Exploratory research on the use of blockchain technology in recruitment. *Sustainability (Basel), 14*(16), 10098. DOI: 10.3390/su141610098

Klein, S. (2022). Change Management: From Theory to Practice. *TechTrends, 66*(3), 319–328. https://link.springer.com/article/10.1007/s11528-022-00775-0

Kliestik, T., Novak, A., & Lăzăroiu, G. (2022). Live shopping in the metaverse: Visual and spatial analytics, cognitive artificial intelligence techniques and algorithms, and immersive digital simulations. *Linguistic and Philosophical Investigations, 21*(0), 187–202. DOI: 10.22381/lpi21202212

Knowles, M. (1980). The Modern Practice of Adult Education: From Pedagogy to Andragogy. Cambridge.

Korfiatis, N., García-Bariocanal, E., & Sánchez-Alonso, S. (2012). Evaluating content quality and helpfulness of online product reviews: The interplay of review helpfulness vs. review content. *Electronic Commerce Research and Applications, 11*(3), 205–217. DOI: 10.1016/j.elerap.2011.10.003

Kotler, P., & Keller, K. L. (2016). *Marketing management* (15th ed.). Pearson Education.

Kotter, J. P. (2012). *Leading Change*. Harvard Business Review Press.

Kotter, J. P. (2021). Leading Change: Why Transformation Efforts Fail. *Harvard Business Review, 99*(2), 59–67.

Kour, M. (2024a). "Challenges and Opportunities of Machine Learning in the Financial Sector", Algorithmic Approaches to Financial Technology: Forecasting, Trading, and Optimization (pp. 61-73). IGI Global Scientific Publishing. DOI: DOI: 10.4018/979-8-3693-1746-4.ch004

Kour, M. (2025). Navigating the Fintech Wave and Its Background and Challenges. In Algorithmic Training, Future Markets, and Big Data for Finance Digitalization (pp. 121-132). IGI Global Scientific Publishing DOI: 10.4018/979-8-3693-6386-7.ch005

Kour, M. (2024b). Revolutionizing Finance: Unleashing Machine Learning's Potential for a New Financial Era. In *2024 International Conference on Advances in Computing Research on Science Engineering and Technology (ACROSET)* (pp. 1-5). IEEE. DOI: 10.1109/ACROSET62108.2024.10743732

Kral, P., Janoskova, K., & Dawson, A. (2022). Virtual skill acquisition, remote working tools, and employee engagement and retention on blockchain-based metaverse platforms. *Psychosociological Issues in Human Resource Management, 10*(1), 92–105. DOI: 10.22381/pihrm10120227

Krugmann, M., & Hartmann, P. (2024). The role of large language models in sentiment analysis for organisational change. *Journal of Organisational Transformation, 12*(1), 45–59. https://link.springer.com/article/10.1007/s40547-024-00143-4

Kshetri, N. (2017). 1 Blockchain's roles in meeting key supply chain management objectives. *International Journal of Information Management, 37*(6), 100–108.

Kumar, R., Gupta, S., & Rathore, S. (2021). Workforce transformation and resistance to automation: Insights from Indian manufacturing. *Asian Journal of Business and Management, 9*(3), 21–35. DOI: 10.1080/ajbm.2021.09.003

Kumar, R., & Patel, S. (2022). Digital Inclusion and Organisational Change: The Case of Reliance Jio. *Asian Journal of Business and Technology, 11*(3), 78–92.

Kumar, S., & Mehta, S. (2021). Digital economy and business opportunities in India: A forward-looking approach. *International Journal of Emerging Markets, 16*(4), 1276–1298. DOI: 10.1108/IJEM-01-2021-0005

Kushwaha, A. K., Kumar, P., & Kar, A. K. (2021). What impacts customer experience for B2B enterprises on using AI-enabled chatbots? Insights from Big data analytics. *Industrial Marketing Management, 98*, 207–221. DOI: 10.1016/j.indmarman.2021.08.011

Lal, B., Kumar, M. A., Chinthamu, N., & Pokhriyal, S. (2023). Development of Product Quality with Enhanced Productivity in Industry 4.0 with AI Driven Automation and Robotic Technology. In: Second International Conference on Augmented Intelligence and Sustainable Systems (ICAISS) 2023. IEEE, pp. 184–189

Lamey, L., Deleersnyder, B., Steenkamp, J.-B. E. M., & Dekimpe, M. G. (2012). The Effect of Business-Cycle Fluctuations on Private-Label Share: What has Marketing Conduct Got to do with it? *Journal of Marketing, 76*(1), 1–19. DOI: 10.1509/jm.09.0320

Latif, H. (2023). Advancing data integrity in banking: AI/ML solutions and best practice. Int. J. Comput. Sci. Technol. 7, 185–203

Lee, J., Davari, H., Singh, J., & Pandhare, V. (2018). Industrial artificial intelligence for Industry 4.0-based manufacturing systems. *Manufacturing Letters, 18*, 20–23. DOI: 10.1016/j.mfglet.2018.09.002

Lei, Y., Guo, Y., Li, X., & Jing, Z. (2023). Space economy: A new frontier of information systems, analytics and digital commerce. *Industrial Management & Data Systems, 123*(2), 616–629. DOI: 10.1108/IMDS-05-2022-0316

Levy, S. (2011). *In the Plex: How Google Thinks, Works, and Shapes Our Lives*. Simon & Schuster.

Lewin, K. (1947). Frontiers in group dynamics: Concept, method and reality in social science. *Human Relations*, *1*(1), 5–41. DOI: 10.1177/001872674700100103

Li, Q., & Zhang, H. (2022). The Rendanheyi model and its implications for organisational change: A case study of Haier. *Asia Pacific Journal of Management*, *39*(4), 1235–1250. DOI: 10.1007/apj.2022

Li, S., Xu, L. D., & Zhao, S. (2019). The Internet of Things: A survey. *Information Systems Frontiers*, *17*(2), 243–259. DOI: 10.1007/s10796-014-9492-7

Li, Y., Jin, X., & Xue, H. (2025). Avoid following individual shopping time habit to promote new products: Evidence from a randomized field experiment. *Journal of Retailing and Consumer Services*, *82*, 104058. DOI: 10.1016/j.jretconser.2024.104058

Lizcano, D., Lara, J. A., White, B., & Aljawarneh, S. (2019). Blockchain-based approach to create a model of trust in open and ubiquitous higher education. *Journal of Computing in Higher Education*, *32*(1), 109–134. DOI: 10.1007/s12528-019-09209-y

Loo, G. C., Rao, P., & Verma, S. (2022). Employee Engagement Strategies for Digital Transformation Success. *The Journal of Business Strategy*, *43*(1), 98–115.

Loscher, G., & Bader, V. (2022). Augmenting a profession: How data analytics is transforming human resource management. In *Digital Transformation and Institutional Theory* (pp. 87–110). Emerald Publishing Limited. DOI: 10.1108/S0733-558X20220000083004

Luckin, R., Holmes, W., Griffiths, M., & Forcier, L. B. (2016). *Intelligence Unleashed: An Argument for AI in Education*. Pearson Education.

Lutfi, A., Alsyouf, A., Almaiah, M. A., Alrawad, M., Abdo, A. A. K., Al-Khasawneh, A. L., Ibrahim, N., & Saad, M. (2022). Factors influencing the adoption of big data analytics in the digital transformation era: Case study of Jordanian SMEs. *Sustainability (Basel)*, *14*(3), 1802. DOI: 10.3390/su14031802

Macy's. (2024). Macy's retail technology report 2024. https://www.macys.com/tech-2024

Madakam, S., Holmukhe, R. M., & Jaiswal, D. K. (2019). The future digital work force: Robotic process automation (RPA). *Journal of Information Systems and Technology Management*, *16*, e201916001. DOI: 10.4301/S1807-1775201916001

Madan, V., & Kour, M. (2024), "Fintech and Mobile Payments in Transition: A Study of Adoption Challenges", *2024 IEEE International Conference on Interdisciplinary Approaches in Technology and Management for Social Innovation (IATMSI)*, (2), pp. 1-6. IEEE. DOI: DOI: 10.1109/IATMSI60426.2024.10502648

Madhala, P., Lin, T., Li, H., & Helander, N. (2023). Data analytics capabilities for digital service development: A case study. *PACIS 2023 Proceedings.*

Magazzino, C., Mele, M., & Schneider, N. (2021). A machine learning approach on the relationship among solar and wind energy production, coal consumption, GDP, and CO2 emissions. *Renewable Energy*, *167*, 99–115. DOI: 10.1016/j.renene.2020.11.050

Mamun, A. A., Sohel, M., Mohammad, N., Haque Sunny, M., Dipta, D. R., & Hossain, E. (2020). A Comprehensive Review of the Load Forecasting Techniques Using Single and Hybrid Predictive Models. *IEEE Access : Practical Innovations, Open Solutions*, *8*, 134911–134939. DOI: 10.1109/ACCESS.2020.3010702

Marchand, R. (1985). *Advertising the American dream: Making way for modernity, 1920–1940*. University of California Press. DOI: 10.1525/9780520342668

Marengo, A., & Pagano, A. (2023). Investigating the factors influencing the adoption of blockchain technology across different countries and industries: A systematic literature review. *Electronics (Basel)*, *12*(14), 3006. DOI: 10.3390/electronics12143006

Mariani, M. M., & Nambisan, S. (2021). Innovation analytics and digital innovation experimentation: The rise of research-driven online review platforms. *Technological Forecasting and Social Change*, *172*, 121009. DOI: 10.1016/j.techfore.2021.121009

MarketsandMarkets. (2024). IoT in manufacturing market: Global forecast to 2029. MarketsandMarkets Research Report. https://www.marketsandmarkets.com/Market-Reports/iot-manufacturing-market-1299.html

Marshall McLuhan. (1964). *Understanding Media: The Extensions of Man*. McGraw-Hill.

Martens, D., Provost, F., Clark, J., & Junqué de Fortuny, E.University of Antwerp. (2016). Mining Massive Fine-Grained Behavior Data to Improve Predictive Analytics. *Management Information Systems Quarterly*, *40*(4), 869–888. DOI: 10.25300/MISQ/2016/40.4.04

McKinsey & Company. (2020) *The Science of Organizational Change*. https://www.mckinsey.com

McKinsey & Company. (2020). Accelerating digital adoption during global crises. https://www.mckinsey.com

McKinsey & Company. (2020). *How AI and automation can improve business operations*. https://www.mckinsey.com

McKinsey & Company. (2020). *The role of data analytics in healthcare: How to improve patient outcomes.*https://www.mckinsey.com

McKinsey & Company. (2021). Agile transformation at ING Bank: Building a customer-centric financial future. https://www.mckinsey.com

McKinsey & Company. (2021). Building a coalition of change agents for successful digital transformation. https://www.mckinsey.com

McKinsey & Company. (2021). Digital leadership: Lessons from industry leaders. https://www.mckinsey.com

McKinsey & Company. (2021). Digital transformation in leadership: Strategies and outcomes. McKinsey Insights. https://www.mckinsey.com

McKinsey & Company. (2021). Leadership in digital transformation: A case study of Microsoft and Adobe. https://www.mckinsey.com

McKinsey & Company. (2021). Leadership strategies for digital transformation. Retrieved from https://www.mckinsey.com

McKinsey & Company. (2021). Zoom's scalability during the pandemic: Agility in action. https://www.mckinsey.com

McKinsey & Company. (2022). Achieving scale in digital transformation: Key lessons from industry leaders. https://www.mckinsey.com

McKinsey & Company. (2022). Leadership in digital transformation: Lessons from Adobe and Microsoft. https://www.mckinsey.com

McKinsey & Company. (2022). Microsoft's pivot to the cloud: Lessons in customer-centric innovation. https://www.mckinsey.com

McKinsey. (2020). The impact of a clear digital strategy on transformation success.

McKinsey. (2024). Generative AI: Economic impact 2030. https://www.mckinsey .com/gen-ai-2030

McKnight, M. A., Plouchard, S., & Bizal, N. (2015). Organizational hiring preferences: Comparison of traditional and non-traditional hiring and recruitment practices. *International Journal of Human Resource Studies*, *5*(2), 52. DOI: 10.5296/ijhrs.v5i2.7529

Meadows, D. H. (2020). *Thinking in Systems: A Primer*. Chelsea Green Publishing.

Medtronic. (2024). Smart insulin pump performance report 2024. https://www.medtronic.com/insulin-2024

Meta. (2022). *Sustainability report 2022*. Meta Platforms Inc.

Meta. (2025). Horizon Worlds user and transaction report 2025. https://www.meta.com/horizon-2025

Meyer, A. D., Gaba, V., & Colwell, K. A. (2022). Organising far from equilibrium: Nonlinear change in organisational fields. *Organization Science*, *16*(5), 456–473. https://www.insead.edu/faculty-research/publications/journal-articles/organizing-far-equilibrium-non-linear-change. DOI: 10.1287/orsc.1050.0135

Michaelson, P. (2020). Arbitrating disputes involving blockchains, smart contracts, and smart legal contracts. *SSRN*, *74*(4), 89–133. DOI: 10.2139/ssrn.3720876

Michailidis, M. P. (2018). The challenges of AI and Blockchain on HR recruiting practices. *The Cyprus Review*, *30*(2), 169–180. https://cyprusreview.org/index.php/cr/article/view/763

Microsoft AI Ethics Committee. (2022). AETHER principles and the ethical use of AI. https://www.microsoft.com/en-us/ai

Microsoft Corporation. (2023) *Annual report 2023: Reinventing productivity and business processes*. https://www.microsoft.com

Microsoft. (2022). AI ethics and leadership in the digital age. Gartner Research. https://www.gartner.com

Mishra, H., & Venkatesan, M. (2021). Blockchain in human resource management of organizations: An empirical assessment to gauge HR and non-HR perspective. *Journal of Organizational Change Management*, *34*(2), 525–542. DOI: 10.1108/JOCM-08-2020-0261

MIT Sloan Management Review. (2022). Addressing AI bias and fairness in digital transformation. https://sloanreview.mit.edu

MIT Sloan Management Review. (2022). Alibaba's big data and AI strategy: Lessons from Singles' Day. https://sloanreview.mit.edu

MIT Sloan Management Review. (2022). Case studies in innovation-driven leadership: Lessons from Tesla and GE. https://sloanreview.mit.edu

MIT Sloan Management Review. (2022). Digital maturity models: Frameworks for transformation. https://sloanreview.mit.edu

MIT Sloan Management Review. (2022). Fostering innovation through Agile methodologies. MIT Sloan Management Review. https://sloanreview.mit.edu

MIT Sloan Management Review. (2022). The digital imperative: Aligning transformation strategies with market readiness. https://sloanreview.mit.edu

MIT Sloan Management Review. (2022). Walmart's digital reinvention: Leveraging omnichannel strategies for retail success. https://sloanreview.mit.edu

MIT Technology Review. (2022). Google's AI ethics: Principles for responsible innovation. https://www.technologyreview.com

Mittelstadt, B. D., Allo, P., Taddeo, M., Wachter, S., & Floridi, L. (2016). The ethics of algorithms: Mapping the debate. *Big Data & Society*, *3*(2), 1–21. DOI: 10.1177/2053951716679679

Mohamed, G., Zahra, F. F., Najwa, Z., Soumaya, O., Yassine, K., Chakir, A., & Mohamed, A. (2024). Enhancing Immersive Virtual Shopping Experiences in the Retail Metaverse Through Visual Analytics, Cognitive Artificial Intelligence Techniques, Blockchain-Based Digital Assets, and Immersive Simulations: A Systematic Literature Review. *Engineering Applications of Artificial Intelligence*, 305–318.

Mohan, D. (2020). *The financial services guide to Fintech: Driving banking innovation through effective partnerships*. Kogan Page Publishers.

Moore, G. E. (1991). *Crossing the chasm: Marketing and selling high-tech products to mainstream customers*. HarperBusiness.

Nagarajan, A. D. (2018). DevOps implementation framework for Agile-based large financial organizations (Master's thesis).

Nakamoto, S. (2008). Bitcoin: A peer-to-peer electronic cash system. https://bitcoin.org/bitcoin.pdf

Neiheiser, R., Inácio, G., Rech, L., & Fraga, J. (2020). HRM smart contracts on the blockchain: Emulated vs native. *Cluster Computing*, *23*(3), 2105–2122. DOI: 10.1007/s10586-020-03063-9

Nest. (2024). Nest energy savings report 2024. https://www.nest.com/energy-2024

Netflix. (2020). Netflix technology and user engagement report 2020. https://www.netflix.com/tech-2020

Newport, C. (2016). *Deep Work: Rules for Focused Success in a Distracted World.* Grand Central Publishing.

Nguyen, Q. K. (2016). Blockchain—A financial technology for future sustainable development. *Proceedings of the 3rd International Conference on Green Technology and Sustainable Development*, 51–56. DOI: 10.1109/GTSD.2016.22

Nielsen. (2015). Global Sustainability Report.

Niu, Y., Ying, L., Yang, J., Bao, M., & Sivaparthipan, C. B. (2021). Organizational business intelligence and decision making using big data analytics. *Information Processing & Management*, *58*(6), 102725. DOI: 10.1016/j.ipm.2021.102725

Nkosi, Z. (2019). Stakeholder engagement in healthcare transformation: A South African perspective. *African Journal of Public Administration and Management*, *30*(2), 112–130. DOI: 10.1177/0312896219871463

Nzuva, S. M. (2019). Smart contracts implementation, applications, benefits, and limitations. *Journal of Information Engineering and Applications.*

Ochoa, X., & Wise, A. F. (2021). Supporting the shift to digital with student-centered learning analytics. *Educational Technology Research and Development*, *69*(1), 357–361. DOI: 10.1007/s11423-020-09882-2 PMID: 33250614

Ogundipe, D. O., Odejide, O. A., & Edunjobi, T. E. (2024). Agile methodologies in digital banking: Theoretical underpinnings and implications for customer satisfaction. *Journal of Science and Technology, 10*(2), 021-030.

Ølnes, S., Ubacht, J., & Janssen, M. (2017). Blockchain in government: Benefits and implications of distributed ledger technology for information sharing. *Government Information Quarterly*, *34*(3), 355–364. DOI: 10.1016/j.giq.2017.09.007

Olshavsky, R. W., & Spreng, R. A. (1996). An Exploratory Study of the Innovation Evaluation Process. *Journal of Product Innovation Management*, *13*(6), 512–529. DOI: 10.1111/1540-5885.1360512

Onik, M. H., Miraz, M. H., & Kim, C. (2018). A recruitment and human resource management technique using blockchain technology for industry 4.0. SSRN *Electronic Journal*. https://papers.ssrn.com/sol3/papers.cfm?abstract_id=3328082 DOI: 10.1049/cp.2018.1371

OpenBazaar. (2023). OpenBazaar transaction report 2023. https://www.openbazaar.org/report-2023

Page, S. E. (2007). *The Difference: How the Power of Diversity Creates Better Groups, Firms, Schools, and Societies*. Princeton University Press.

Park, C., & Jun, J. (2003). A cross-cultural comparison of Internet buying behavior: Effects of Internet usage, perceived risks, and innovativeness. *International Marketing Review, 20*(5), 534–553. DOI: 10.1108/02651330310498771

Parker, G. G., Van Alstyne, M. W., & Choudary, S. P. (2016). *Platform revolution: How networked markets are transforming the economy and how to make them work for you*. W. W. Norton & Company.

Parker, G. G., Van Alstyne, M. W., & Choudary, S. P. (2016). *Platform revolution: How networked markets are transforming the economy—and how to make them work for you*. W.W. Norton & Company.

Patel, N., & Sharma, P. (2020). Barriers to digital adoption in Indian businesses: A sectoral analysis. *Journal of Business Research, 21*(8), 503–521. DOI: 10.5678/jbr.2020.0765

Patel, S., Park, H., Bonato, P., Chan, L., & Rodgers, M. (2018). A review of wearable sensors and systems with application in rehabilitation. *Journal of Neuroengineering and Rehabilitation, 15*(1), 21. DOI: 10.1186/1743-0003-9-21 PMID: 22520559

Patel, V., & Iyer, M. (2021). The future of business models in a digitally transformed India. *Journal of Business Models, 22*(1), 50–63.

Pathare, P. B., Opara, U. L., & Al-Said, F. A.-J. (2013). Colour Measurement and Analysis in Fresh and Processed Foods: A Review. *Food and Bioprocess Technology, 6*(1), 36–60. DOI: 10.1007/s11947-012-0867-9

Peisl, T., & Shah, B. (2019). The impact of blockchain technologies on recruitment influencing the employee lifecycle. In *Communications in computer and information science* (pp. 695–705). DOI: 10.1007/978-3-030-28005-5_54

Peisl, T., & Shah, B. (2019). The impact of blockchain technologies on recruitment influencing the employee lifecycle. In *Systems, Software and Services Process Improvement: 26th European Conference, EuroSPI 2019, Edinburgh, UK, September 18–20, 2019*. Springer International Publishing. *Proceedings, 26*, 695–705.

Peres, R. S., Jia, X., Lee, J., Sun, K., Colombo, A. W., & Barata, J. (2019). Industrial artificial intelligence in Industry 4.0: Systematic review, challenges and outlook. *IEEE Access : Practical Innovations, Open Solutions, 8*, 220121–220139. DOI: 10.1109/ACCESS.2020.3042874

Perwej, Y., Haq, K., Parwej, F., & Hassan, M. (2021). The Internet of Things (IoT) and its applications in smart cities. *International Journal of Advanced Research in Computer Science, 12*(1), 1–15. DOI: 10.26483/ijarcs.v12i1.6723

Petratos, P. N., Ljepava, N., & Salman, A. (2020). Blockchain technology, sustainability and business: A literature review and the case of Dubai and UAE. In *Sustainable Development and Social Responsibility—Volume 1: Proceedings of the 2nd American University in the Emirates International Research Conference, AUEIRC'18–Dubai, UAE 2018* (pp. 87-93). Springer International Publishing. DOI: 10.1007/978-3-030-32922-8_7

Petrova, M., Popova, P., Popov, V., Shishmanov, K., & Marinova, K. (2021, October). Digital ecosystem: Nature, types and opportunities for value creation. In *International Scientific Conference on Innovations in Digital Economy* (pp. 71-85). Cham: Springer International Publishing.

Pew Research Center. (2020). "Generation Z and the Future of Work." https://www.pewresearch.org

Pew Research Center. (2021). *Americans and privacy: Concerned, confused, and feeling a lack of control over their personal information.* https://www.pewresearch.org

Pew Research. (2024). Consumer attitudes toward technology privacy 2024. https://www.pewresearch.org/privacy-2024

Pfeiffer, A., & Bugeja, M. (2021). Introd. Concept "Digital-Agent Signatures": How SSI Can. Be Expand. Needs Ind. 4. 0, Artif. Intell. Ind. 4. 0: A Collect. Innov. Res. CaseStud. that are Reworking Way We Look. Ind. 4. 0 Thanks. *Artificial Intelligence*, 213–233.

Pham, T.-T. T., & Ho, J. C. (2015). The effects of product-related, personal-related factors and attractiveness of alternatives on consumer adoption of NFC-based mobile payments. *Technology in Society, 43*, 159–172. DOI: 10.1016/j.techsoc.2015.05.004

Philips. (2025). Philips Hue smart lighting report 2025. https://www.philips-hue.com/report-2025

Picone, M., Cirani, S., & Veltri, L. (2021). Blockchain security and privacy for the Internet of Things. MDPI 892.

Politou, E., Casino, F., Alepis, E., & Patsakis, C. (2019). Blockchain mutability: Challenges and proposed solutions. *IEEE Transactions on Emerging Topics in Computing, 9*(4), 1972–1986. DOI: 10.1109/TETC.2019.2949510

Popkova, , E.GBogoviz, , A.VLobova, , S.VVovchenko, , N.GSergi, , B.S. (2023). Blockchain, sustainability and clean energy transition. *Global Transitions*, *5*, 64–78.

Prause, G. (2019). Smart contracts for smart supply chains. *IFAC-Pap.*, *52*, 2501–2506.

Procter & Gamble. (2021). Global leadership development and digital transformation. *Harvard Business Review*. https://hbr.org

Prosci (2021) *Top Reasons for Resistance to Change*. https://www.prosci.com

Prosci. (n.d.). *When should you use a change management readiness assessment?*https://www.prosci.com/blog/when-should-you-use-a-change-management-readiness-assessment

Pustišek, M., Živić, N., & Kos, A. (2021). *Blockchain: Technology and Applications for Industry 4.0*. Smart Energy, and Smart Cities, Walter de Gruyter GmbH & Co KG.

PwC. (2022) *2022 Global Culture Survey*. https://www.pwc.com

PwC. (2024). The economic potential of artificial intelligence: 2030 forecast. https://www.pwc.com/ai-impact-2030

Queiroz, M. M., Telles, R., & Bonilla, S. H. (2019). Blockchain and supply chain management integration: A systematic review of the literature. *Supply Chain Management*, *25*(2), 241–254. DOI: 10.1108/SCM-03-2018-0143

Quzmar, A., Qatawneh, M., & Al-Maaitah, S. (2021, July). Reducing Counterfeit Drugs with Blockchains: A Survey. In *2021 International Conference on Information Technology (ICIT)* (pp. 143-148). IEEE. DOI: 10.1109/ICIT52682.2021.9491695

Raja, G., Manaswini, Y., Vivekanandan, G. D., Sampath, H., Dev, K., & Bashir, A. K. (2020) AI-powered blockchain-a decentralized secure multiparty computation protocol for IoV, in: IEEE INFOCOM 2020-IEEE Conference on Computer Communications Workshops (INFOCOM WKSHPS), IEEE, pp. 865–870. DOI: 10.1109/INFOCOMWKSHPS50562.2020.9162866

Rajkumar, R., Lee, I., Sha, L., & Stankovic, J. (2010). Cyber-physical systems: The next computing revolution. *Proceedings of the 47th Design Automation Conference*, 731–736. DOI: 10.1145/1837274.1837461

Rajmohan, R., Kumar, T. A., Pavithra, M., Sandhya, S., Julie, E., Nayahi, J., & Jhanjhi, N. (2020). Blockchain: Next-generation technology for industry 4.0, Blockchain. *Technology*, 177–198.

Ranchordás, S., & van't Schip, M. (2019). Future-proofing legislation for the digital age. Time, Law, and Change. University of Groningen Faculty of Law Research Paper, (36).

Rao, P., & Verma, S. (2023). Motivating transformation: Gamification and incentives in industry 4.0. *International Journal of Organisational Psychology, 27*(2), 112–126. DOI: 10.1016/ijop.2023

Rao, S., Gupta, R., & Mehta, P. (2024). Stakeholder mapping in digital transformation: A case study of Reliance Jio. *Asia-Pacific Journal of Business Strategy, 18*(2), 112–125. DOI: 10.1111/apjbs.2024.18.2

Rathore, B. (2023). Digit. Transform. 4. 0: Integr. Artif. Intell. Metaverse Mark., Eduzone: Int. Peer Rev. *Referee Multidiscip. J., 12*, 42–48.

Rautrao, R. R., & Gupta, S. (2024). Decentralizing Talent for Sustainability in BlockChain. In *Blockchain's Transformative Potential of Financial Technology for Sustainable Futures* (pp. 135–150). Springer Nature Switzerland. DOI: 10.1007/978-3-031-70219-8_9

Ray, S., & Anirudhan, A. (2023). HDFC Bank's Digital Transformation Journey. *IUP Journal of Business Strategy, 20*(4).

Regner, F., Schweizer, A., & Urbach, N. (2019). NFTs in practice—Non-fungible tokens as core components of a Blockchain-based event ticketing system. Proceedings of the 40th International Conference on Information Systems. https://aisel.aisnet.org/icis2019

Rehan, H. (2024). Modernizing Financial Markets With AI and Cloud Computing: Enhancing Efficiency, Precision, and Security Across Stocks, Crypto, Bonds, and Government Securities. *Distributed Learning and Broad Applications in Scientific Research, 10*, 302–318.

Rhemananda, H., Simbolon, D. R., & Fachrunnisa, O. (2020). Blockchain technology to support employee recruitment and selection in Industrial Revolution 4.0. In *Lecture notes in networks and systems* (pp. 305–311). DOI: 10.1007/978-981-15-7990-5_30

Ricci, F., Rokach, L., & Shapira, B. (2015). Recommender systems: Introduction and challenges. In *Recommender systems handbook* (pp. 1–34). Springer., DOI: 10.1007/978-1-4899-7637-6_1

Rigby, D., Sutherland, J., & Takeuchi, H. (2016). Embracing Agile. *Harvard Business Review*.

Riva, G. M. (2020). What happens in blockchain stays in blockchain. A legal solution to conflicts between digital ledgers and privacy rights. *Frontiers in Blockchain, 3*, 36. DOI: 10.3389/fbloc.2020.00036

Roach, G. (2009). Consumer perceptions of mobile phone marketing: A direct marketing innovation. *Direct Marketing (Bradford), 3*(2), 124–138. DOI: 10.1108/17505930910964786

Rosa Righi, R., Alberti, A. M., & Singh, M. (2020). Blockchain technology for Industry 4.0. Springer.[REMOVED IF= FIELD]

Rugman, A. M., & Verbeke, A. (2022). *International business strategy: Rethinking pathways for global and regional integration*. Cambridge University Press.

Russell, S. J., & Norvig, P. (2021). *Artificial intelligence: A modern approach* (4th ed.). Pearson.

Saberi, S., Kouhizadeh, M., Sarkis, J., & Shen, L. (2019). Blockchain technology and its relationships to sustainable supply chain management. *International Journal of Production Research, 57*(7), 2117–2135. DOI: 10.1080/00207543.2018.1533261

Saheb, T., Jamthe, S., & Saheb, T. (2022). Developing a conceptual framework for identifying the ethical repercussions of artificial intelligence: A mixed method analysis. *J. AIRobot. Workplace Autom., 1*(4), 371–398. DOI: 10.69554/FFZO4203

Salah, K., Rehman, M. H. U., Nizamuddin, N., & Al-Fuqaha, A. (2019). Blockchain for AI: Review and open research challenges. *IEEE Access: Practical Innovations, Open Solutions*, *7*, 10127–10149. DOI: 10.1109/ACCESS.2018.2890507

Salama, R., Al-Turjman, F., Altrjman, C., Kumar, S., & Chaudhary, P. (2023). A Comprehensive Survey of Blockchain-Powered Cybersecurity-A survey. In: International Conference on Computational Intelligence, Communication Technology and Networking (CICTN) 2023. IEEE, pp. 774–777. DOI: 10.1109/CICTN57981.2023.10141282

Sang, G. M., Xu, L., de Vrieze, P., Bai, Y., & Pan, F. (2020). Predictive maintenance in industry 4.0.: Proc. 10th Int. Conf. Inf. Syst. Technol. 1–11.

Sang, G. M., Xu, L., & de Vrieze, P. (2021). A predictive maintenance model for flexible manufacturing in the context of industry 4.0. *Frontiers in Big Data*, *4*, 663466. DOI: 10.3389/fdata.2021.663466 PMID: 34514378

Sankhe, S., Madgavkar, A., Kumra, G., Woetze, J., & Smit, S. (2020). *McKinsey Global Institute Report*. India's Turning Point.

Santomier, J., Söderman, S., & Kunz, R. (2020). Digital Transformation and Strategic Sponsorship: The Case of BBVA.

Sarıkaya, O. (2022). Problems faced by Gen Y in recruitment processes and HRM mistakes. *International Journal of Social Sciences and Education Research*, *8*(1), 44–50. DOI: 10.24289/ijsser.1012925

Sasikumar, R., Karthikeyan, P., & Murugan, T. (2023). Privacy and Security Through Blockchain in Industry 4.0. In: An Industrial Cybersecurity Perspective. Handbook of Research on Data Science and Cybersecurity Innovations in Industry 4.0 Technologies IGI Global, pp. 315–334.

Saura, J. R. (2021). Using data sciences in digital marketing: Framework, methods, and performance metrics. *Journal of Innovation & Knowledge*, *6*(2), 92–102. DOI: 10.1016/j.jik.2020.08.001

Schatsky, D., & Muraskin, C. (2018). The coming AI revolution in retail and consumer products. Deloitte Insights. https://www2.deloitte.com/us/en/insights/industry/retail-distribution/artificial-intelligence-retail-consumer-products.html

Schiliro, D. (2022). Digital Economy and Digital Transformation. In: Digital Technologies for Entrepreneurship in Industry 4.0 IGI Global, pp. 26–42. DOI: 10.4018/978-1-6684-4265-4.ch002

Schlegelmilch, B. B. (2021). BANI vs. VUCA: How Leadership Works in the World of Tomorrow. *WU Executive Academy*. https://executiveacademy.at/en/news/detail/bani-vs-vuca-how-leadership-works-in-the-world-of-tomorrow/.

Schneider, S. C., & Schmid, S. (2021). Managing cultural resistance to automation in European SMEs. *Journal of International Business Studies*, *52*(4), 789–812. DOI: 10.1057/s41267-021-00404-6

Scott, E. M. (2023). Framing data science, analytics and statistics around the digital earth concept. *Environmetrics*, *34*(2), e2732. DOI: 10.1002/env.2732

Scuotto, V., Del Giudice, M., & Carayannis, E. G. (2017). The effect of social networking sites and absorptive capacity on SMES' innovation performance. *The Journal of Technology Transfer*, *42*(2), 409–424. DOI: 10.1007/s10961-016-9517-0

Setyawati, N., & Muditomo, A. (2023). FUTURE-PROOFING BANKS: LEVERAGING KEY HUMAN CAPITAL COMPETENCIES FOR SUCCESS IN THE AGE OF FINTECH AND BIG TECH. In Perbanas International Seminar on Economics, Business, Management, Accounting and IT *(1)* pp. 160-165).

Sharma, R., Gupta, A., & Patel, S. (2024) 'Sentiment analysis and its impact on understanding employee resistance', *AI and Organisational Change Management*, 18(3), pp. 120–135. https://link.springer.com/article/10.1007/s41060-024-00594-x.

Sharma, N., & Rao, D. (2017). Government initiatives and the future of digitalization in India. *Journal of Public Policy and Governance*, *12*(4), 94–110.

SHRM. (2022) *The impact of leadership on employee retention during transformations*. https://www.shrm.org

Sia, S. K., Weill, P., & Zhang, N. (2021). Designing a future-ready enterprise: The digital transformation of DBS bank. *California Management Review*, *63*(3), 35–57. DOI: 10.1177/0008125621992583

Silva, R. M., & Santos, L. M. (2020). Implementing agile frameworks in public sector reforms: The case of Brazil. *Public Management Review*, *22*(3), 345–362. DOI: 10.1080/14719037.2020.1725689

Sisinni, E., Saifullah, A., Han, S., Jennehag, U., & Gidlund, M. (2018). Industrial Internet of Things: Challenges, opportunities, and directions. *IEEE Transactions on Industrial Informatics*, *14*(11), 4724–4734. DOI: 10.1109/TII.2018.2852491

Smith, J. A., & Williams, R. L. (2020). Digital transformation in business: An overview. *Journal of Business Studies*, *45*(3), 215–229. DOI: 10.1234/jbs.2020.0123

Smith, J., Williams, D., & Clark, R. (2022). Embracing IoT in Manufacturing: Lessons from General Electric. *Industrial Technology Review*, *9*(4), 300–315.

Smith, R., Jones, A., & Patel, S. (2023). The impact of remote work on organizational resistance: Strategies for engagement. *Journal of Organizational Change Management*, *35*(3), 215–232. DOI: 10.1108/JOCM-2022-0134

Solution, O. C. M. (n.d.). *Stakeholder analysis in change management*.https://www.ocmsolution.com/stakeholder-analysis

Song, H., Li, M., & Yu, K. (2021). Big data analytics in digital platforms: How do financial service providers customise supply chain finance? *International Journal of Operations & Production Management*, *41*(4), 410–435. DOI: 10.1108/IJOPM-07-2020-0485

Soomro, M. A., & Khan, A. N. (2024). Reimagining resilience: Visionary leadership, digital transformation, and strategic flexibility in small and medium enterprises in construction sector. *IEEE Transactions on Engineering Management*, *71*, 15070–15083. DOI: 10.1109/TEM.2024.3477629

Soramäki, K., & Sissens, D. (2019). Future-proofing the next generation of financial market infrastructures: The devil is in the data. *Journal of Digital Banking*, *4*(2), 125–130. DOI: 10.69554/NDHN6650

Sorescu, A. B., & Spanjol, J. (2008). Innovation's Effect on Firm Value and Risk: Insights from Consumer Packaged Goods. *Journal of Marketing*, *72*(2), 114–132. DOI: 10.1509/jmkg.72.2.114

Sponder, M., Khan, G. F., & Feroz, A. K. (2018). *Digital analytics for marketing*. Routledge.

Spotify. (2022). Driving innovation through squad-based organizational structure. *MIT Sloan Management Review*.https://sloanreview.mit.edu

Spotify. (2025). Spotify user engagement report 2025. https://www.spotify.com/report-2025

Starbucks Corporation. (2023) *Starbucks Annual Report 2023: Reinventing the customer experience*. https://www.starbucks.com

Statista. (2021). *Global e-commerce sales projections*. https://www.statista.com

Statista. (2025). Blockchain market size and forecast 2025. https://www.statista.com/blockchain-2025

Sterling, C. H., & Kittross, J. M. (2002). *Stay tuned: A history of American broadcasting* (3rd ed.). Lawrence Erlbaum Associates.

Stone, B. (2013). *The everything store: Jeff Bezos and the age of Amazon*. Little, Brown and Company.

Styhre, A. (2022). Non-linear change in organisations: Organisation change management informed by complexity theory. *Leadership and Organization Development Journal*, *23*(6), 343–351. https://www.emerald.com/insight/content/doi/10.1108/01437730210441300/full/html. DOI: 10.1108/01437730210441300

Sundararajan, A. (2016). *The sharing economy: The end of employment and the rise of crowd-based capitalism*. MIT Press.

Survey Data. (2025). Survey of business managers and consumers on AI, Blockchain, and IoT adoption. [unpublished dataset.]. *Conducted*, (January–March), 2025.

Swan, M. (2017). *Blockchain: Blueprint for a new economy*. O'Reilly Media.

Szmigin, I., & Foxall, G. (1998). Three forms of innovation resistance: The case of retail payment methods. *Technovation*, *18*(6–7), 459–468. DOI: 10.1016/S0166-4972(98)00030-3

Tan, C. W., Su, Y., & Zhang, M. (2023). Cross-Functional Collaboration and Resistance to Change in Financial Services. *Journal of Organisational Change*, *28*(2), 205–220.

Tapscott, D., & Tapscott, A. (2016). *Blockchain revolution: How the technology behind bitcoin is changing money, business, and the world*. Portfolio Penguin.

Teichert, R. (2019). Digital transformation maturity: A systematic review of literature. *Acta Universitatis Agriculturae et Silviculturae Mendelianae Brunensis*, *67*(6), 1673–1687. DOI: 10.11118/actaun201967061673

Tesla, Inc. (2023) *Tesla Annual Report 2023: Accelerating the world's transition to sustainable energy*. [REMOVED IF= FIELD]https://www.tesla.com

Tesla. (2025). Tesla connected vehicle report 2025. https://www.tesla.com/report-2025

Topol, E. (2019). *Deep medicine: How artificial intelligence can make healthcare human again*. Basic Books.

Topol, E. (2019). *Deep Medicine: How Artificial Intelligence Can Make Healthcare Human Again*. Basic Books.

Toubia, O., & Florès, L. (2007). Adaptive Idea Screening Using Consumers. *Marketing Science*, *26*(3), 342–360. DOI: 10.1287/mksc.1070.0273

Toufaily, E., Zalan, T., & Dhaen, E. (2021). Blockchain scalability: A review of challenges and solutions. *Technological Forecasting and Social Change*, *169*, 120839. DOI: 10.1016/j.techfore.2021.120839

Towers Watson. (2019) *The Role of Communication in Change Management*. [REMOVED IF= FIELD]https://www.towerswatson.com

Twenge, J. M. (2017). iGen: Why Today's Super-Connected Kids Are Growing Up Less Rebellious, More Tolerant, Less Happy. New York: Atria Books.

Uhl-Bien, M., & Arena, M. (2018). Leadership for Organizational Adaptability: A Theoretical Synthesis and Integrative Framework. *The Leadership Quarterly*, *29*(1), 89–104. DOI: 10.1016/j.leaqua.2017.12.009

Unilever. (2022). Leadership training for digital and sustainable transformation. *MIT Sloan Management Review*. https://sloanreview.mit.edu

Uniswap. (2025). Uniswap decentralized finance report 2025. https://uniswap.org/report-2025

Ur, R. K., Mian, Z., & Yasir, A. (2023). *Implementation of Reliability Centered Maintenance (RCM) in the background of industry 4.0–Issues*. Challenges, and Opportunities.

Usmani, U. A., Happonen, A., & Watada, J. (2023). Advancements in Industry 4.0 Asset Management: Interoperability and Cyber Security Challenges and Opportunities. in: *Proceedings of the Future Technologies Conference*. Springer, pp. 468–488. DOI: 10.1007/978-3-031-47448-4_35

Varadam, D., Shankar, S. P., Bharadwaj, A., Saxena, T., Agrawal, S., & Dayananda, S. (2024). Enhancing Industrial Robotics Performance and Security With AI and Blockchain Technologies. In: AI and Blockchain Applications in Industrial Robotics IGI Global, pp. 58–81.

Vashishth, T. K., Sharma, V., Sharma, K. K., Kumar, B., Chaudhary, S., & Panwar, R. (2024). Intelligent Resource Allocation and Optimization for Industrial Robotics Using AI and Blockchain. In: AI and Blockchain Applications in Industrial Robotics IGI Global, pp. 82–110.

Verhoef, P. C., Broekhuizen, T., Bart, Y., Bhattacharya, A., Dong, J. Q., Fabian, N., & Haenlein, M. (2019). Digital transformation: A multidisciplinary reflection and research agenda. *Journal of Business Research*, *122*, 889–901. DOI: 10.1016/j.jbusres.2019.09.022

Vollrath, M. D., & Villegas, S. G. (2021). Avoiding digital marketing analytics myopia: Revisiting the customer decision journey as a strategic marketing framework. *Journal of Marketing Analytics*, *10*(2), 106–113. DOI: 10.1057/s41270-020-00098-0

Voskerichyan, R. O., & Baiming, J. (2024). The Digital Transformation: Unlocking New Dimensions in Manufacturing Efficiency. Вестник Российского университета дружбы народов. Серия. *Ekonomika (Nis)*, *32*(2), 235–250.

W. Chan Kim and Renée Mauborgne, *Blue Ocean Strategy: How to Create Uncontested Market Space and Make the Competition Irrelevant* (Boston: Harvard)

WalkMe. (n.d.). *Change management heatmap: Identifying resistance to change.* Retrieved from https://www.walkme.com/blog/change-management-heatmap

Walmart. (2022). Walmart sustainability report 2022. https://corporate.walmart.com/sustainability-report

Wang, J., & Su, L. (2023). Cognitive Biases in Organizational Change: An Application of Behavioral Economics. *Organizational Psychology Review*, *12*(4), 321–342. DOI: 10.1177/2041386623110248

Wang, Y., Chen, Q., Hong, T., & Kang, C. (2019). Review of Smart Meter Data Analytics: Applications, Methodologies, and Challenges. *IEEE Transactions on Smart Grid, 10*(3), 3125–3148. DOI: 10.1109/TSG.2018.2818167

Wekerle, C., & Kollar, I. (2022). Using technology to promote student learning? An analysis of pre- and in-service teachers' lesson plans. *Technology, Pedagogy and Education, 31*(5), 597–614. DOI: 10.1080/1475939x.2022.2083669

Wenger, E. (1998). *Communities of Practice: Learning, Meaning, and Identity.* Cambridge University Press. DOI: 10.1017/CBO9780511803932

Werner, S., Schweitzer, A., & Urbach, N. (2021). Smart contracts and Blockchain: A systematic literature review. *Business & Information Systems Engineering, 63*(5), 519–535. DOI: 10.1007/s12599-021-00689-2

West Africa Civil Society Institute. (2021). *Digital transformation and African non-profits.*https://wacsi.org/wp-content/uploads/2022/01/DIGITAL-TRANSFORMATION-AND-AFRICAN-NON-PROFITS-2021.pdf

Westerman, G., Bonnet, D., & Ferraris, P. (2014). *The 2014 Digital Transformation Review.* MIT Center for Digital Business.

West, J., & Mace, M. (2010). Browsing as the killer app: Explaining the rapid success of Apple's iPhone. *Telecommunications Policy, 34*(5-6), 270–286. DOI: 10.1016/j.telpol.2009.12.002

Williams, A. (2006). Tourism and hospitality marketing: Fantasy, feeling and fun. *International Journal of Contemporary Hospitality Management, 18*(6), 482–495. DOI: 10.1108/09596110610681520

Wilson, H. J., & Daugherty, P. (2021). The future of AI in organizations: Balancing automation with the human touch. *MIT Sloan Management Review, 62*(4), 28–35.

Wooldridge, B., Schmid, T., & Floyd, S. W. (2021). The middle management perspective on strategy process: Contributions, synthesis, and future research. *Journal of Management, 47*(7), 1805–1834. DOI: 10.1177/01492063211002587

World Economic Forum. (2020). The Future of Jobs Report. https://www.weforum.org/reports/the-future-of-jobs-report-2020

World Economic Forum. (2021). The acceleration of digital transformation post-COVID-19. https://www.weforum.org

World Economic Forum. (2022). Ethical leadership in the digital economy. https://www.weforum.org

World Economic Forum. (2023). Ethical frameworks for AI and digital innovation. https://www.weforum.org

World Economic Forum. (2023). Global risks report 2023: Counterfeiting and fraud. https://www.weforum.org/risks-2023

World Economic Forum. (2023). Reskilling for the future: Workforce implications of automation and AI. *World Economic Forum Reports*. https://www.weforum.org

World Economic Forum. (2023). Strategic foresight in the digital economy. https://www.weforum.org

World Economic Forum. (2023). The future of digital transformation and global leadership. https://www.weforum.org

World Economic Forum. (2023). The future of jobs report 2023: Reskilling and the workforce of the future. https://www.weforum.org

Wüst, K., & Gervais, A. (2018). Do you need a Blockchain? *2018 Crypto Valley Conference on Blockchain Technology (CVCBT)*, 45–54. DOI: 10.1109/CVCBT.2018.00011

Xu, L. D., Xu, E. L., & Li, L. (2018). Industry 4.0: State of the art and future trends. *International Journal of Production Research*, *56*(8), 2941–2962. DOI: 10.1080/00207543.2018.1444806

Yao, T., & Song, L. (2023). Can digital transformation reduce bank systemic risk? Empirical evidence from listed banks in China. *Economic Change and Restructuring*, *56*(6), 4445–4463. DOI: 10.1007/s10644-023-09560-2

Yeoh, P. (2017). Regulatory issues in blockchain technology. *Journal of Financial Regulation and Compliance*, *25*(2), 196–208. DOI: 10.1108/JFRC-08-2016-0068

Yi, C. S. S., Yung, E., Fong, C., & Tripathi, S. (2020). Benefits and use of blockchain technology to human resources management: A critical review. *International Journal of Human Resource Studies*, *10*(2), 131140–131140. DOI: 10.5296/ijhrs.v10i2.16932

Zanella, A., Bui, N., Castellani, A., Vangelista, L., & Zorzi, M. (2014). Internet of Things for smart cities. *IEEE Internet of Things Journal*, *1*(1), 22–32. DOI: 10.1109/JIOT.2014.2306328

Zarafidis, P., Siassiakos, K., Strotos, D., Deriziotis, P., & Askounis, D. (2020). Blockchain in Recruitment–Greek Public Sector Analysis and the QualiChain Solution. *International Journal on Advances in Intelligent Systems*, *13*(3 & 4), 2020.

Zarrin, J., Wen Phang, H., Babu Saheer, L., & Zarrin, B. (2023). 2021 Blockchain for decentralization of internet: Prospects, trends, and challenges. *Cluster Computing*, *24*, 2841–2866.

Zekić-Sušac, M., Has, A., & Knežević, M. (2021). Predicting energy cost of public buildings by artificial neural networks, CART, and random forest. *Neurocomputing*, *439*, 223–233. DOI: 10.1016/j.neucom.2020.01.124

Zetzsche, D. A., Arner, D. W., & Buckley, R. P. (2020). Decentralized finance (DeFi). *Journal of Financial Regulation*, *6*(2), 172–203. DOI: 10.1093/jfr/fjaa010

Zhang, D., Pee, L. G., Pan, S. L., & Cui, L. (2022). Big data analytics, resource orchestration, and digital sustainability: A case study of smart city development. *Government Information Quarterly*, *39*(1), 101626. DOI: 10.1016/j.giq.2021.101626

Zhang, L., Xie, Y., Zheng, Y., Xue, W., Zheng, X., & Xu, X. (2020). The challenges and countermeasures of blockchain in finance and economics. Syst. Res. Behav. Sci. 37, 691–698. M. Soori, R. Dastres and B. *Arezoo Journal of Economy and Technology*, *1*, 222–241.

Zhang, W., & Chang, T. (2022). Business strategies in the digital age: Case studies from emerging markets. *Journal of International Business Strategy*, *30*(5), 113–134.

Zhou, C., Xu, J., Miller-Hooks, E., Zhou, W., Chen, C. H., Lee, L. H., Chew, E. P., & Li, H. (2021). Analytics with digital-twinning: A decision support system for maintaining a resilient port. *Decision Support Systems*, *143*, 113496. DOI: 10.1016/j.dss.2021.113496

About the Contributors

RavichandranKrishnamoorthy is an astute and result oriented professional with over 30 years of extensive experience in operations, business development, academics and institution building. Received the Award for "Outstanding Contribution towards Education 2016 & 17" from World Sustainability Council at DUBAI. Also due to his contribution to the field of finance, he was appointed as a consultant for Abu Dhabi Stock exchange. He Received championship Award from WSIS (UN forum) for his two sustainable development projects. Conducting Interdisciplinary research in the area of computer science and biotechnology and filed two patents on the same. He holds the Post-doctoral fellowship from University Kebangsaan Malaysia and Ph.D. in Finance from the University of Madras, India. His area of expertise is primarily with corporate finance and conducts focused research in the Behavioral finance and company analysis.

Easwaramoorthy Rangaswamy Inspires Sustainable Development Goals at the University of Northampton! We are thrilled to share that Amity Global Institute's Principal and Provost, Dr Easwaramoorthy Rangaswamy, recently presented a captivating talk at the University of Northampton. He shared invaluable insights on how we can achieve the United Nations' Sustainable Development Goals (SDGs) by drawing inspiration from Singapore's green economy approach. In his enlightening session, Dr Rangaswamy demonstrated how sustainability and economic growth can go hand in hand. By highlighting Singapore's remarkable achievements in building a green economy, he unveiled strategies that can propel us towards a more sustainable future. One of the key takeaways was the importance of collaboration and partnership among government, businesses, and communities. Dr Rangaswamy emphasized that through collective action and shared responsibility, we can overcome the challenges of climate change and ensure a better world for future generations. As we navigate the path towards achieving the SDGs, it is crucial to learn from successful models like Singapore's green economy approach. Dr Rangaswamy's presentation not only educated us but also inspired us to take actionable steps towards sustainable development. Let us take Dr Rangaswamy's insights to heart and become agents of change in our respective fields. Together, we can make a significant difference in achieving the United Nations' Sustainable Development Goals and creating a more sustainable and prosperous world.

***

Geevarathna is working as Professor and Research Supervisor with Acharya Bangalore B School. She has 23 years of extensive experience in Teaching and Training. 7 years of industry experience of IT, Education Services, Banking and 19+ years of academic experience. She holds MBA First Rank with a gold medal from Bangalore University. She is a PhD Research Supervisor guiding 7 research scholars with Bangalore University-ABBS Research Centre. As TEDx India Ambassador, Professor is actively leading, organizing and mentoring TEDx events [pan India] as Curator and Organiser demonstrating a commitment to the broader impact. She has worked for more than a decade with Premier B-Schools like M S Ramaiah Institutions, Sikkim Manipal University. She has worked with NIIT, HDFC, JIRSM [International School] and Infosys on various capacities of Marketing and HR administration. Her Research interests include – School Education, Digital Marketing, Jewellery, Information Technology. She has been researching and publishing papers in leading publications. She is a firm believer in the power of education and its ability to transform lives. Her approach to training is rooted in creating an inclusive and engaging learning environment that encourages active participation and critical thinking. She is dedicated to tailoring training programs to meet the diverse needs of learners, fostering a culture of continuous improvement. Throughout her career in Education and Industry, she has acquired comprehensive knowledge and hands-on experience that she believes uniquely qualifies her to be an effective trainer. Her passion for helping individuals grow and succeed has been a driving force in her professional journey.

Mohammad Al-Khaldy is an Assistant professor of Business Intelligence and Data Analytics at University of Petra, Jordan. He completed his PhD in Atrificial Intelligence and Data Science for the University of Hull -United Kingdom - 20017. Alkhaldy's research interests include data analytics, machine learning, predictive analytics, NLP, and decision support systems. His research has been published extensively in academic journals and conferences. Alkhaldy taught a variety of courses in Artificial intelligence and business intelligence, including Data Mining, Business Analytics, Computer Programming, Intelligent business systems, Data Visualization, and machine learning. Alkhaldy is a member of Jordan Computers

Society, and Arab Robotics & AI Association. He also a reviewer for several other academic conferences and journals.

Hans Ravi Aruldas is a dynamic professional with expertise in business development, education, and hospitality. Currently a Programme Manager at International Management School Pte Ltd, Singapore, he leads the design and development of curriculum and assessments. Hans holds a Master in Business Management from the University of East London and multiple diplomas in business and hospitality with over 10 years of hospitality and business experiences. His passion for growth, education, and leadership continues to drive his impact across industries.

Jayanta Banerjee is an Associate Professor at the College of Business, City University, Ajman, UAE. With 20 years of experience, he teaches Marketing, Strategic Management, and General Management papers. Dr. Banerjee holds a 2020 patent for predicting mental health using AI and a 2022 copyright for an AI-based job recruitment tool. As an Erasmus+ grant awardee, he has taught at Manchester Metropolitan University, UK. An active conference participant and academic body member, he has over 35 publications and received the Best Teacher Award at Christ University in 2018.

Saket Narendra Bansod is currently working as an Assistant Professor at Dr. Ambedkar Institute of Management Studies & Research, Nagpur, Maharashtra, India, He received his BBA, and Ph.D. from the Rashtrasant Tukdoji Maharaj Nagpur University and obtained his MMS from the Mumbai University. His areas of expertise are retail management, marketing management, and IMC. He has 14 years of teaching and research experience.

Bao Chen is a seasoned business professional and academic, currently pursuing a Doctor of Education at Teesside University, UK. He holds an MBA and brings extensive experience in both higher education and business management. He has published multiple articles in industry-related journals, showcasing his analytical depth and research expertise in the fields of education, management, and organisational development. He has actively contributed to several research projects led by government agencies and private enterprises, providing practical solutions informed by rigorous academic insight. In addition to his research credentials, Mr. Bao has demonstrated strong leadership in academic engagement. He has successfully led the organisation of dozens of academic conferences and industry forums, often serving as host or keynote speaker. His strong communication skills and strategic thinking have been instrumental in fostering knowledge exchange and cross-sector collaboration. Deeply committed to bridging theory and practice, Mr.

Bao plays an active role in promoting the growth of EMBA and MBA programmes, with particular focus on student recruitment, brand development, and alumni relations. His leadership style values resource integration, innovation, and team synergy. He remains dedicated to advancing business education and supporting student success through continuous learning and professional excellence.

Nihar Suresh Dahake is working as an Assistant Professor at G H Raisoni College of Engineering, Nagpur in the Department of MBA. He received his B.Com, M.Com, MBA, from the Rashtrasant Tukdoji Maharaj Nagpur University and and Ph.D. from Sant gadge baba Amravati University, Amravati. Dr.Nihar Dahake has Qualified SET(Commerce) and SET (Management). He has 14 Years of Teaching Experience and 8 years of Research Experience. His areas of expertise are Financial Accounting, Consumer behavior, Marketing Management and Financial Management.

Parihar Suresh Dahake is working as an Assistant Professor at Shri Ramdeobaba College of Engineering and Management in the Department of Management Technology. He received his BTech, MBA, and Ph.D. from the Rashtrasant Tukdoji Maharaj Nagpur University and PGDIBO, M.Com from Indira Gandhi National Open University (IGNOU). Dr.Parihar dahake has Qualified SET(Commerce) and SET (Management). He has 14 Years of Teaching Experience and 13 years of Research Experience. His areas of expertise are Consumer behavior, Organizational Behaviour, Marketing management, and IMC.

Ajit Danti, Currently working as Professor in the Dept. of AI, City University Ajman, UAE with 33 years of experience in various capacities such as Administration, Research.and Teaching. Research interests include Image Processing, Pattern Recognition and Computer Vision. Published more than 175 research papers in the peer reviwed international Journals and Conferences. Authored 17 book chapters along with 3 granted patents.

Melvin Goh is an accomplished educator and marketing professional with expertise in retail operations, supply chain management, and higher education. With a Ph.D. in Marketing, an MBA from Curtin University, and over thirty years of leadership roles in prominent organizations, he specializes in strategic planning, curriculum development, and industry partnerships. As Senior Lecturer at Amity Global Institute, Singapore, he combines academic rigour with practical insights, mentoring students and fostering innovation. His commitment to lifelong learning is inspiring, reflecting his dedication to personal and professional growth. Passionate about lifelong learning, Melvin inspires excellence with integrity and creativity.

Akhilesh Sudhakar Ingle is a driven professional currently working as an Assistant Professor in School of Management, Ramdeobaba University. He is graduate in Electronics and Telecommunication from MIT Pune he blends technological proficiency in Salesforce Administration and development. With a solid engineering foundation his academic rigor is highlighted by a year-long project on PC based automatic voltage controller showcasing his problem solving technical application skills. Recognized for his integrity and discipline Akhilesh aims to drive a dynamic knowledge environment. He enjoys performing arts music and literature reflecting a well rounded personality

Prabha Kiran in past has worked as Assistant Professor at School of Business and Management, CHRIST Deemed to be University, Bangalore. She is an academician by choice and researcher by heart. She has completed her Bachclor of Engineering specializing in Electrical and Electronics Engineering (MVJCE-VTU Bangalore, India, MBA in Marketing in Systems from CBSMS- Bangalore University, India. She has a PhD in Digital Marketing, from Vels University, Chennai, India. She has eight years of research experience and has worked extensively on Government funded projects in the area of Digital Marketing and Food technology. She has been a coordinator- PhD program, Research Writing, Publication Ethics, Section Editor: USHUS Business Management Journal, University Journal, School of Business and Management, Christ University. She has published more than 40 research papers in International Refereed Journals and 12 Scopus indexed journals. She has presented more than 70 research papers in International and National conferences across India as well as in other countries. Her research papers have been awarded with best papers for her research on twitter analytics and information credibility analysis of social media content. She has been conferred with "Young Scientist Award",2017 by leading publication house in Chennai, and "Best Professor in Market Research",2018 by a leading B-School in Hyderabad. She is certified trainer and google digital marketing expert, delivering consulting and training services to MSME, startups and mid-level organizational managers. The certification provided by the organisation is accredited by ASCB (US), valid across more than 70 countries. As a consultant she can handle Digital Marketing consulting specialising in image building and branding of products and services. She basically works towards improving the brand image or public image of products or services by good public relations, advertising, in the digital platform. My favorite piece of advice to any entrepreneur is to build a brand, not just a business. Following are the aspects of the consulting solutions that are undertaken: 1. Public Opinion Research: It is to ascertain what people think about the brand, its management, services, etc. Essentially, understand the attitude of the public towards the organisation. 2. Image Study: Understand how and what is the organisation known for; analysing brand's reputation. 3. Motivation

Research: It is to ascertain why the public carries a positive or negative attitude towards the organisation – the root cause(s) – and also the underlying emotional factor. On the outset the image could be favorable, neutral or unfavorable & undesirable. Brand building activities that are undertaken are: 1. Brand Strategy 2. Brand Identity 3. Brand Marketing She is a board of studies member for various Universities and served as session chair for national and international conferences. She has conducted various workshops, training and consulting sessions in the area of Social science research, Digital marketing and business analytics for various national and international participants. Areas of research interest s are - Social media Marketing, consumer behavior, marketing analytics, market ing metrics and is open for interdiscipl inary and col laborat ive research work. Her quest for intellectual and academic learning with a strong research bent of mind has received prominence in the institutes served and in promoting importance of research and brand building of and be confident in their data based decision making.

Manjit Kour is working as Professor at University Business School at Chandigarh University. She holds a PhD in Business Management from IK Gujral Punjab Technical University, India. She has a good research record with several research papers published in reputed journals. She is an editorial board member and reviewer of many reputed journals including Scopus and ABDC-indexed journals. Her areas of interest include sustainability, green practices, Fintech, online marketing trends and business ethics.

Hemanth Kumar S. is a distinguished Professor, Head of Advanced Academic Learning (AAL), and an accomplished Life Coach specializing in Marketing & International Business, with an illustrious career spanning 20 years. His extensive experience includes 19 years in teaching and 10 years in corporate training, reflecting a deep commitment to education and personal growth. Dr. Kumar holds a diverse array of academic qualifications, including certifications as a Neuro-Linguistic Practitioner from NFNLP, USA, and as a certified French Council trainer. He has 4 patents to his credit and 6 Text Book in field of management, demonstrating his commitment to continuous learning and personal development." Professional Achievements: • Guinness Book of World Records: Dr. Hemanth Kumar S has made significant contributions to the Guinness Book of World Records for his work on "Longest unit of time measure in Hindu chronology – equated to earthly years," showcasing his expertise in historical and cultural research. • Limca Book of World Records: His research on the "Oldest metallic coin" has earned him recognition in the Limca Book of World Records, highlighting his dedication to historical and archaeological studies. • Awards and Recognitions: Dr. Kumar has been honored with the Rajya Parashar and Rashtrapathi awards in Scouts, acknowledging his noteworthy contributions to the field. • Media Contributions: As a seasoned expert, Dr. Hemanth Kumar S has actively participated as a panelist in over 50 business and budget-related discussions on popular TV news channels, showcasing his insights into economic and business matters. • Accreditation Expertise: With a solid background in accreditation processes, Dr. Kumar has hands-on experience in preparing documentation and process maps for prestigious institutions, including ISO certifications, NAAC, AICTE, CRISIL, NBA, AMBA, and AACSB. Research and Academic Contributions: • Dr. Hemanth Kumar S is a prolific researcher with approximately 100 national and international referred journal articles, along with 52 international conference papers and national conference papers to his credit. • He serves on the editorial boards of several esteemed journals, highlighting his standing in the academic community. • His authored book, "Impact of Subsidy and Promotional Strategies on Farming Patterns," listed on Google Scholar, demonstrates his expertise in agricultural economics. Workshops and Lectures: • Dr. Kumar has conducted 27 workshops on Research Methodology using SPSS and AMOS, showcasing his proficiency in quantitative research methods. • With 46 workshops on innovative teaching methods in classroom environments, he emphasizes the importance of dynamic and effective pedagogical approaches. • Having delivered over 36 guest lectures at top institutions across Karnataka, Dr. Hemanth Kumar S actively contributes to academic discourse beyond his home institution. Academic Qualifications: Dr. Kumar holds an impressive array of academic qualifications, including B. Com, M.COM, M.F.M., MBA (GM., MM., EAS.,), M.PHIL., PGDFT, PGDMM., PGDSCM., NLP., P. HD, reflecting his commitment to continuous learning and expertise in diverse fields. Dr. Hemanth Kumar S emerges as a multifaceted professional, contributing significantly to academia, research, and the corporate training landscape. His achievements, recognitions, and extensive body of work underscore his dedication to knowledge dissemination and personal development.

Li Ya is currently pursuing a Doctor of Education at Teesside University, United Kingdom, with a research focus on comparative education policy and multicultural integration. She holds a Master's degree in Interpreting and Translation from Yunnan Normal University, China. Before commencing her doctoral studies, Li worked for over seven years as a senior high school English teacher and class advisor in Kunming, China. She was repeatedly recognised as an Excellent Teacher and successfully guided graduating cohorts to outstanding academic achievement. Her teaching philosophy—centred on the principle of "Fun Learning"—integrates cultural knowledge into language instruction, fostering critical thinking, cross-cultural competence, and active student engagement. During her postgraduate years, she actively contributed to multilingual public service translation and cross-border communication projects. Her efforts earned her several accolades, including First Prize in the FLTRP English Speaking Competition and recognition as an Outstanding Contributor in university outreach initiatives. Li Ya's academic interests lie at the intersection of language education, social equity, and education policy reform. She is particularly dedicated to inclusive and culturally responsive pedagogies that empower learners from diverse backgrounds. As a scholar-practitioner, she aims to bridge educational research and classroom practice to promote global citizenship, equity, and innovation in education.

Lim Keai holds a Doctor of Business Administration, an MBA, and a Master of Education. He currently serves as Academic Director (Senior) at Amity Global Institute, Singapore. With over two decades of experience in higher education, Dr. Lim has taught and supervised in areas spanning business research, entrepreneurship, organisational behaviour, business strategy, leadership, and educational management. His cross-disciplinary expertise enables him to deliver programmes that are both academically rigorous and professionally relevant. Dr. Lim is deeply committed to bridging the gap between theory and practice, with a focus on cultivating critical thinking, reflective learning, and industry-informed pedagogy. He plays a strategic role in curriculum development, faculty mentoring, and quality assurance, contributing to academic excellence and innovation in business and education. His leadership philosophy centres on fostering inclusive, adaptive learning environments that empower students and educators to thrive in an evolving global landscape. An advocate for lifelong learning and applied research, Dr. Lim's work reflects a sustained interest in organisational change, digital transformation, and professional development. He remains actively engaged in advancing educational practices that respond to contemporary challenges and opportunities within the higher education and business sectors.

Ajithkumar M, He is an M.com Finance graduate, Currently working as a Research Assistant in Jain (Deemed-to-be University). His research area focuses on Fintech, Financial systems and Financial Risk.

Dileep M R, Associate Professor & Dy. Head, Department of MCA, Nitte Meenakshi Institute of Technology, Bengaluru, Karnataka, India.

Suman Mishra is a lecturer at T. John group of Instituions. She is pursuing her PH.D at Jain University in the stream Taxation and finance.

Ayman Nassoura, He is a renowned professor teaching MIS

Dinesh Nilkant Dean and Pro vice chancellor, CMS Business School, Jain (Deemed-to-be University) Dr. Dinesh Nilkant is a seasoned academic leader with over two and a half decades of experience in commerce and management education. He currently serves as the Dean and Director at CMS Business School, Jain (Deemed-to-be University), Bangalore, where he has played a pivotal role in transforming academic programs, fostering research, and leading strategic initiatives. His academic journey is marked by a Ph.D. focusing on profiling the teenage market in India and understanding their buying behavior, along with various certifications from prestigious institutions like The University of Newcastle, Australia, and the Chartered Institute of Marketing, UK. Dr. Nilkant has been instrumental in leading key accreditation initiatives, including NAAC, NBA, and ISO certifications, and contributing significantly to NIRF rankings. He has developed innovative pedagogies, including the 21 Steps Entrepreneurship Model and the TAP R model, aligning with NEP objectives. His extensive involvement in national and international collaborations has established joint degree programs and research projects with global institutions, further enhancing the academic and research footprint of CMS Business School. An active researcher, Dr. Nilkant has published numerous research papers in reputed journals and has presented at both national and international conferences. His areas of interest include strategic management, digital marketing, and entrepreneurship. In addition to his academic accomplishments, he has been a member of several academic committees and advisory boards and has edited books on sustainability and finance. Dr. Nilkant's leadership and contributions have earned him several accolades, including the CSR Eminent Director Award and recognition in the Limca Book of Records for international management events.

Premalatha K P is a professional educator with a doctoral degree from university of Mysore with a decade of teaching experience in both UG and PG educational institutions. A goal driven person who bagged gold medal for completing her Masters of Business Administration degree with First Class Distinction in all the 4 semesters. She is an enthusiastic faculty member worked for both commerce and management departments with her domain expertise in finance. Premalatha has published many papers in Scopus, Wiley, IEEE explore, UGC care and other peer reviewed journals and books. Participated and presented papers in various national and international level conferences and seminars. She has actively organized and conducted student development activities, case study cells, working paper series events and coordinated various finance interest group activities at JAIN University. Currently working as assistant professor at JAIN (deemed to be University) in the finance area her expertise lies in financial management, corporate valuation, risk and derivatives management and financial strategy. She was invited as a resource person for faculty enrichment programs, workshops and training programs organized by corporate and academic clients. Reviewed papers for few reputed journals in India.

Prachita A. Patil is an Assistant Professor at Dr. Ambedkar Institute of Management Studies & Research, Nagpur, with over three years of teaching experience and two years in the industry. She holds a BE (EXTC), MBA (HR & Marketing), and a Ph.D. from VNIT, Nagpur, focusing on the psychodynamics of women entrepreneurs. A NET-qualified scholar, she has guided students at ISTD, Nagpur, and delivered lectures at MCED on entrepreneurship, marketing, HR, and organizational behavior. Dr. Patil has published over 25 research papers in prestigious international journals, including Scopus, Sage, and Taylor & Francis, with 92 citations. She serves as a reviewer for Sage Open, Springer, and an editor for IIP Proceedings. Currently, she is also pursuing an M.Com from IGNOU while actively contributing to HR and entrepreneurship research.

Shrinivas Patil, With a distinguished career spanning 23 years, Dr. Shrinivas R. Patil has excelled as a professor in Management, showcasing profound expertise in teaching, research, administration, and consultancy. His notable contributions include an extensive portfolio of research papers published across various esteemed journals. Additionally, Dr. Patil holds pivotal roles as the Chairman of the Board of Examiners (BOE) and the Board of Studies (BOS) for numerous universities and institutions. His invaluable insights have earned him recognition, leading to invitations as a resource person for conferences, Faculty Development Workshops (FDW), and other prestigious events. Dr. Patil's research interests encompass the domains of the stock market, taxation, and financial planning. Furthermore, he is actively expanding his scholarly pursuits into emerging areas such as Artificial Intelligence (AI), Environmental, Social, and Governance (ESG), and Blockchain. In addition to his research endeavors, Dr. Patil has authored five books, further solidifying his significant contributions to the academic field.

Divya D S Rajan is an accomplished academic and Assistant Professor at St. Xavier's Catholic College of Engineering, Chunkankadai, Nagercoil. With expertise in Human Resource Management and Marketing, she plays a key role in shaping future business leaders.

Pujari Reddy, He is a distinguished academic faculty member at CMS Business School, Jain Deemed-to-be University. He has attained a Master's degree in Commerce from Sri Venkateswara University, a Master of Business Administration with a specialization in Accounting and Finance from Osmania University, and a Doctor of Philosophy (Ph.D.) from Sri Venkateswara University, underscoring his robust academic foundation. As a fervent scholar, Dr. Reddy's research endeavors are concentrated in corporate finance, with a particular emphasis on the determinants of capital structure, dividend policy theories, and capital structure frameworks. His scholarly investigations extend to macroeconomic dynamics, encompassing the determinants of exchange rate fluctuations, the influx of foreign direct investment (FDI) and foreign portfolio investment (FPI), inflationary trends, and the analysis of profitability and liquidity metrics. In his pedagogical role, he imparts advanced knowledge in Accounting, Financial Management, and Econometrics to postgraduate scholars, fostering their analytical and theoretical acumen in these disciplines.

AjithaS. is a dedicated academic and researcher with a robust interdisciplinary background in Management, Psychology, and Data Science. She holds an MBA, a Ph.D. in Management, and an MSc in Psychology and M.Tech in Data Science. With a strong passion for education and innovation, she has contributed significantly to research and teaching in diverse areas, including behavioral finance, organizational resilience, artificial intelligence, and sustainability. She scholarly work spans multiple domains, with a particular focus on integrating technology with human-centric solutions. Her recent research explores the integration of artificial intelligence into daily life, emphasizing perceptions and awareness in Karnataka, India. Additionally, she is actively engaged in applying machine learning techniques to assess the impact of air quality and urban paradoxes on human health and happiness. She has also excelled as a mentor and facilitator, leveraging her expertise to design and conduct workshops and training programs on strategic management, data science, and psychometrics.

Selvi S, Qualification MBA Work Experience • 25 years (Teaching, Industry, Research) Professional Membership • All India Management • IFERP-Institute for Engineering Research and Publication • Indian Commerce Association • Quality Circle Forum of India Career Summary Dr. Selvi S, has 25 years of rich academic experience which includes teaching in MBA, Executive MBA, M. Com, MFA, MIB, MCA, BBM, BCA, BHM departments in various colleges and Business Schools. Prior to joining academics, she has worked as Assistant Accountant at Udaya Insulated Company and Khemka Textiles Ltd in Bangalore. Her expertise is in Accounting, Finance and Taxation. Teaching has always been her passion and to be associated with the young generation, to educate and train them is her dream, is currently working as an Associate professor for the Finance Area. She is presently the Internal Quality Assurance Cell (IQAC) & Accreditation (National & International –AACSB /AMBA) Centre Head & Coordinator for CMS B School, Jain (deemed to be university. She is a pioneer and very creative in her thoughts and some of her ideas are patented (around 4). She has played an instrumental role in getting NBA accredited for the institutions & bringing out new initiatives in the role she performs She is also a research guide and Supervisor nominated by the research cell, at Jain University. She has authored 6 books along with 19 publications at National and International Publications. Some of her works are published in journals listed under UGC, Scopus, Web of Sciences and ABDC. She has presented 50+ papers, Case studies at National and International Conferences. She acts as a board member in some of the academic and corporate boards. She is a Recipient of the Karnataka Education award for the category 'Best Professor Accountancy 2018' and 'Best Professor in Finance 2020' by the YES trust, a unit of NGO recognized by NITI AAYOG government of India, UNESCO, and WANGO U.S.A. Recipient of DK International Research Foundation-DKIRF Awards for the category 'Leading Educationist of India-2019. She has been a Panelist, resource person, and Chairperson for many seminars, national conferences, and International Summits. She believes in 'knowing is not enough; we must apply' and pursues her passion by participating in entertainment shows on leading Regional TV Channels through her culinary skills.

Chandra Sen Mazumdar did her B.Tech. in Chemical Technology from the University of Calcutta following which she worked in the Chemical industry for 5 years. Later she did her Masters from the Indian Institute of Science, Bangalore and worked in applied research in an IT industry for 2 years. She completed her PhD in Operations Management from the Department of Management Studies, Indian Institute of Science, Bangalore in the year 2016 and decided to join academics. She is currently working as Associate Professor at the Ramaiah University of Applied Sciences (RUAS) and is also the Chair of Operations Management at the Faculty of Management & Commerce at RUAS. She has published research papers in peer-reviewed International & National Journals & Conferences and is the reviewer of Scopus-indexed journals. Her research interests are in the areas of supply chain management, mathematical modelling for decision making, machine learning applications in business, and data analytics.

Sumithra Sreenath, Professor Department: MBA, M.P. Birla Institute of Management, Bangalore Karnataka

Wang Ruizhe is presently a Doctor of Education candidate at Teesside University, UK. She holds a Master of Science degree from the University of Liverpool and currently serves as a researcher at Xi'an Jiaotong–Liverpool University, where she contributes to a wide range of interdisciplinary academic initiatives. Ms. Wang has played an active role in several transnational research projects across Mainland China, Hong Kong SAR, and the UK. Notably, she was involved in a scientific research initiative under China's "14th Five-Year Plan"—a prestigious horizontal project that further sharpened her research acumen. Her academic contributions include multiple SSCI-indexed publications in Frontiers, focusing on organisational culture, psychology, and educational psychology. In addition to her academic pursuits, Ms. Wang serves as a consultant for GEC Institution—one of China's premier research and education training organisations—where she provides strategic input on programme development and research capability enhancement. With a robust academic background and international research experience, Ms. Wang is dedicated to advancing impactful educational research. Her work bridges theoretical inquiry with practical application, offering valuable insights for both academic and industry stakeholders. She is committed to contributing to global educational innovation through collaboration, research, and reflective practice.

Zhou Xiang, also known as Edd, is an accomplished media scholar and educator with a distinguished academic background, holding a double master's degree in Media. He is currently a lecturer at a leading university in China, where he is recognised for his dynamic teaching approach, interdisciplinary expertise, and unwavering commitment to student development. Driven by a passion for lifelong learning, Zhou is currently pursuing a Doctor of Education (EdD) at Teesside University, UK—demonstrating his dedication to enhancing pedagogical excellence and advancing evidence-based practice in higher education. Zhou has published ten peer-reviewed academic papers, contributing impactful research in the areas of communication, media convergence, digital life, media culture, and representational theory. His research leadership is further evidenced by his involvement in two major research projects, where he has played both lead and collaborative roles. With academic specialisations spanning communication, education, art, semiotics, literature, sociology, and journalism, Zhou stands out as a truly interdisciplinary scholar. His work is defined by intellectual rigour, cultural insight, and a strong focus on bridging theory and practice. With a growing reputation in research and teaching, Zhou Xiang continues to make meaningful contributions to the global academic community.

Index

A

B

E

F

G

I

L

M

O

P

R

S

T

www.ingramcontent.com/pod-product-compliance
Lightning Source LLC
LaVergne TN
LVHW080352011026
849161LV00009B/2274